THE PILGRIM'S PROGRESS

JOHN BUNYAN (1628–88) was born at Elstow, near Bedford, the eldest son of a tinker. His schooling was slight. In 1644 he was mustered in the Parliamentarian army and stationed at Newport Pagnell. During the early 1650s he underwent the prolonged spiritual crisis he later graphically recorded in one of the classics of Puritan spirituality, *Grace Abounding to the Chief of Sinners* (1666). Following his conversion he joined the Bedford Baptist church, and the congregation quickly recognized his gift for preaching. This led, in 1656, to the beginning of a literary career in the course of which he would publish some sixty works of controversial, expository and practical divinity, marked by an uncompromising zeal, a trenchant directness of style and a particular concern for the spiritual welfare of common people. During the twelve-year imprisonment for nonconformity which followed the Restoration (1660–72), writing became the chief means by which to fulfil his vocation. It was not, however, until the publication in 1678 of *The Pilgrim's Progress* that his genius declared itself. The imaginative persuasiveness and realistic authenticity of this allegory have earned for it an unprecedently extensive readership and for its author his unique place in literary history. It was followed in 1680 by its sequel, *The Life and Death of Mr. Badman*, by the elaborate multi-level allegory *The Holy War* in 1682 and by Part II of *The Pilgrim's Progress* in 1684, works which substantiate Bunyan's claim to be the founder of the English novel.

N. H. KEEBLE lectures in English at the University of Stirling. He has edited J. M. Lloyd Thomas's abridgement of *The Autobiography of Richard Baxter* (1974) and published *Richard Baxter: Puritan Man of Letters* (Oxford, 1982) as well as introductory guides to *Romeo and Juliet* and *Richard II* (1980) and a number of articles on late medieval, Renaissance and seventeenth-century English literature. He is currently working on a study of the literary culture of nonconformity 1660–1702.

THE WORLD'S CLASSICS

====

JOHN BUNYAN

The Pilgrim's Progress

====

Edited with an Introduction by
N. H. KEEBLE

Oxford   New York

OXFORD UNIVERSITY PRESS

Oxford University Press, Walton Street, Oxford OX2 6DP

Oxford New York Toronto
Delhi Bombay Calcutta Madras Karachi
Petaling Jaya Singapore Hong Kong Tokyo
Nairobi Dar es Salaam Cape Town
Melbourne Auckland

and associated companies in
Beirut Berlin Ibadan Nicosia

Oxford is a trade mark of Oxford University Press

This edition first published 1984 as a World's Classics paperback
Reprinted 1986, 1988

British Library Cataloguing in Publication Data

Bunyan, John
The pilgrim's progress.—(The World's classics)
I. Title II. Keeble, N. H.
823'.4[F] PR3330.A2
ISBN 0-19-281607-1

Library of Congress Cataloging in Publication Data

Bunyan, John, 1628-1688.
The pilgrim's progress.
(The World's classics)
Bibliography: p.
Includes index.
I. Title.
PR 3330.A2K43 1984 828'.407 83-11455
ISBN 0-19-281607-1 (pbk.)

Printed in Great Britain by
Hazell Watson & Viney Limited
Aylesbury, Bucks

$6.50

ACKNOWLEDGEMENTS

IN my annotation I several times refer to Professor Roger Sharrock's Oxford English Texts edition of *The Pilgrim's Progress*, but I should like here to acknowledge a more general indebtedness to his commentary, which has served me as a model in formulating my own less comprehensive notes. It is from Professor Sharrock's edition that the index is taken. My chronology derives chiefly from John Brown's biography, and for my glossary I turned most often to *The Oxford English Dictionary*.

For their kind help with specific queries I thank Dr. Geoffrey F. Nuttall, Dr. John Drane of the Department of Religious Studies at the University of Stirling, Dr. David Bebbington of the Department of History, Dr. Lance Butler and Dr. David Reid of the Department of English Studies. To Dr. Reid I am especially grateful for offering what was invariably good advice on an earlier draft of the introduction.

Some passages in the introduction derive from previously published articles on '*The Pilgrim's Progress*: a Puritan Fiction', *The Baptist Quarterly*, xxviii (1980), and 'Christiana's Key: The Unity of *The Pilgrim's Progress*' in Vincent Newey (ed.), *The Pilgrim's Progress: Critical and Historical Views* (Liverpool, 1980), and are here reprinted by permission.

University of Stirling   N. H. KEEBLE

To Oliver, Owen and Sophie

CONTENTS

INTRODUCTION

I

The Pilgrim's Progress holds a unique place in the history of our literature. No other seventeenth-century text save the King James Bible, nothing from the pen of a writer of Bunyan's social class in any period, and no other Puritan, or, indeed, committed Christian work of any persuasion, has enjoyed such an extensive readership. How it was that a poorly educated Bedfordshire tinker and sectarian preacher could manage so to overcome barriers of class, culture and dogma has surprised and perplexed many.[1] The circumstances of the book's composition open the way to one answer, for, it seems, the author was himself as surprised as anyone else by what he produced: *The Pilgrim's Progress* was not what Bunyan had intended to write. The prefatory 'Apology' to Part I tells that, 'writing of the Way/ And Race of Saints' and falling 'suddenly into an Allegory', he was forced by the rapid multiplication of ideas 'Like sparks that from the coals of Fire do flie' to put aside the treatise on which he was engaged in order to give free rein to his imagination (p. 1).[2] In the 'Apology' Bunyan is apprehensive lest his readers fail to recognize in the product of this unpremeditated surrender to inspiration an edifying piece of theological exposition. Indeed, his alternating tone of defiance and anxiety suggests that he himself had misgivings about the result, misgivings which probably explain why the book was withheld from publication for several years.[3] And its eventual publication in 1678 appears to have been not a confident resolve but the expedient of a man who could neither quite trust what he had

[1] Bunyan's life and circumstances are summarized below in the Chronology. For a fuller account, see the standard biography by John Brown, *John Bunyan (1628–88): His Life, Times and Work*, rev. F. M. Harrison, tercentenary edn (1928) and Roger Sharrock, *John Bunyan*, corrected reissue (1968), pp. 9–50.

[2] References in the text are to this edition of *The Pilgrim's Progress*.

[3] For identification of the treatise on the 'Race of Saints', the date of the composition of *The Pilgrim's Progress* and the delay before publication, see below, nn. to p. 1.

created nor yet bear to put aside what had involved him so
intimately in its writing: confronted with conflicting advice from
friends, Bunyan had it printed 'To prove then who advised for
the best' (p. 2).

This diffidence is remarkable for Bunyan was no novice setting
out on his literary career. The threat posed to his ministry by
what he took to be erroneous Christian doctrines had first
prompted him into print over twenty years before. The controversial impulse has always been a spur to Christian authorship.
St. Paul himself wrote what may have been his first epistle to
rescue the Galatians from the influence of those judaising
Christians who had 'bewitched' them. Bunyan's combativeness,
evident in his glances at the Churches of Rome and of England in
The Pilgrim's Progress, is not a peculiarly Puritan attribute which
witnesses to narrow-mindedness but the result of a pastoral
concern which, like Paul's, would save men from the 'Ditch' of
doctrinal error as well as from the 'Quagg' of moral turpitude
(p. 51). The challenge Bunyan met came from the Quakers.
They began evangelizing in Bedfordshire in 1654, and their
leader, George Fox, was there in 1655, the year Bunyan moved
from Elstow to Bedford to begin his preaching ministry. Con-
troversies with their converts led to Bunyan's first publications,
Some Gospel-Truths Opened (1656) and its *Vindication* (1657).[4]
These were followed by such homiletic pieces as *A Few Sighs
from Hell* (1658), the theological treatise *The Doctrine of the Law
and Grace Unfolded* (1659), the verses of *Prison Meditations*
(1663), and, in 1666, Bunyan's autobiographical *Grace Abounding to the Chief of Sinners*. Hence, by the mid-1660s he had
behind him a considerable body of work. Although expository in
manner and not distinctively different in kind from the mass of
Puritan practical divinity, it already displayed an aptitude for
striking analogies ('a Similitude', he wrote in *Gospel-Truths*, 'is
warrantable; for both Christ and his Apostles did sometimes
use them to the end soules might be the better informed'),[5] a

---

[4] For the circumstances in which Bunyan began his literary career
see the introduction to *The Miscellaneous Works of John Bunyan*,
general editor Roger Sharrock, 12 vols. in progress (Oxford, 1976–),
i, pp. xv–xxx (hereafter cited as *Misc. Wks.*).   [5] *Misc. Wks.*, i. 75–6.

trenchant style, and a fine scorn for hypocrisy detected, often, in the rich and lordly. *Grace Abounding* had traced in sensitive detail and with considerable emotional intensity just those stages of the Christian experience which would be represented in *The Pilgrim's Progress*,[6] and we catch, too, in these earlier works, our first glimpses of Bunyan's skill in combining the incisively realistic with the typically representative. Such later allegorical figures as Mr. By-ends are adumbrated here, and there is a presentation in dialogue of the arguments to be given to Ignorance, sometimes in his very phrases.[7] Bunyan had, then, handled this material before and was accustomed to carry through successfully preconceived literary plans. No wonder he was disconcerted to find his work being 'eaten out' by an unforeseen allegory.

The apparent implications of this history were drawn out by Coleridge in a famous note which ascribed the achievement of *The Pilgrim's Progress* entirely to Bunyan's own genius:

> in that admirable Allegory, the first Part of Pilgrim's Progress, which delights every one, the interest is so great that spite of all the writer's attempts to force the allegoric purpose on the Reader's mind by his strange names – Old Stupidity of the Town of Honesty, &c. &c. – his piety was baffled by his genius, and the Bunyan of Parnassus had the better of the Bunyan of the Conventicle – and with the same illusion as we read any tale known to be fictitious, as a novel – we go on with his characters as real persons.[8]

Bunyan's imagination, that is to say, was so fired as to defeat his didactic purpose. Although it may have been intended but to illustrate and impress a particular conception of the Christian life, and so apparently of interest to but a limited audience, *The Pilgrim's Progress* is lifted above the body of seventeenth-century Puritan writing precisely because its inspired author was liberated from the constraints of his theology. This train of thought leads to the view of the book as the work of a 'transcendent genius', 'as

[6] On the relationship between *Grace Abounding* and *The Pilgrim's Progress* see Roger Sharrock, 'Spiritual Autobiography in *The Pilgrim's Progress*', *Review of English Studies*, xxiv (1948), 102–20.

[7] *Misc. Wks.*, i. 301, ii. 172–6.

[8] *Coleridge on the Seventeenth Century*, ed. Roberta Florence Brinkley (New York, 1968), p. 475.

original as anything in literature can be',[9] its innovative fictional realism inexplicable save by reference to its author's own daemon.

However, to hold that 'No one has done so much with so little help from predecessors or contemporaries'[10] is to set Bunyan in the context of a tradition foreign to him. To be sure, he owes almost nothing to the literary culture of his time. Not only did he make no secret of his ignorance, glossing 'ex Carne & Sanguine Christi', 'The Lattine I borrow' (p. 190), but he defiantly asserted it as a strength: 'I have not writ at a venture, not borrowed my Doctrine from Libraries. I depend upon the sayings of no man: I found it in the Scriptures of Truth, among the true sayings of God.' That Christian's education is not by a course of reading prescribed by Evangelist but by the Bible and the Holy Spirit in the House of the Interpreter accords with Bunyan's dismissal of the value of mere 'humane' learning as compared to the experiential knowledge and illumination of faith:

Reader, if thou do finde this book empty of Fantastical expressions, and without light, vain, whimsical Scholar-like terms, thou must understand, it is because I never went to School to *Aristotle* or *Plato*, but was brought up at my fathers house, in a very mean condition, among a company of poor Countrey-men. But if thou do finde a parcel of plain, yet sound, true and home sayings, attribute it but to the Lord Jesus, his gifts and abilities, which he hath bestowed upon such a poor Creature, as I am, and have been.[11]

But if to be brought up 'among a company of poor Countrey-men' was to be denied the learning of a Donne and to be no part of the courtly tradition of Spenser, Sidney, Jonson and Dryden, it was, in seventeenth-century Puritan East Anglia, to be possessed of another kind of knowledge and to be part of another tradition quite as vital.

Puritanism was a pre-eminently social movement whose con-

---

[9] Walter Allen, *The English Novel* (Harmondsworth, 1958), p. 32.

[10] J. W. Hales, *Folia Litteraria* (1893), in Roger Sharrock (ed.), *The Pilgrim's Progress: a Casebook* (1976), p. 78 (hereafter cited as *Casebook*).

[11] *Misc. Wks.*, viii. 51, ii. 16. On the extent of Bunyan's reading, see Richard L. Greaves, *John Bunyan* (Appleford, Berks., 1969), esp. pp. 153–60.

siderable literature was characterized by a fascinated interest in the actual experiences of men. Milton's oft-quoted rejection of the spiritual ease and sloth of monasticism testifies to the Puritan conviction that it is by playing a full part in this world that salvation is won: 'I cannot praise a fugitive and cloister'd vertue, unexercis'd & unbreath'd, that never sallies out and sees her adversary, but slinks out of the race, where that immortall garland is to be run for, not without dust and heat'.[12] Hence, Puritan divinity was above all practical, or, as we should say, moral and casuistical, concerned with problems of daily living. Scholasticism was rejected as firmly as monasticism. Furthermore, despite our modern sense of the word (a legacy of seventeenth-century anti-Puritan satire), the Puritan did not conceive the way to salvation to consist in abstinence or asceticism but in a right use of our physical natures. Eating, drinking, singing, music and dancing are celebrated throughout *The Pilgrim's Progress* as divine gifts. It is precisely Christiana's human affection for her husband which sets her on the road to the Celestial City.

If the Puritan was responsive to the external world and concerned with his everyday social, domestic and commercial behaviour, the doctrine of justification by faith encouraged in him an equal preoccupation with inner, personal experience, for it was upon the validity of his experience of grace, the sincerity of his faith, that his claim to election rested. It is an 'experimental' or experiential faith which Ignorance lacks and which Faithful sets against Talkative's opinionative religion. Just such an experimental confession of Faith' as Faithful here mentions (p. 68) was a condition of membership of such congregational churches as the Bedford church to which Bunyan belonged. But Puritans were habitually given to exchanging experiences in every context, just as Bunyan's pilgrims do. The classics of Puritan spirituality are predominantly autobiographical: Bunyan's own *Grace Abounding*, Richard Baxter's *Reliquiae Baxterianae* (1696), George Fox's *Journal* (1694), perhaps, too, Milton's *Samson Agonistes* (1672). Although our evidence for this bias is necessarily literary, it was not a peculiarly literary phenomenon, nor confined to the literate.

---

[12] John Milton, *Areopagitica* (1644), in *Selected Prose*, ed. C. A. Patrides (Harmondsworth, 1974), p. 213.

The crucial incident in *Grace Abounding* of the 'poor women' of Bedford 'sitting at a door in the Sun, and talking about the things of God', which led, in time, to Bunyan's own conversion, illustrates how efficacious and potent such unashamed spoken communication of commitment could be.[13]

By virtue of that conversion, Bunyan entered into this tradition of realistic plain dealing, and through it reinterpreted the older folk culture of originally pagan story to which he was also heir. It was his genius to realize artistically this tradition. It was not that his genius was baffled by his piety but that it startled both author and first readers by imaginatively representing Puritan experience rather than explicating it with anecdotal illustration. It is, in the sense that the realistic method of his allegory was almost inevitable, 'not fantastic to assert that it was the Puritan culture as much as Bunyan that produced *The Pilgrim's Progress*.'[14]

II

It is that realism which is the most immediately striking feature of *The Pilgrim's Progress*. Everything, as C. S. Lewis remarked, 'is visualized in terms of the contemporary life that Bunyan knew'.[15] We discern this in the circumstantial details of the journey, which takes us along a muddy, poorly signposted seventeenth-century road, over hills, through dark valleys and across bleak moors, exposed to inclement weather and at risk from floods, sobered by the sight of criminals' skeletons hanging from a wayside gibbet, apprehensive of footpads, fearful of being benighted and grateful for the refuge of an inn or house hospitable to travellers, for sheer physical exhaustion, the weariness of aching legs, is never far away. The various incidents on this journey are often mundanely inconsequential: travellers who misdirect; a dog barking when a door is knocked; boys scrumping apples; the longings of a pregnant woman. This 'strong sense...of an actual reality'[16] is enforced by the conversations we overhear. They

---

[13] John Bunyan, *Grace Abounding*, ed. Roger Sharrock (Oxford, 1962), § 37.

[14] Q. D. Leavis, *Fiction and the Reading Public* (1932), p. 97.

[15] *Casebook*, p. 197.

[16] F. R. Leavis, in *Casebook*, pp. 213–14.

concern 'the actual, unimaginary problems of living besetting the average man and woman of the time'.[17] Newly-met travellers compare notes of their experiences on the road – when Christian and Faithful meet they have 'sweet discourse of all things that had happened to them in their Pilgrimage' (p. 55) – and tell each other of their wives, family and background, as does By-ends to Christian (p. 81). Wayfarers are anxious to glean from passers-by information about the road ahead, and may disagree between themselves as to which is the better route, as do Christian and Hopeful over the meadow-path which 'lay along by the way on the other side of the fence' (p. 91). Conversations often take a gossipy turn: talk of neighbours and acquaintances, of where they live, how they earn their living, of their standing in the community, and rumours that they are not all they seem, occupy a good deal of the time. Christian is very ready to put Faithful straight about Talkative, who, though a 'pretty man' to 'them that have not through acquaintance with him', is yet 'near home...ugly enough' (p. 64).

It is, indeed, remarkable how little we rely upon the narrator for our information. Much of the immediacy of *The Pilgrim's Progress* derives from the fact that, as in drama, we hear the characters directly in dialogue. And, again as in drama, this makes for ironic misunderstandings and misapprehensions as the limited points of view of these characters lead them to misjudge both themselves and others. *The Pilgrim's Progress* is built up of a series of such ironic encounters. Beneath their superficial exchanges, new acquaintances are engaged in weighing each other up, trying to determine whether or not they will make suitable travelling companions. Christian is constantly taxed to tell his friends from his foes, and it is a measure of what he (and the reader with him) learns by experience that, while at the outset he falls an easy prey to Mr. Worldly Wiseman, he is later able to see through Talkative and to lay bare the fallacies of Mr. Money-love, until he achieves the independence of judgement evidenced in the wisdom and generosity of his comments on Little-faith. To attain

[17] Arnold Kettle, *An Introduction to the English Novel*, 2nd edn, vols. (1967), i. 40.

and maintain this discerning perception demands a constant mental alertness. Christian takes not merely physical ease in the Chamber Peace after what Christiana later calls the 'breathing' Hill Difficulty (p. 178) but psychological and spiritual rest. The Plain of Ease is, however, 'but *narrow*' (p. 87) and the struggle must go on.

Bunyan's 'genius for realistic observation'[18] is still more evident in his characterization of those travellers who so tax Christian. Bunyan has an uncommonly sharp eye for the appearances, mannerisms and behaviour of ordinary people, and vivifies his characters, as Henri Talon has put it, by the 'slightest flick of the brush',[19] such as Ignorance's 'briskness' (p. 101), Atheist's 'very great laughter' (p. 110) or Madam Bubble's 'Swarthy Complexion' (p. 252). Such details, however, do not merely vivify and particularize: they suggest moral states – Ignorance's unreflecting complacency, Atheist's cynical derisiveness, Madam Bubble's rather tawdry worldliness. That they do suggest, rather than embody, is important. U. Milo Kaufmann has well cautioned us against speaking of Bunyan's characters as 'types'. Certainly, they do have a symbolic function within the allegory, but they fulfil it as examples: they 'do not incarnate but exemplify a particular quality'. Honest's remark, 'Not Honesty in the *Abstract*, but *Honest* is my Name' (p. 205), leads Kaufmann to point out that generally the names of Bunyan's characters 'are adjectival in nature rather than substantival, and hence hint at attribute rather than essence'.[20] Even such a minor character as Pliable illustrates the consequence. He *is* pliable, in his readiness first to join and then to abandon Christian when the going gets tough. Our later glimpses of him 'sneaking among' his neighbours and 'leering away' from acquaintances (pp. 14, 56) are, however, of a man whose behaviour declares guilt and embarrassment at what he has done. Guilt and embarrassment are no part of the abstract quality of pliability: they are the experiences of a pliable man. Bunyan,

[18] Sharrock, *Bunyan*, p. 89.
[19] Henri Talon, *John Bunyan: the Man and his Works*, transl. Barbara Wall (1951), p. 218.
[20] U. Milo Kaufmann, *The Pilgrim's Progress and Traditions in Puritan Meditation* (New Haven, 1966), p. 90.

that is to say, recognizes that moral qualities have their reality in people's lives, but he appreciates further that, though people may be virtuous or vicious to a degree, no human personality is ever wholly subsumed into a particular virtue or vice. His moral realism is the source of his artistic realism.

So it is that, despite his Calvinism, that most discriminatory of theologies, Bunyan never rigidly categorizes. On the one hand, his villains are never impressive in their villainy, never wholly evil. Evil in *The Pilgrim's Progress* is a pervasive meanness, pettiness and selfishness, which yet allows a kind of friendliness: it is not a defiant amorality, nor a deliberate policy, nor a demonic power (though that is its source). We shall find nothing to compare with Marlowe's Dr. Faustus and Mephostophilis, Shakespeare's Iago, Tourneur's Vendice, Webster's Vittoria Corombona or Milton's Satan. What we do find is more familiar: the narrow-mindedness of Obstinate which has no patience with '*Craz'd-headed Coxcombs*' (p. 10); the smugness of Formalist and Hypocrisy who look at each other and laugh at Christian's strict adherence to the way (p. 34); the snobbery of Shame who, anxious to be in the social swim, objects to religion as a 'pitiful, low, sneaking business' unbecoming 'the brave spirits of the times' (p. 59). On the other hand, our saints are not rare exemplars: Evangelist discerns 'many weaknesses' in Christian and Faithful (p. 71). Anyone who doubts that Bunyan was keenly aware that faith coexists with failings need only read over the stories of Little-faith and Mr. Fearing, true pilgrims both. Indeed, in Part II Bunyan deliberately associates with the women Christiana and Mercy the young, the old, the infirm, the despondent, the tiresomely scrupulous, the fearful and the anxious. These seem improbable saints and unlikely literary heroes. Yet, that they hardly look any different from ordinary people is both the theological point and the source of the fictional verisimilitude. The regenerate are distinguished from the unregenerate not by any exceptional abilities or virtue but by their faith: they keep on going. As Christian says, 'all the Kings Subjects are not his Champions: nor can they, when tried, do such feats of War as he' (p. 106). That Bunyan's pilgrims are people like us implies we, like them, might have glory if we will: that they are people like us is what fascinates us as we read. We cannot here distinguish realism from evangelism, story from point, Parnassus from Conventicle.

The same holds true of Christian himself. Although many commentators have distinguished Part I from Part II as 'epic',[21] Christian is not really cast in the traditional heroic mould. It is only in Part II, when the pilgrims have the benefit of his example to guide them, that he is likened to Hercules (p. 199): in Part I itself, Christian trembles for fear of the lions before the Palace Beautiful. Bunyan offers us no simple progress in the sense of 'steady improvement'.[22] Christian confesses to Prudence that 'when I would be doing of that which is best, that which is worst is with me' and that the 'Golden hours' of pure devotion come 'but seldom' (p. 41); even at the end, he nearly drowns in the river. Left to his own devices, Christian never would have achieved the quest: and yet, quite against the odds, he does achieve it. Any man may, through grace, become a Christian hero. That is the point which Bunyan, like every Puritan preacher before him, would impress upon us. Certainly, the imaginative skill with which he does so surpasses that of his predecessors and creates, incidentally, the hero of realistic story: it is Christian's fallibility which arouses both our sympathy for him in his predicaments and our anxiety about their outcome. This, however, is the fortunate literary consequence of a theological conviction held by a man who was also a writer of genius; it is not the consequence of that writer's imaginative escape from his theology.

That there is tension in the book might appear strange in view of the predestinarian tenets of Calvinism. It derives again from Bunyan's presentation of events from the point of view of those

[21] e.g. E. M. W. Tillyard, *The English Epic and its Background* (1954), p. 393, and, in slightly different terms, Sharrock, *Bunyan*, p. 138 and Talon, *Bunyan*, p. 219.

[22] This is not to go as far as Stanley E. Fish, who, in *Self-Consuming Artifacts* (Berkeley and Los Angeles, 1972), pp. 224–64, argued against any kind of progress; he has been ably answered by John R. Knott, 'Bunyan's Gospel Day', in *Casebook*, pp. 221–43. For discussion of the senses of 'progress' applicable to *The Pilgrim's Progress*, see Philip Edwards, 'The Journey in *The Pilgrim's Progress*', in Vincent Newey (ed.), *The Pilgrim's Progress: Critical and Historical Views* (Liverpool, 1980), pp. 111–17.

who undergo them, from his concentration upon the human experience of grace. For the saints, assurance of salvation is neither easily won nor sustained. While it may be doctrinally true that the elect necessarily persevere and are saved, this dogma is not confidently applied to themselves by those whose self-evident weaknesses give grounds for doubting their election. Hence, Christian can lose the roll which 'was the assurance of his life, and acceptance at the desired Haven' (p. 37) and be imprisoned in Doubting Castle because the experiential fact of uncertainty is incontrovertible. The saint only gains final assurance of perseverence when he has persevered. As Coleridge annotated *The Pilgrim's Progress*, the argument from predestination is 'a mere identical proposition followed by an assertion of God's prescience. Those who will persevere, will persevere, and God foresees'.[23] Or, as the Shepherd answers Christian's query, '*How far is it thither?*', 'Too far for any, but those that *shall* get thither indeed' (p. 98). The discovery of their identity must await the end of the story. And so tension and suspense are inevitable, as James, to Great-heart's approval, comments upon the case of Mr. Fearing, who 'was always afraid that he should come short of whither he had a desire to go', 'No fears, no Grace' (p. 211).

Bunyan captures the linguistic mannerisms of ordinary people as accurately as he depicts their behaviour and appearance and as firmly as he realizes their inner emotional and spiritual life. In this regard, his very unfamiliarity with literary tradition worked to his advantage, for his model was not any convention of pastoral rusticity but the actual language as spoken on the roads and in the towns of East Anglia. Since at least 1578 Puritan divines had been prompted by their pastoral concern for common people to call for a plain style of preaching in contrast to the 'metaphysical' ingenuity of such court preachers as Lancelot Andrewes.[24] This was a moral and practical, rather than aesthetic,

---

[23] *Coleridge on the Seventeenth Century*, pp. 483-4.
[24] Harold Fisch, 'The Puritans and the Reform of Prose Style', *Journal of English Literary History*, xix (1952), 231-2, finds the earliest example in a Paul's Cross sermon by Laurence Chaderton of this date. For some account of Puritan style and stylistics, see this article and Perry Miller, *The New England Mind: the Seventeenth Century* (1939; rpt. Boston, 1961), pp. 331-62; L. A. Sasek, *The*

preference, which Bunyan shared. He rebuked learned ministers who, instead of preaching directly 'the simplicity of the Gospel', 'do tickle the ears of their hearers with vain Philosophy' and 'nuzzle up your people in ignorance with *Aristotle, Plato*, and the rest of your heathenish Philosophers',[25] and in *The Pilgrim's Progress* fine speaking is repeatedly associated with hypocrisy, and both, often, with elevated social status: By-ends comes from Fair-speech and has married into high society. Yet it is not the plain-ness of Bunyan's style, the absence of such carefully wrought rhetorical periods as Jeremy Taylor can supply, so much as its authentically colloquial tone, quite unlike the poised urbanity we find in Dryden or Halifax, which persuades us of the reality of his characters. By 1684, Bunyan had himself come to realize how remote from contemporary literary fashion this 'native Language' was (p. 136), but, though Part II, unlike Part I, was a premedi-tated undertaking written by a man now enjoying considerable standing, Bunyan did not refine his style. This cost his reputation dear for over a century after his death. Measured by Augustan canons, Bunyan's style was a 'prophanation', the work of a Grub Street hack so wanting in 'sublimity' as to be quite unacceptable to men of quality. Regard for Bunyan was, pronounced David Hume in 1757, próof merely of bad taste.[26] Not until the Romantic age could Coleridge point to Bunyan's 'lowest style of English' as essential to 'the reality of the vision' and Southey commend his 'homespun style' as 'a clear stream of current English'.[27] The idiomatic phraseology, colloquial usages and provincial forms of

Literary Temper of the English Puritans (1961; rpt. New York, 1969), pp. 39–56; Kenneth B. Murdock, *Literature and Theology in Colonial New England* (1949; rpt. New York, 1963), pp. 34–65; N. H. Keeble, *Richard Baxter: Puritan Man of Letters* (Oxford, 1982), pp. 48–68.

25 *Misc. Wks.*, i. 345.
26 Edward Smith, *A Poem on the Death of Mr. John Philips* ([1708]), p. 8; Joseph Addison, *The Spectator*, ed. Donald F. Bond, 5 vols. (Oxford, 1965), iv. 365; John Dennis, *The Critical Works*, ed. E. N. Hooker, 2 vols. (Baltimore, 1939–43), ii. 29–30; John Arbuthnot, *The History of John Bull*, ed. A. W. Bower and R. A. Erickson (Oxford, 1976), pp. 63, 92; David Hume, *The Philosophical Works*, ed. T. H. Green and T. H. Grose, 4 vols. (1882; rpt. 1964), iii. 268–9.
27 *Coleridge on the Seventeenth Century*, p. 475; *Casebook*, p. 57.

this style root us firmly in the no-nonsense world of proverbial wisdom, physical reality and emotional intensity: Hill Difficulty puts the pilgrims into a 'pelting heat' (p. 178); they walk 'a good stitch' (p. 228); Mrs. Timorous leaves Mercy 'to go a fooling' with the curt common sense of 'while we are out of danger we are out; but when we are in, we are in' (p. 151); Christian 'snibbeth his fellow' with the brusque remark 'Thou talkest like one, upon whose head is the Shell to this very day' (p. 104).

III

Yet there are phrases with quite another cadence: when Discretion hears of Christian's gratitude for the security of the Palace Beautiful, 'she smiled, but the water stood in her eyes' (p. 39); when Christian wounds Apollyon, 'he did smile, and look upward: but twas the dreadfullest sight that ever I saw' (p. 50). These adversative conjunctions set against the observed scene a world of spiritual knowledge and power which, in its very nature, is not susceptible of precise definition. For all its concreteness, there are mysteries in *The Pilgrim's Progress*: Much-afraid sings crossing the river, 'but none could understand what she said' (p. 259). Bunyan's landscapes of spiritual torment and spiritual ease take us far from Bedfordshire. And the setting itself has a disconcerting way of dissolving before our eyes: the rooms of the Interpreter's House are as often open to the sky as enclosed.[28] When Bunyan's realism is most pointed, the context can be from quite another literary mode: Christian faces his foe for the very good reason 'he had no Armour for his back' but that foe is a monster (p. 46); the lock hindering the fugitives' escape 'went *damnable* hard' but it is a giant's castle they flee (p. 96). *The Pilgrim's Progress* is, in truth, an ambivalent work: not only our first novel, but our last allegory, a bridge between two worlds, the medieval and the modern. In this, it exactly reflects the paradoxical temper of the Puritan, whose religious tradition encouraged him to be not merely a shrewd observer of men and their environment but also to detect significance in what he observed. His keen sense of

---

[28] This is more fully discussed in James Turner, 'Bunyan's Sense of Place', in Newey (ed.), pp. 91–110.

Providence led him to look at the world in order to interpret it, to discern through it God's will. 'Ther is no object that we see, no action that we doe, no evill that we feele, or fear, but we may make some spiritual advantage of all,' wrote the New England Puritan Anne Bradstreet.[29] Bunyan entirely agreed, as the emblems in the Interpreter's House or Prudence's answers to Matthew's emblematical catechism illustrate (pp. 191–2). It is this sensitivity to the interpenetration of the visible by the invisible, this consciousness of the immanence of ultimate realities, which explains, and should mute ridicule of, the Puritan habit of setting children on life's journey with just such names as Bunyan's pilgrims bear.[30]

This 'spiritualization' of mortal experience is the source of the allegorical method of *The Pilgrim's Progress*. 'Realistic allegory' may appear an oxymoron, if not a contradiction, but there is in Bunyan's execution no indecorousness and little tension because he was habituated to treating the sublunary world as an allegory of translunary truth. And, however engaging the realism, the allegory, as its marginalia remind us, requires of its reader the same interpretative effort as Bunyan believed the natural order demanded: both were 'Metaphors' to be 'Turned up' (p. 134). And he clearly expected his reader to share his own delight in the challenge this presented. He openly entices the would-be reader of *The Pilgrim's Progress* with the promise of 'pleasant' and teasing 'Riddles' ('Wouldst thou divert thy self from Melancholly?') and speaks of his work in terms which recall the 'metaphysicals' and their penchant for cunning and obscurity: it is 'something rare' whose 'Riddles' that 'lie couch't within' its 'misterious lines' 'nimble Fancies' are invited to solve. To accept the invitation is not, however, to engage in the merely enjoyable: it is a 'profitable' exercise; to 'loose thy self' in his allegory is 'to find thy self' (pp. 6, 7, 141). Small wonder, then, that the after-supper entertainment at Gaius' house is riddling or that the progress of

[29] *Poems of Anne Bradstreet*; ed. R. Hutchinson (New York, 1969), p. 187. This attitude is discussed in Kaufmann, esp. pp. 175–95, Miller, pp. 207–35 and Keeble, pp. 108–13.

[30] Charles W. Bardsley, *Curiosities of Puritan Nomenclature* (1880), pp. 117–201; E. G. Withycombe, *The Oxford Dictionary of English Christian Names*, 3rd edn (Oxford, 1977), pp. xxxvii–xl.

the pilgrims to the full self-knowledge of sanctity is measured in part by their skill in solving riddles. Christian's smile at the House of the Interpreter is a smile not merely of admiration but of satisfaction and pleasure at fathoming the emblem of the valiant man (p. 28). When Mercy and the boys blush for failing to solve the 'Riddle' of the spider, it is not merely the application which affects them: it is also their want of percipience, of wit. Christiana, by contrast, is a 'Woman quick of apprehension' (p. 165). Her companions have been caught out, and that ill becomes a saint, for the godly are an alert and curious people:

> Things that seem to be hid in words obscure,
> Do but the Godly mind the more alure;
> To Study what those Sayings should contain,
> That speak to us in such a Cloudy strain. (p. 139)

In just the same way Bunyan read the Bible, not, like Archbishop Tillotson, as an eminently reasonable and lucid handbook of ethics, but as 'some curious riddles of secrets'.[31] It, too, was a 'Cabinet' to be unlocked with a 'Key', its *Hard* Texts 'Nuts' to be cracked (pp. 4, 139, 219). It is, of course, from the Bible that the central image of *The Pilgrim's Progress* derives. The way 'as straight as a Rule can make it' (p. 23), in contrast to such a 'crooked Lane' as that by which Ignorance arrives (p. 101), recalls the distinction drawn in the Old Testament wisdom literature and prophets between the straight way of the righteous and the crooked ways of the wicked, an image which was given a specifically Christian significance in the dominical saying 'I am the way' (John 14:6). And the association of spiritual progress with a physical journey in the Genesis story of Abraham's setting-out from Mesopotamia recurs repeatedly thereafter in Old Testament narratives. These narratives Bunyan read typologically, that is, as prefiguring the New Testament, their characters 'types' or symbols of Christ,[32] but he also discerned in them patterns of experience applicable more generally to believers. It had 'happened to *Israel*'

---

[31] John Bunyan, *The Holy War*, ed. Roger Sharrock and James F. Forrest (Oxford, 1980), p. 116.

[32] For an account of Puritan typology see Thomas M. Davis, 'The Exegetical Traditions of Puritan Typology', *Early American Literature*, v (1970), 11–50.

as it happens to Christian and to all Christians (p. 36). The frequent references to the Exodus saga, for example, hinge on a parallel between the Israelites' experiences on their physical journey to Canaan and the Christian's spiritual experience: the believer must forsake the Egypt of carnality through the Red Sea of conversion and baptism to remain firm in the wilderness of trial in hope of the Canaan of salvation. The interpretation of Old Testament story in Hebrews (especially chapters 3 and 11), which saw the patriarchs as 'strangers and pilgrims in the earth' journeying to a 'heavenly country', a 'continuing city' (11: 13–16, 13:14), encouraged such allegorical readings.[33] Hence, when Bunyan appealed to the 'Dark Figures, Allegories' of the Bible to justify his method (p. 4) he had in mind far more than those passages we should recognize as figurative: he was appealing to what he took to be the literary technique of the Holy Spirit. If the Conventicle sharpened Bunyan's keen eye, and so facilitated the realism of his narrative, it fostered in him no less a hermeneutical habit which enabled him to invest that narrative with the significance only expository didacticism had hitherto handled. It was, indeed, the Conventicle which enabled him to scale Parnassus.

[33] This is discussed more at large in Brainerd P. Stranahan, 'Bunyan and the Epistle to the Hebrews', *Studies in Philology*, lxxix (1982), 279–96.

NOTE ON THE TEXT

PROFESSOR Roger Sharrock established the definitive text of *The Pilgrim's Progress* in his revision (1960) of James Blanton Wharey's Oxford English Texts edition (1928). It is based upon the first editions of Parts I and II (1678; 1684) and, for passages subsequently added by Bunyan, upon the earliest editions in which they appeared, with all significant textual variants recorded in footnotes. This choice of copy text is explained in Professor Sharrock's introduction, whose discussion of the circumstances of composition, detailed description of the seventeenth-century editions and thorough analysis of their relationship should be consulted on all bibliographical and textual questions. This text is here reprinted from its *editio minor* in the Oxford Standard Authors series (1966), with the alteration only of marginal page references to Part I in Part II (to correspond with the new pagination), and the omission of footnotes.

Asterisks in the text indicate some of Bunyan's own marginal glosses. Daggers (†) refer the reader to the present editor's explanatory notes at the end of the book, where there is also a Glossary.

SELECT BIBLIOGRAPHY

*(The place of publication of the works cited is London,
unless otherwise stated.)*

THE most fully annotated and standard text of *The Pilgrim's Progress*
is Roger Sharrock's revision of James Blanton Wharey's edition (2nd
edn, 1960), part of the Clarendon Press's Oxford English Texts edition
of Bunyan's corpus which will, when complete, supersede George
Offor's three volumes of *The Works of John Bunyan* (1853; rpt. 1862).
It is planned to comprise twelve volumes of *Miscellaneous Works*
(1976–) and separate editions of *Grace Abounding to the Chief of
Sinners* (1962), *The Holy War* (1980) and *The Life and Death of Mr
Badman* (forthcoming). Part I of a facsimile edition of *The Pilgrim's
Progress*, first published in 1875, was reprinted by Gresham Books in
1978. Roger Sharrock introduces and lightly annotates a modernized
text in the Penguin English Library series (1965). There is a Rinehart
edition, edited by Louis Martz (New York, 1949), and editions in
Everyman's Library, edited by G. B. Harrison (1954) and in Signet
Books, with an afterword by F. R. Leavis (New York, 1964).

The authoritative guide to Bunyan's publications is F. M. Harrison,
A Bibliography of the Works of John Bunyan (Oxford, 1932). Roger
Sharrock surveys the secondary material in *The English Novel: Select
Bibliographical Guides*, ed. A. E. Dyson (1974) and Richard L. Greaves
has compiled *An Annotated Bibliography of John Bunyan Studies*
(Pittsburgh, 1972).

The historical and ecclesiastical context of Bunyan's life is most
readily learned from Michael R. Watts, *The Dissenters: from the
Reformation to the French Revolution* (Oxford, 1978) and the experience of Puritanism best understood from Geoffrey F. Nuttall, *The Holy
Spirit in Puritan Faith and Experience*, 2nd edn (Oxford, 1947). The
standard biography is John Brown, *John Bunyan: his Life, Times and
Work*, tercentenary edn revised by F. M. Harrison (1928), who wrote
his own *John Bunyan: a Story of his Life* (1928; rpt. 1964). Also biographical are G. O. Griffith, *John Bunyan* (1937), G. B. Harrison, *John
Bunyan: a Study in Personality* (1928), O. E. Winslow, *John Bunyan*
(New York, 1961) and, with a psychological bias, Monica Furlong,
Puritan's Progress (1975). W. Y. Tindall, *John Bunyan: Mechanick
Preacher* (New York, 1934) relates Bunyan to mid-seventeenth century
radical Puritanism, as does Christopher Hill in the second appendix to
his *The World Turned Upside Down* (1972; rpt. Harmondsworth,

1975). Richard L. Greaves, *John Bunyan* (Appleford, Berks., 1969) is an admirably lucid exposition of Bunyan's theology.

The literary consequences of Puritanism are the subject of William Haller, *The Rise of Puritanism* (1938; rpt. New York, 1957), Lawrence A. Sasek, *The Literary Temper of the English Puritans* (Baton Rouge, 1961) and Owen Watkins, *The Puritan Experience* (1972). U. Milo Kaufmann's analysis of Bunyan's literary techniques in the light of Puritan hermeneutics and devotion in *The Pilgrim's Progress and Traditions in Puritan Meditation* (New Haven, 1966) can be thoroughly recommended. The allegory is set in the wider Christian tradition in R. M. Frye, *God, Man and Satan: Patterns of Christian Thought and Life in Paradise Lost and The Pilgrim's Progress* (Princeton, 1960).

J. A. Froude's *Bunyan* in the English Men of Letters series (1880) has been superseded as an introduction to Bunyan's literary career by Roger Sharrock, *John Bunyan* (1954; corrected reissue 1968) and Henri Talon, *John Bunyan: the Man and his Works*, translated by Barbara Wall (1951). Bunyan's work is related to different aspects of literary history in: Walter Allen, *The English Novel* (1954; rpt. Harmondsworth, 1958); Arnold Kettle, *An Introduction to the English Novel*, 2 vols., 2nd edn (1967); Q. D. Leavis, *Fiction and the Reading Public* (1932; rpt. 1965); G. R. Owst, *Literature and Pulpit in Medieval England*, 2nd edn (Oxford, 1961); James Sutherland, *English Literature of the Late Seventeenth Century* (Oxford, 1969); E. M. W. Tillyard, *The English Epic and its Background* (1954).

Roger Sharrock, who has written on *The Pilgrim's Progress* in Arnold's Studies in English Literature series (1966), has collected together examples of critical comment on Bunyan's allegory in *The Pilgrim's Progress: a Casebook* (1976). Vincent Newey (ed.), *The Pilgrim's Progress: Critical and Historical Views* (Liverpool, 1980) is a collection of original essays. Coleridge has some characteristically suggestive notes in *Coleridge on the Seventeenth Century*, ed. Roberta Florence Brinkley (1955; rpt. New York, 1968). There are appreciative essays in F. R. Leavis, *The Common Pursuit* (1952) and by Maurice Hussey in *From Donne to Marvell*, ed. Boris Ford (Harmondsworth, 1956), and critical analyses of Bunyan's narrative and fictional technique in Stanley E. Fish, *Self-Consuming Artifacts* (1972; rpt. Berkeley and Los Angeles, 1974), Wolfgang Iser, *The Implied Reader* (1974) and Dorothy Van Ghent, *The English Novel* (1953; rpt. New York, 1961).

Specific aspects of *The Pilgrim's Progress* are discussed in the following essays: David J. Alpaugh, 'Emblem and Interpretation in *The Pilgrim's Progress*', *English Literary History*, xxxiii (1966); James

F. Forrest, 'Bunyan's Ignorance and The Flatterer', *Studies in Philology*, lx (1963); Daniel Gibson Jr., 'On the Genesis of *Pilgrim's Progress*', *Modern Philology*, xxxii (1935); Harold Golder, 'Bunyan's Giant Despair', *Journal of English and Germanic Philology*, xxx (1931); Dayton Haskin, 'The Burden of Interpretation in *The Pilgrim's Progress*', *Studies in Philology*, lxxix (1982); N. H. Keeble, '*The Pilgrim's Progress*': a Puritan Fiction', *The Baptist Quarterly*, xxviii (1980); Geoffrey F. Nuttall, 'The Heart of *The Pilgrim's Progress*', in *Reformation Principle and Practice: Essays in Honour of Arthur Geoffrey Dickens*, ed. Peter Brooks (1980); J. G. Patrick, 'The Realism of *The Pilgrim's Progress*', *The Baptist Quarterly*, xiii (1949–50); Roger Pooley, ' "The Wilderness of this World" – Bunyan's *Pilgrim's Progress*', *The Baptist Quarterly*, xxvii (1978); Roger Sharrock, 'Personal Vision and Puritan Tradition in Bunyan', *The Hibbert Journal*, lvi (1957) and 'Spiritual Autobiography in *The Pilgrim's Progress*', *Review of English Studies*, xxiv (1948); Brainerd P. Stranahan, 'Bunyan and the Epistle to the Hebrews: his source for...*The Pilgrim's Progress*', *Studies in Philology*, lxxix (1982).

1628   John Bunyan ('Bunyon' or 'Bunnian') is born (baptised 30 November), the eldest of three children of Thomas Bunyan (1603–76) by his second wife Margaret Bentley (1603–44), whom he had married on 23 May 1627. Bunyan described himself as 'of a low and inconsiderable generation; my fathers house being of that rank that is meanest, and most despised of all the families in the Land', but his father, a 'braseyer' or tinker, though poor, was by no means destitute. The family home was in the parish of Elstow, just over a mile from both the village of Elstow and the county town of Bedford.

1630s   'Notwithstanding the meanness and inconsiderableness of my Parents, it pleased God to put it into their heart, to put me to School.' Bunyan attends either the grammar school at Bedford or the school in the neighbouring parish of Houghton Conquest, but, it seems, for only a short period. Thereafter, he follows his father's trade.

1644   The death of Bunyan's mother (20 June) is followed a month later by that of his sister Margaret and his father's marriage to his third wife Anne (d. 1680). These bereavements and his father's prompt remarriage may have a bearing on the 'wildness' of his youth which Bunyan later described in *Grace Abounding*. In November, Bunyan is mustered in a county levy for the Parliamentarian army and is stationed at the garrison at Newport Pagnell.

1645–6   What active service Bunyan sees is uncertain: it is almost certainly very little, perhaps none. No significant battles were fought near Newport Pagnell, early claims that Bunyan was at the seige of Leicester in 1645 are without evidence, Bunyan himself mentions no specific engagement, and the battles of *The Holy War* do not show indisputable proof of first-hand military experience. But the garrison was visited by notable Parliamentarians and Bunyan would have had there the opportunity to see detachments of the New Model Army in transit and to hear radical Puritan preachers.

1647   Bunyan's company is disbanded and he returns to Elstow.

1648/9?   Bunyan marries his first wife (her name is unknown), 'whose Father was counted godly'. 'Though we came together as poor as poor might be ... yet this she had for her part, *The Plain Mans Path-way to Heaven*, and *The Practice of Piety*.' These were two of the century's most popular works of practical and devotional divinity, by Arthur Dent (1601) and Lewis Bayly (1612) respectively. Four children are born to this marriage (the eldest, a blind daughter, Mary, in 1650).

1649/50–3 Bunyan undergoes the prolonged spiritual crisis later recounted in *Grace Abounding*.

post 1650 Bunyan meets, and is counselled by, John Gifford, minister to the open communion Baptist church at Bedford.

1654 Bunyan joins John Gifford's church.

1655 Bunyan moves to Bedford and begins, 'in private' and 'with much weakness and infirmity', to 'discover my Gift' to the Bedford congregation. John Gifford dies, and is succeeded by John Burton.

1656 Bunyan is 'appointed' to a more ordinary and publick preaching the Word'. Disputes with Quakers result in his first publication, *Some Gospel-Truths Opened*, with a commendatory preface by John Burton.

1658 Bunyan's first wife dies. *A Few Sighs from Hell*, his first evangelistic piece, is published.

1659 Bunyan marries his second wife, Elizabeth, who will bear him three children. *The Doctrine of the Law and Grace Unfolded*, his most substantial theological treatise, is published. The reputation of his preaching is by now considerable.

1660 Bunyan is one of the first Puritans to experience the consequences of the reinstatement of pre-1642 legislation and the reimposition of episcopal authority following the Restoration. He is arrested in November for holding a conventicle (that is, an illegal religious meeting). John Burton dies; the Bedford church becomes a separatist (or nonconformist) congregation, meeting in secret during the 'Great Persecution' promoted by the 'Clarendon Code' of penal legislation enacted against nonconformists following the re-establishment of the episcopal Church of England by the Act of Uniformity (1662).

1661 In January Bunyan is sentenced to imprisonment in Bedford jail, initially for a term of three months, but his continued refusal to give assurance that he would refrain from preaching if released prolongs the term to twelve years. A petition by his wife for his release in August is unsuccessful.

1661–72   During his imprisonment Bunyan is allowed occasional freedom and attends some meetings of the Bedford church. In prison, he supports his family by making shoe laces, preaches to the inmates and spends much time writing.

1661   *Profitable Meditations*, Bunyan's first prison book, is published.

1663   Samuel Fenn and John Whiteman elected pastors of the Bedford church. *Christian Behaviour* published.

1665   *The Holy City* published.

1666   Bunyan publishes his spiritual autobiography, *Grace Abounding*.

1667–72   Bunyan is probably working on *The Pilgrim's Progress* (see below, p. 1 and n.).

1672   John Whiteman dies. On 21 January, Bunyan is elected pastor of the Bedford church, in March he is released and on 9 May he is licensed to preach under Charles II's Declaration of Indulgence. (This release may have interrupted the composition of *The Pilgrim's Progress*: see below, p. 101 and n.) The Bedford church purchases a barn, which is licensed as a Congregational meeting place, and looks forward to a more settled corporate life.

1672–88   Bunyan's pastoral diligence and evangelistic zeal, and the spiritual authority he exercises in Hertforshire, Cambridgeshire and Bedfordshire, earn him the nickname 'Bishop Bunyan'. He preaches still further afield (including London) but declines invitations to move from Bedford, where, it seems, he still works as a tinker, at least occasionally.

mid 1670s   Bunyan seeks friends' advice on the advisability of publishing *The Pilgrim's Progress* (see below, p. 2).

1676   Bunyan's father dies.

1677   Bunyan is imprisoned during the first six months of the year.

1678   *The Pilgrim's Progress* is published.

1680   *The Life and Death of Mr. Badman* is published, as a sequel to *The Pilgrim's Progress*.

1682   *The Holy War* is published. *The Second Part of The Pilgrim's Progress* is published by 'T.S.' (the Baptist Thomas Sherman), possibly one of the works which prompts Bunyan to write his own continuation (see below, p. 136 and n.).

1684   *The Second Part of The Pilgrim's Progress* is published.

1685	Bunyan makes a deed of gift of all his property to his wife, a protection for his family against the consequences of possible fining, confiscation and imprisonment during the 'Tory Revenge', a period of renewed persecution of nonconformists in the early 1680s (in reaction to the Exclusion Crisis) which continued into James II's reign.

1686	*A Book for Boys and Girls* is published.

1688	Riding on horseback in heavy rain from Reading to London, Bunyan contracts a fever, and dies on 31 August at the home of his London friend John Strudwick. He is buried in Bunhill Fields, Finsbury, on 3 September.

THE PILGRIM'S PROGRESS

FROM THIS WORLD
TO THAT WHICH IS
TO COME

Joseph Middleton
1784

THE
Pilgrim's Progress
FROM
THIS WORLD,
TO
That which is to come:

Delivered under the Similitude of a

DREAM

Wherein is Discovered,
The manner of his setting out,
His Dangerous Journey; And safe
Arrival at the Desired Countrey.

I have used Similitudes, Hos. 12. 10.

By *John Bunyan.*

Licensed and Entred according to Order.

LONDON,
Printed for *Nath. Ponder* at the *Peacock*
in the *Poultrey* near *Cornhil,* 1678.

THE AUTHOR'S *APOLOGY*[†]
FOR HIS BOOK

WHEN at the first I took my Pen in hand,[†]
Thus for to write; I did not understand
That I at all should make a little Book
In such a mode; Nay, I had undertook
To make another, which when almost done,
Before I was aware, I this begun.

And thus it was: I writing of the Way
And Race of Saints in this our Gospel-Day,[†]
Fell suddenly into an Allegory
About their Journey, and the way to Glory,
In more than twenty things, which I set down;
This done, I twenty more had in my Crown,
And they again began to multiply,
Like sparks that from the coals of Fire do flie.
Nay then, thought I, if that you breed so fast,
I'll put you by your selves, lest you at last
Should prove ad infinitum, and eat out
The Book that I already am about.

Well, so I did; but yet I did not think
To shew to all the World my Pen and Ink
In such a mode; I only thought to make
I knew not what: nor did I undertake
Thereby to please my Neighbour; no not I,
I did it mine own self to gratifie.

Neither did I but vacant seasons spend
In this my Scribble; Nor did I intend
But to divert my self in doing this,
From worser thoughts, which make me do amiss.

Thus I set Pen to Paper with delight,
And quickly had my thoughts in black and white.
For having now my Method by the end;
Still as I pull'd, it came;[†] and so I penn'd

It down, until it came at last to be
For length and breadth the bigness which you see.

Well, when I had thus put mine ends together,
I shew'd them others, that I might see whether
They would condemn them, or them justifie:
And some said, let them live; some, let them die:
Some said, John, print it; others said, Not so:
Some said, It might do good; others said, No.

Now was I in a straight, and did not see
Which was the best thing to be done by me:
At last I thought, Since you are thus divided,
I print it will, and so the case decided.

For, thought I; Some I see would have it done,
Though others in that Channel do not run;
To prove then who advised for the best,
Thus I thought fit to put it to the test.

I further thought, if now I did deny
Those that would have it thus, to gratifie,
I did not know, but hinder them I might,
Of that which would to them be great delight.

For those that were not for its coming forth;
I said to them, Offend you I am loth;
Yet since your Brethren pleased with it be,
Forbear to judge, till you do further see.

If that thou wilt not read, let it alone;
Some love the meat, some love to pick the bone:
Yea, that I might them better palliate,
I did too with them thus Expostulate.

May I not write in such a stile as this?
In such a method too, and yet not miss
Mine end, thy good? why may it not be done?
Dark Clouds bring Waters, when the bright bring none;
Yea, dark, or bright, if they their silver drops
Cause to descend, the Earth, by yielding Crops,
Gives praise to both, and carpeth not at either,
But treasures up the Fruit they yield together:
Yea, so commixes both, that in her Fruit

None can distinguish this from that, they suit
Her well, when hungry: but if she be full,
She spues out both, and makes their blessings null.

You see the ways the Fisher-man doth take
To catch the Fish; what Engins doth he make?
Behold! how he ingageth all his Wits;
Also his Snares, Lines, Angles, Hooks and Nets:
Yet Fish there be, that neither Hook, nor Line,
Nor Snare, nor Net, nor Engin can make thine;
They must be grop'd for, and be tickled too,
Or they will not be catcht, what e're you do.

How doth the Fowler seek to catch his Game,
By divers means, all which one cannot name?
His Gun, his Nets, his Lime-twigs, light and bell:
He creeps, he goes, he stands; yea, who can tell
Of all his postures? Yet there's none of these
Will make him master of what Fowls he please.
Yea, he must Pipe, and Whistle to catch this;
Yet if he does so, that Bird he will miss.

If that a Pearl may in a Toads-head dwell,[†]
And may be found too in an Oister-shell;
If things that promise nothing, do contain
What better is then Gold; who will disdain,
(That have an inkling of it,) there to look,
That they may find it? Now my little Book,
(Tho void of all those paintings that may make
It with this or the other man to take,)
Is not without those things that do excel,
What do in brave, but empty notions dwell.

Well, yet I am not fully satisfied,
That this your Book will stand, when soundly try'd.

Why, what's the matter? It is dark, *what tho?*
But it is feigned, what of that I tro?
Some men by feigning words as dark as mine,
Make truth to spangle, and its rayes to shine.

But they want solidness: Speak man thy mind:
They drown'd the weak; Metaphors make us blind.

Solidity, indeed becomes the Pen
Of him that writeth things Divine to men:
But must I needs want solidness, because
By Metaphors I speak; was not Gods Laws,
His Gospel-laws in older time held forth
By Types,† Shadows and Metaphors? Yet loth
Will any sober man be to find fault
With them, lest he be found for to assault
The highest Wisdom. No, he rather stoops,
And seeks to find out what by pins and loops,
By Calves, and Sheep; by Heifers, and by Rams;
By Birds and Herbs, and by the blood of Lambs;†
God speaketh to him: And happy is he
That finds the light, and grace that in them be.

Be not too forward therefore to conclude,
That I want solidness; that I am rude:
All things solid in shew, not solid be;
All things in parables despise not we,
Lest things most hurtful lightly we receive;
And things that good are, of our souls bereave.

My dark and cloudy words they do but hold
The Truth, as Cabinets inclose the Gold.

The Prophets used much by Metaphors
To set forth Truth; Yea, who so considers
Christ, his Apostles too, shall plainly see,
That Truths to this day in such Mantles be.

Am I afraid to say that holy Writ,
Which for its Stile, and Phrase, puts down all Wit,
Is every where so full of all these things,
(Dark Figures, Allegories,) yet there springs
From that same Book that lustre, and those rayes
Of light, that turns our darkest nights to days.

Come, let my Carper, to his Life now look,
And find There darker Lines, then in my Book
He findeth any. Yea, and let him know,
That in his best things there are worse lines too.

May we but stand before impartial men,

To his poor One, I durst adventure Ten,
That they will take my meaning in these lines
Far better then his lies in Silver Shrines.†
Come, Truth, although in Swadling-clouts, I find
Informs the Judgement, rectifies the Mind,
Pleases the Understanding, makes the Will
Submit; the Memory too it doth fill
With what doth our Imagination please;
Likewise, it tends our troubles to appease.

Sound words I know Timothy is to use;†
And old Wives Fables he is to refuse,
But yet grave Paul him no where doth forbid
The use of Parables; in which lay hid
That Gold, those Pearls, and precious stones that were
Worth digging for; and that with greatest care.

Let me add one word more, O Man of God!
Art thou offended? dost thou wish I had
Put forth my matter in another dress,
Or that I had in things been more express?
Three things let me propound, then I submit
To those that are my betters, (as is fit.)

1. I find not that I am denied the use
Of this my method, so I no abuse
Put on the Words, Things, Readers, or be rude
In handling Figure, or Similitude,
In application; but, all that I may,
Seek the advance of Truth, this or that way:
Denyed did I say? Nay, I have leave,
(Example too, and that from them that have
God better pleased by their words or ways,
Then any Man that breatheth now adays,)
Thus to express my mind, thus to declare
Things unto thee that excellentest are.

2. I find that men (as high as Trees) will write
Dialogue-wise;†yet no Man doth them slight
For writing so: Indeed if they abuse
Truth, cursed be they, and the craft they use

To that intent; but yet let Truth be free
To make her Salleys upon Thee, and Me,
Which way it pleases God. For who knows how,
Better then he that taught us first to Plow,†
To guide our Mind and Pens for his Design?
And he makes base things usher in Divine.

   3. *I find that holy Writ in many places,*
Hath semblance with this method, where the cases
Doth call for one thing to set forth another:
Use it I may then, and yet nothing smother
Truths golden Beams; Nay, by this method may
Make it cast forth its rayes as light as day.

   And now, before I do put up my Pen,
I'le shew the profit of my Book, and then
Commit both thee, and it unto that hand
That pulls the strong down, and makes weak ones stand.

   This Book it chaulketh out before thine eyes,
The man that seeks the everlasting Prize:†
It shews you whence he comes, whither he goes,
What he leaves undone; also what he does:
It also shews you how he runs, and runs,
Till he unto the Gate of Glory comes.

   It shews too, who sets out for life amain,
As if the lasting Crown they would attain:
Here also you may see the reason why
They loose their labour, and like fools do die.

   This Book will make a Travailer of thee,
If by its Counsel thou wilt ruled be;
It will direct thee to the Holy Land,
If thou wilt its Directions understand:
Yea, it will make the sloathful, active be;
The Blind also, delightful things to see.

   Art thou for something rare, and profitable?
Wouldest thou see a Truth within a Fable?
Art thou forgetful? wouldest thou remember
From New-years-day to the last of December?
Then read my fancies, they will stick like Burs,

And may be to the Helpless, Comforters.
   This Book is writ in such a Dialect,
As may the minds of listless men affect:
It seems a Novelty, and yet contains
Nothing but sound and honest Gospel-strains.

   Wouldst thou divert thy self from Melancholly?
Would'st thou be pleasant; yet be far from folly?
Would'st thou read Riddles, and their Explanation,
Or else be drownded in thy Contemplation?
Dost thou love picking-meat? or would'st thou see
A man i' the Clouds, and hear him speak to thee?
Would'st thou be in a Dream, and yet not sleep?
Or would'st thou in a moment Laugh and Weep?
Wouldest thou loose thy self, and catch no harm?
And find thy self again without a charm?
Would'st read thy self, and read thou know'st not what
And yet know whether thou art blest or not,
By reading the same lines? O then come hither,
And lay my Book, thy Head and Heart together.

JOHN BUNYAN

THE PILGRIM'S PROGRESS:
IN THE SIMILITUDE
OF A DREAM

As I walk'd through the wilderness of this world, I lighted on a
certain place, where was a *Denn; And I laid me down in that place
to sleep: And as I slept I dreamed a Dream. I dreamed, and behold
I saw a Man *clothed with Raggs standing in a certain place, with his face
from his own House, a Book in his hand, and a great burden upon his Back.
I looked, and saw him open the Book, and Read therein; and as he
read, he wept and trembled: and not being able longer to contain,
he brake out with a lamentable cry; saying, *what shall I do?

In this plight therefore he went home, and refrained himself as
long as he could, that his Wife and Children should not perceive his
distress; but he could not be silent long, because that his trouble
increased: wherefore at length he brake his mind to his Wife and
Children; and thus he began to talk to them, O my dear Wife, said
he, and you the Children of my bowels, I your dear friend am in my self
undone, by reason of a burden that lieth hard upon me: moreover, I am for
certain informed, that this our City will be burned with fire from Heaven,
in which fearful overthrow, both my self, with thee, my Wife, and you my
sweet babes, shall miserably come to ruine; except (the which, yet I see not)
some way of escape can be found, whereby we may be delivered. At this his
Relations were sore amazed; not for that they believed, that what
he said to them was true, but because they thought, that some
frenzy distemper had got into his head: therefore, it drawing towards night, and they hoping that sleep might settle his brains,
with all hast they got him to bed; but the night was as troublesome
to him as the day: wherefore instead of sleeping, he spent it in
sighs and tears. So when the morning was come, they would know
how he did; and he told them worse and worse. He also set to
talking to them again, but they began to be hardened; *they also
thought to drive away his distemper by harsh and surly carriages

The *Gaol.

* Isa. 64. 6.
Lu. 14. 33.
Psal. 38. 4.
Hab. 2. 2.
Act. 16. 31.

* His Out-
cry.

* Carnal
Physick for
a Sick Soul.

to him: sometimes they would deride, sometimes they would chide, and sometimes they would quite neglect him: wherefore he began to retire himself to his Chamber to pray for, and pity them; and also to condole his own misery: he would also walk solitarily in the Fields, sometimes reading, and sometimes praying: and thus for some days he spent his time.

Now, I saw upon a time, when he was walking in the Fields, that he was (as he was wont) reading in his Book, and greatly distressed in his mind; and as he read, he burst out, as he had done before, crying, *What shall I do to be saved?*

I saw also that he looked this way, and that way, as if he would run; yet he stood still, because, as I perceived, he could not tell which way to go. I looked then, and saw a man named *Evangelist* coming to him, and asked, *Wherefore dost thou cry?* He answered, Sir, I perceive, by the Book in my hand, that I am Condemned to die, and *after that to come to Judgment; and I find that I am not *willing to do the first, nor *able to do the second. • Heb. 9. 27. • Job 16. 21, 22. • Ezek. 22. 14.

Then said *Evangelist,*† Why not willing to die? since this life is attended with so many evils? The Man answered, Because I fear that this burden that is upon my back, will sink me lower then the Grave; and I shall fall into *Tophet. And Sir, if I be not fit to go to Prison, I am not fit (I am sure) to go to Judgment, and from thence to Execution; and the thoughts of these things make me cry. • Isa. 30. 33.

Then said *Evangelist*, If this be thy condition, why standest thou still? He answered, Because I know not whither to go, Then he gave him a *Parchment-Roll, and there was written within, *Fly from the wrath to come. • *Conviction of the necessity of flying.* • Mat. 3. 7.

The Man therefore Read it, and looking upon *Evangelist* very carefully; said, Whither must I fly? Then said *Evangelist*, pointing with his finger over a very wide Field, Do you see yonder *Wicket-gate?*†The Man said, No. Then said the other, Do you see yonder *shining light? He said, I think I do. Then said *Evangelist*, Keep that light in your eye, and go up directly thereto,*so shalt thou see the Gate; at which when thou knockest, it shall be told thee what thou shalt do. • Mat. 7. Psal. 119. 105. • 2 Pet. 1. 19. • *Christ and the way to him cannot be found without the Word.*

So I saw in my Dream, that the Man began to run; Now he had not run far from his own door, but his Wife and Children perceiving

• Luke 14.
26.
• Gen. 19.
17.
• *They that
fly from the
wrath to
come, are a
Gazing-
Stock to the
world.*
Jer. 20. 10.
• Obstinate
and Pliable
follow him.
it, began to cry after him to return: *but the Man put his fingers in his Ears, and ran on crying, Life, Life, Eternal Life: so he looked not behind him, *but fled towards the middle of the Plain.

The Neighbours also came out to *see him run, and as he ran, some mocked, others threatned; and some cried after him to return: Now among those that did so, there were two that were resolved to fetch him back by force. *The name of the one was *Obstinate*, and the name of the other *Pliable*. Now by this time the Man was got a good distance from them; But however they were resolved to pursue him; which they did and in little time they over-took him. Then said the Man, Neighbours, *Wherefore are you come?* They said, To perswade you to go back with us; but he said, That can by no means be: You dwell, said he, in the City of *Destruction*,†(the place also where I was born) I see it to be so; and dying there, sooner or later, you will sink lower then the Grave, into a place that burns with Fire and Brimstone: Be content good Neighbours, and go along with me.

• *Obstinate.*
**What!* said *Obstinate*, and leave our Friends, and our Comforts behind us!*

• *Christian.*
• 2 Cor. 4.
18.
• Luke 15.
**Yes, said *Christian*,† (for that was his name) because, that all, which you shall forsake, is not *worthy to be compared with a little of that that I am seeking to enjoy, and if you will go along with me, and hold it, you shall fare as I my self; for there where I go, is *enough, and to spare; Come away, and prove my words.

Obst. *What are the things you seek, since you leave all the world to find them?*

• 1 Pet. 1.
4.
• Heb. 11.
16.
Chr. I seek an *Inheritance, incorruptible, undefiled, and that fadeth not away*; and it is laid up in Heaven, *and fast there, to be bestowed at the time appointed, on them that diligently seek it. Read it so, if you will, in my Book.

Obst. *Tush*, said *Obstinate*, *away with your Book; will you go back with us, or no?*

• Luke 9.
62.
Chr. No, not I, said the other; because I have laid my hand to the *Plow.

Obst. *Come then, Neighbour Pliable, let us turn again, and go home without him; there is a company of these Craz'd-headed Coxcombs, that when they take a fancy by the end, are wiser in their own eyes then seven men that can render a reason.*

Pli. Then said *Pliable*, Don't revile; if what the good *Christian* says is true, the things he looks after are better then ours; my heart inclines to go with my Neighbour.

Obst. What! more Fools still? be ruled by me and go back; who knows whither such a brain-sick fellow will lead you? Go back, go back, and be wise.

Chr. *Come with me Neighbour *Pliable*, there are such things to be had which I spoke of, and many more Glories besides; If you believe not me, read here in this Book; and for the truth of what is exprest therein, behold, all is confirmed by the *blood of him that made it.

Pli. * *Well Neighbour* Obstinate *(said* Pliable*) I begin to come to a point; I intend to go along with this good man, and to cast in my lot with him: But my good Companion, do you know the way to this desired place?*

Chr. I am directed by a man whose name is *Evangelist*, to speed me to a little Gate that is before us, where we shall receive instruction about the way.

Pli. Come then, good Neighbour, let us be going. Then they went both together.

Obst. And I will go back to my place, said *Obstinate:* I will be no Companion of such mis-led fantastical Fellows.

Now I saw in my Dream, that when *Obstinate* was gon back, *Christian* and *Pliable* went *talking over the Plain; and thus they began their discourse,

Chr. Come Neighbour *Pliable*, how do you do? I am glad you are perswaded to go along with me; and had even *Obstinate* himself, but felt what I have felt of the Powers, and Terrours of what is yet unseen, he would not thus lightly have given us the back.

Pliable. Come Neighbour Christian, *since there is none but us two here, tell me now further, what the things are: and how to be enjoyed, whither we are going.*

Chr. I can better conceive of them with my Mind, then speak of them with my Tongue: But yet since you are desirous to know, I will read of them in my Book.

Pli. And do you think that the words of your Book are certainly true?

Chr. Yes verily, for it was made by him that *cannot lye.

Pli. Well said; what things are they?

Marginal notes:
*Christian and Obstinate pull for Pliable's Soul.
* Heb. 13. 20, 21.
* Pliable concented to go with Christian.
* Talk between Christian, and Pliable.
* Tit. 1. 2.

*Isa. 45.
17.
John 10.
27, 28, 29.

Chr. There is an *endless Kingdom to be Inhabited, and ever-lasting life to be given us;.that we may Inhabit that Kingdom for ever.

Pli. Well said, and what else?

*2 Tim. 4.
8.
Rev. 3. 4.
Matth. 13.

Chr. There are Crowns of Glory to be given us; *and Garments that will make us shine like the Sun in the Firmament of Heaven.

Pli. This is excellent; And what else?

*Isa. 25. 8.
Rev. 7. 16,
17.

Chr. There shall be no more crying, *nor sorrow; For he that is owner of the place, will wipe all tears from our eyes.

Chap. 21.
4.

Pli. And what company shall we have there?

*Isa. 6. 2.
1 Thess. 4.
16, 17.
Rev. 5. 11.

Chr. There we shall be with *Seraphims*, *and *Cherubins*, Creatures that will dazle your eyes to look on them: There also you shall meet with thousands, and ten thousands that have gone before us to that place; none of them are hurtful, but loving, and holy; every one walking in the sight of God; and standing in his presence with

*Rev. 4. 4.
*Chap. 14.
1, 2, 3, 4, 5.
*John 12.
25.

acceptance for ever: In a word, there we shall see the *Elders with their Golden Crowns: There we shall see the Holy *Virgins with their Golden Harps. There we shall see *Men that by the World were cut in pieces, burnt in flames, eaten of Beasts, drownded in the Seas, for the love that they bare to the Lord of the place; all well,

*2 Cor. 5.
2, 3, 5.

and cloathed with *Immortality, as with a Garment.

Pli. The hearing of this is enough to ravish ones heart; but are these things to be enjoyed? how shall we get to be Sharers hereof?

*Isa. 55.
12.
John 7. 37.

Chr. The Lord, the Governour of that Countrey, hath Recorded that *in this Book: The substance of which is, If we be truly willing to have it, he will bestow it upon us freely.

Chap. 6.
37.
Rev. 21. 6.
Chap. 22.
17.

Pli. Well, my good Companion, glad am I to hear of these things: Come on, let us mend our pace.

Chr. I cannot go so fast as I would, by reason of this burden that is upon my back.

Now I saw in my Dream, that just as they had ended this talk, they drew near to a very *Miry Slow* that was in the midst of the Plain, and they being heedless, did both fall suddenly into the bogg. The name of the Slow was *Dispond.*† Here therefore they wallowed for a time, being grieviously bedaubed with the dirt; And *Christian*, because of the burden that was on his back, began to sink in the Mire.

Pli. Then said Pliable, *Ah, Neighbour* Christian, *where are you now?*

Chr. Truly, said *Christian,* I do not know.

Pli. At that *Pliable* began to be offended; and angerly, said to his Fellow, *Is this the happiness you have told me all this while of? if we have such ill speed at our first setting out, What may we expect, 'twixt this and our Journeys end?* *May I get out again with my life, you shall possess the brave Country alone for me.* And with that he gave a desperate struggle or two, and got out of the Mire, on that side of the Slow which was next to his own House: So away he went, and *Christian* saw him no more.

• *It is not enough to be pliable.*

Wherefore *Christian* was left to tumble in the Slow of *Dispond* alone; but still he endeavoured to struggle to that side of the Slow, that was still further *from his own House, and next to the Wicket-gate; the which he did, but could not get out, because of the burden that was upon his back: But I beheld in my Dream, that a Man came to him, whose name was *Help,* and asked him, *What he did there?*

• *Christian in trouble, seeks still to get further from his own House.*

Chr. Sir, said *Christian,* I was bid go this way, by a Man called *Evangelist,* who directed me also to yonder Gate, that I might escape the wrath to come: And as I was going thither, I fell in here.

*Help. But why did you not look for *the steps?*

• *The Promises.*†

Chr. Fear followed me so hard, that I fled the next way, and fell in.

Help. Then, said he, *Give me thy hand;* so he gave him his hand, and *he drew him out, and set him upon sound ground, and bid him go on his way.

• *Help lifts him out.*
• *Psal. 40. 2.*

Then I stepped†to him that pluckt him out, and said; Sir, Wherefore (since over this place is the way from the City of *Destruction,* to yonder *Gate*) is it, that *this* Plat is not mended, that poor Travellers might go thither with more security? And he said unto me, this *Miry slow,* is such a place as cannot be mended: It is the descent whither the *scum and filth that attends conviction for sin doth continually run, and therefore is it called the *Slow of Dispond*: for still as the sinner is awakened about his lost condition, there ariseth in his soul many fears, and doubts, and discouraging apprehensions, which all of them get together, and settle in this place: And this is the reason of the badness of this ground.

• *What makes the Slow of Dispond.*

It is not the *pleasure of the King that this place should remain so bad; his Labourers also, have by the direction of His Majesties

• *Isa.* 35. 3, 4.

Surveyors, been for above this sixteen hundred years, imploy'd about this patch of ground, if perhaps it might have been mended: yea, and to my knowledge, saith he, *Here* hath been swallowed up, at least, Twenty thousand Cart Loads; yea Millions of wholesom Instructions, that have at all seasons been brought from all places of the Kings Dominions; (and they that can tell, say, they are the best Materials to make good ground of the place,) If so be it might have been mended, but it is the *Slow of Dispond* still; and so will be, when they have done what they can.

True, there are by the direction of the Law-giver, certain good and substantiall *steps, placed even through the very midst of this *Slow*; but at such time as this place doth much spue out its filth, as it doth against change of weather, these steps are hardly seen; or if they be, Men through the diziness of their Heads, step besides; and then they are bemired to purpose, notwithstanding the steps be there; but the ground is *good when they are once got in at the Gate.

Now I saw in my Dream, that by this time *Pliable was got home to his House again. *So his Neighbours came to visit him; and some of them called him wise Man for coming back; and some called him Fool for hazarding himself with *Christian*; others again did mock at his Cowardliness; saying, Surely since you began to venture, I would not have been so base to have given out for a few difficulties. So *Pliable* sat sneaking among them. But at last he got more confidence, and then they all turned their tales, and began to deride poor *Christian* behind his back. And thus much concerning *Pliable.*

Now as Christian was walking solitary by himself, he espied one afar off, come crossing over the field *to meet him; and their hap was to meet just as they were crossing the way of each other. The Gentleman's name was, Mr. *Worldly-Wiseman,* he dwelt in the Town of *Carnal-Policy,* a very great Town, and also hard by, from whence Christian came. This man then meeting with Christian, and having some inckling of him, for Christians setting forth from the City of *Destruction,* was much noised abroad, not only in the Town, where he dwelt, but also it began to be the *Town*-talk in some other places, Master *Worldly-Wiseman* therefore, having some guess of him, by beholding his laborious going, by observing his sighs and groans, and the like; began thus to enter into some talk with *Christian.*

• The Promises of forgiveness and acceptance to life by Faith in Christ.

• 1 Sam. 12.
23.

• Plyable got home, and is visited of his Neighbours.

• His entertainment by them at his return.

• Mr. Worldly-Wiseman meets with Christian.

Worl. *How now, good fellow, whither away after this burdened manner?*   *Talk betwixt Mr. Worldly-Wiseman and Christian.*

Chr. A burdened manner indeed, as ever I think poor creature had. And whereas you ask me, *Whither away*, I tell you, Sir, I am going to yonder Wicket-gate before me; for there, as I am informed, I shall be put into a way to be rid of my heavy burden.

Worl. *Hast thou a Wife and Children?*

Chr. Yes, but I am so laden with this burden, that I cannot take that pleasure in them as formerly: methinks, I am as *if I had none.   • I Cor. 7. 29.

Worl. *Wilt thou hearken to me, if I give thee counsel?*

Chr. If it be good, I will; for I stand in need of good counsel.

Worl. **I would advise thee then, that thou with all speed get thy self rid of thy burden; for thou wilt never be settled in thy mind till then: nor canst thou enjoy the b∷n fits of the blessing which God hath bestowed upon thee till then.*   • Mr. Worldly-Wiseman's Counsel to Christian.

Chr. That is that which I seek for, even to be rid of this heavy burden; but get it off my self I cannot: nor is there a man in our Country that can take it off my shoulders; therefore am I going this way, as I told you, that I may be rid of my burden.

Worl. *Who bid thee go this way to be rid of thy burden?*

Chr. A man that appeared to me to be a very great and honorable person; his name, as I remember is *Evangelist*.

Worl. **I beshrow him for his counsel; there is not a more dangerous and troublesome way in the world, than is that unto which he hath directed thee; and that thou shalt find, if thou wilt be ruled by his counsel: Thou hast met with something (as I perceive) already; for I see the dirt of the Slow of Dispond is upon thee; but that Slow is the beginning of the sorrows that do attend those that go on in that way: hear me, I am older than thou! thou art like to meet with in the way which thou goest, Wearisomness, Painfulness, Hunger; Perils, Nakedness, Sword, Lions, Dragons, Darkness; and in a word, death,† and what not? These things are certainly true, having been confirmed by many testimonies. And why should a man so carelessly cast away himself, by giving heed to a stranger.*   • Mr. Worldly-Wiseman Condemned Evangelists Counsel.

Chr. Why, Sir, this burden upon my back is more terrible to me than are all these things which you have mentioned: *nay, methinks I care not what I meet with in the way, so be I can also meet with deliverance from my burden.   • The frame of the heart of young Christians.

Worl. *How camest thou by thy burden at first?*

Chr. By reading this Book in my hand.

• Worldly-
Wiseman
*does not like
that Men
should be
Serious in
reading the
Bible.*

Worl. **I thought so; and it is happened unto thee as to other weak men, who meddling with things too high for them, do suddenly fall into thy distractions; which distractions do not only unman men, (as thine I perceive has done thee) but they run them upon desperate ventures, to obtain they know not what.*

Chr. I know what I would obtain; it is ease for my heavy burden.

Worl. *But why wilt thou seek for ease this way, seeing so many dangers attend it, especially, since (hadst thou but patience to hear me) I could direct thee to the obtaining of what thou desirest, without the dangers that thou in this way wilt run thy self into: yea, and the remedy is at hand. Besides, I will add, that instead of those dangers, thou shalt meet with much safety, friendship, and content.*

Chr. Pray Sir open this secret to me.

• *Whether
Mr.
Worldly
prefers
Morality
before the
Straight
Gate.*

Worl. **Why in yonder Village, (the Village is named* Morality*) there dwells a Gentleman, whose name is* Legality*, a very judicious man (and a man of a very good name) that has skill to help men off with such burdens as thine are, from their shoulders: yea, to my knowledge he hath done a great deal of good this way:* Ai*, and besides, he hath skill to cure those that are somewhat crazed in their wits with their burdens. To him, as I said, thou mayest go, and be helped presently. His house is not quite a mile from this place; and if he should not be at home himself, he hath a pretty young man to his Son, whose name is* Civility*, that can do it (to speak on) as well as the old Gentleman himself: There, I say, thou mayest be eased of thy burden, and if thou art not minded to go back to thy former habitation, as indeed I would not wish thee, thou mayest send for thy wife and Children to thee to this Village, where there are houses now stand empty, one of which thou mayest have at reasonable rates: Provision is there also cheap and good, and that which will make thy life the more happy, is, to be sure there thou shalt live by honest neighbors, in credit and good fashion.*

• Christian
*Snared by
Mr.
Worldly
Wisemans
Word.*

*Now was *Christian* somewhat at a stand, but presently he concluded; if this be true which this Gentleman hath said, my wisest course is to take his advice, and with that he thus farther spoke.

Chr. Sir, which is my way to this honest man's house?

• *Mount
Sinai.*

Worl. *Do you see yonder* **high hill?*†

Chr. Yes, very well.

Worl. By that *Hill* you must go, and the first house you come at is his.

So *Christian* turned out of his way to go to Mr. *Legality's* house for help: but behold, when he was got now hard by the *Hill*, it seemed so high, and also that side of it that was next the way side, did hang so much over, that Christian was *afraid to venture fur- ther, lest the *Hill* should fall on his head: wherefore there he stood still, and wotted not what to do. Also his burden, *now*, seemed heavier to him, than while he was in his way. There came also *flashes of fire out of the Hill, that made *Christian afraid that he should be burned: here therefore he swet, and did quake for *fear. And now he began to be sorry that he had taken Mr. *Worldly- Wisemans* counsel; and with that he saw *Evangelist coming to meet him; at the sight also of whom he began to blush for shame. So *Evangelist* drew nearer, and nearer, and coming up to him, he looked upon him with a severe and dreadful countenance: and thus began to reason with *Christian*.

Evan. *What doest thou here? said he: at which word *Christian* knew not what to answer: wherefore, at present he stood speechless before him. Then said *Evangelist* farther, *Art not thou the man that I found crying, without the walls of the City of* Destruction?

Chr. Yes, dear Sir, I am the man.

Evan. *Did not I direct thee the way to the little Wicket-gate?*

Chr. Yes, dear Sir said *Christian*.

Evan. *How is it then that thou art so quickly turned aside, for thou art now out of the way?*

Chr. I met with a Gentleman, so soon as I had got over the *Slow of Dispond*, who perswaded me, that I might in the *Village* before me, find a man that could take off my burden.

Evan. *What was he?*

Chr. He looked like a Gentleman, and talked much to me, and got me at last to yield; so I came hither: but when I beheld this Hill, and how it hangs over the way, I suddenly made a stand, lest it should fall on my head.

Evan. *What said that Gentleman to you?*

Chr. Why, he asked me whither I was going, and I told him.

Evan. *And what said he then?*

Chr. He asked me if I had a Family, and I told him: but, said I,

Christian afraid that Mount Sinai would fall on his head.

* Exod. 19. 18.
* Ver. 16.
* Heb. 12. 21.
* *Evangelist findeth Christian under Mount Sinai and looketh severely upon him.*
* *Evangelist reasons afresh with Christian.*

I am so loaden with the burden that is on my back, that I cannot take pleasure in them as formerly.

Evan. And what said he then?

Chr. He bid me with speed get rid of my burden, and I told him 'twas ease that I sought: And said I, I am therefore going to yonder *Gate* to receive further direction how I may get to the place of deliverance. So he said that he would shew me a better way, and short, not so attended with difficulties, as the way, Sir, that you set me: which way, said he, will direct you to a Gentleman's house that hath skill to take off these burdens: So I believed him, and turned out of that way into this, if haply I might be soon eased of my burden: but when I came to this place, and beheld things as they are, I stopped for fear, (as I said) of danger: but I now know not what to do.

Evan. Then (said Evangelist) *stand still a little, that I may shew thee the words of God.* So he stood trembling. *Then* (said Evangelist) **See that ye refuse not him that speaketh; for if they escaped not who refused him that spake on Earth, *much more shall not we escape, if we turn away from him that speaketh from Heaven. He said moreover, *Now the just shall live by faith; but if any man draws back, my soul shall have no pleasure in him.* He also did thus apply them, *Thou art the man that art running into this misery, thou hast began to reject the counsel of the most high, and to draw back thy foot from the way of peace, even almost to the hazarding of thy perdition.*

Then *Christian* fell down at his foot as dead, crying, Woe is me, for I am undone: at the sight of which *Evangelist* caught him by the right hand, saying, All manner of sin and blasphemies shall be forgiven unto men; be not faithless, but believing; then did *Christian* again a little revive, and stood up trembling, as at first, before *Evangelist*.

Then *Evangelist* proceeded, saying, *Give more earnest heed to the things that I shall tell thee of.* I will now shew thee who it was that deluded thee, and who 'twas also to whom he sent thee. *The man that met thee, is one *Worldly-Wiseman*, and rightly is he so called; partly, *because he favoureth only the Doctrine of this World (therefore he always goes to the Town of *Morality* to Church) and partly *because he loveth that Doctrine best, for it saveth him from the

Margin notes:
• Heb. 12. 25.
• Evangelist Convinces Christian of his Error.
• Chap. 10. 38.

Matth. 12. Mark 3.

• Mr. Worldly-Wiseman discribed by Evangelist.
• 1 John 4. 5.
• Gal. 6. 12.

Cross; and because he is of this carnal temper, therefore he seeketh to prevent my ways, though right. *Now there are three things in this mans counsel that thou must utterly abhor.†

• Evangelist *discovers the deceit of* Mr. Worldly Wiseman.

1. His turning thee out of the way.

2. His labouring to render the Cross odious to thee.

3. And his setting thy feet in that way that leadeth unto the administration of Death.

First, Thou must abhor his turning thee out of the way; yea, and thine own consenting thereto: because this is to reject the counsel of God, for the sake of the counsel of a *Worldly-Wiseman*. The Lord says, *Strive to enter in at the strait gate*, the gate to which I sent thee; *for strait is the gate that leadeth unto life, and few there be that find it*. From this little wicket-gate, and from the way thereto hath this wicked man turned thee, to the bringing of thee almost to destruction; hate therefore his turning thee out of the way, and abhor thy self for harkening to him.

• Luke 13. 24. • Mat. 7. 13, 14.

Secondly, Thou must abhor his labouring to render the Cross odious unto thee; for thou art to *prefer it before the treasures in Egypt*: besides the King of glory hath told thee, *that he that will save his life shall lose it*: and *he that comes after him, and hates not his father and mother, and wife, and children, and brethren, and sisters; yea, and his own life also, he cannot be my Disciple*. I say therefore, for a man to labour to perswade thee, that that shall be thy death, without which the truth hath said, thou canst not have eternal life, this Doctrine thou must abhor.

• Heb. 11. 25, 26. • Mark 8. 34. John 13. 25. Mat. 10. 39. • Luke 14. 26.

Thirdly, thou must hate his setting of thy feet in the way that leadeth to the ministration of death. And for this thou must consider to whom he sent thee, and also how unable that person was to deliver thee from thy burden.

He to whom thou wast sent for ease, being by name *Legality*, is the Son of the *Bond woman which now is, and is in bondage with her children, and is in a mystery this *Mount *Sinai*, which thou hast feared will fall on thy head. Now if she with her children are in bondage, how canst thou expect by them to be made free? This *Legality* therefore is not able to set thee free from thy burden. No man was as yet ever rid of his burden by him, no, nor ever is like to be: ye cannot be justified by the Works of the Law; for by the

• Gal. 4. 21, 22, 23, 24, 25, 26, 27. • *The Bond-Woman.*

deeds of the Law no man living can be rid of his burden: therefore
Mr. *Worldly-Wiseman* is an alien, and Mr. *Legality* a cheat: and for
his Son *Civility*, notwithstanding his simpering looks, he is but an
hypocrite, and cannot help thee. Believe me, there is nothing in all
this noise, that thou hast heard of this sottish man, but a design to
beguile thee of thy Salvation, by turning thee from the way in
which I had set thee. After this *Evangelist* called aloud to the Heavens
for confirmation of what he had said; and with that there came
words and fire out of the Mountain under which poor Christian
stood, that made the hair of his flesh stand. The words were thus
pronounced, *As many as are of the works of the Law, are under the curse;
for it is written, Cursed is every one that continueth not in all things which
are written in the Book of the Law to do them.*

•Gal. 3. 10.

Now *Christian* looked for nothing but death, and began to cry out
lamentably, even cursing the time in which he met with Mr.
Worldly-Wiseman, still calling himself a thousand fools for hearkening to his counsel: he also was greatly ashamed to think that this
Gentlemans arguments, flowing only from the flesh, should have
that prevalency with him as to cause him to forsake the right way.
This done, he applied himself again to *Evangelist* in words and sense
as follows.

• *Christian*
Enquired if
he may yet
be Happy.

Chr. *Sir, what think you? is there hopes? may I now go back
and go up to the *Wicket-gate*, shall I not be abandoned for this, and
sent back from thence ashamed. I am sorry I have hearkened to this
man's counsel, but may my sin be forgiven.

Evang. Then said *Evangelist* to him, Thy sin is very great, for by
it thou hast committed two evils; thou hast forsaken the way that
is good, to tread in forbidden paths: *yet will the man at the Gate
receive thee, for he has *good will* for men; only, said he, take heed
that thou turn not aside again, lest thou perish from the way when
his wrath is *kindled but a little. Then did *Christian* address himself
to go back, *and *Evangelist*, after he had kist him, gave him one
smile, and bid him God speed; so he went on with haste, neither
spake he to any man by the way; nor if any man asked him, would
he vouchsafe them an answer. He went like one that was all the
while treading on forbidden ground, and could by no means think
himself safe, till again he was got into the way which he left to

• *Evangel-*
ist comforts
him.

• Psal. 2
last.

follow Mr. *Worldly-Wiseman*'s counsel: so in process of time *Christian* got up to the Gate. Now over the Gate there was Written, *Knock and it shall be opened unto you.* *He knocked therefore, more then once • Mat. 7. 8. or twice, *saying,*

> *May I now enter here? will he within*
> *Open to sorry me, though I have bin*
> *An undeserving Rebel? then shall I,*
> *Not fail to Sing his lasting praise on high.*

At last there came a grave Person to the Gate: named *Good-will,*† who asked, *Who was there? and whence he came? and what he would have?*

Chr. Here is a poor burdened sinner, I come from the City of *Destruction*, but am going to Mount *Zion*,† that I may be delivered from the wrath to come; I would therefore, Sir, since I am informed that by this Gate is the way thither, know if you are *willing* to let me in.

Good-Will. *I am *willing* with all my heart, said he; and with that he opened the Gate.

• *The Gate will be opened to broken-hearted sinners.*

So when *Christian* was stepping in, the other gave him a pull: Then said *Christian*, what means that? The other told him, a little distance from this Gate, there is erected a strong Castle, of which * *Belzebub*† is the Captain: from thence both he, and them that are with him, Shoot Arrows at those that come up to this Gate; if happily they may die before they can enter in. Then, said *Christian*, *I rejoyce and tremble. So when he was got in, the man of the Gate asked him, Who directed him thither?

• *Satan envies those that enter the straight Gate.*
• *Christian Entred the Gate with Joy and trembling.*
• *Talke between Good Will and Christian.*

Chr. *Evangelist bid me come hither and knock, (as I did;) And he said, that you, Sir, would tell me what I must do.

Good Will. An open Door is set before thee, and no man can shut it.

Chr. Now I begin to reap the benefits of my hazzards.

Good Will. But how is it that you came alone?

Chr. Because none of my Neighbours saw their danger as I saw mine.

Good Will. Did any of them know of your coming?

Chr. Yes, my Wife and Children saw me at the first, and called after me to turn again: Also some of my Neighbours stood crying, and calling after me to return; but I put my Fingers in mine Ears, and so came on my way.

Good Will. But did none of them follow you to perswade you to go back?

Chr. Yes, both *Obstinate*, and *Pliable*: But when they saw that they could not prevail, *Obstinate* went railing back; but *Pliable* came with me a little way.

Good Will. But why did he not come through?

Chr. We indeed came both together, until we came at the Slow of *Dispond*, into the which, we also suddenly fell. And then was my Neighbour *Pliable* discouraged, and would not adventure further. *Wherefore getting out again, on that side next to his own House; he told me, I should possess the brave Countrey alone for him: So he went his way, and I came mine. He after *Obstinate*, and I to this Gate.

Good Will. Then said *Good Will*, Alas poor man, is the Cœlestial Glory of so small esteem with him, that he counteth it not worth running the hazards of a few difficulties to obtain it?

Chr. Truly, said *Christian*, I have said the truth of *Pliable*, and if I should also say all the truth of my self, it will appear there is *no betterment 'twixt him and my self. 'Tis true, he went back to his own house, but I also turned aside to go in the way of death, being perswaded thereto by the carnal arguments of one Mr. *Worldly Wiseman*.

Good Will. Oh, did he light upon you! what, he would have had you a sought for ease at the hands of Mr. *Legality*; they are both of them a very cheat: but did you take his counsel?

Chr. Yes, as far as I durst, I went to find out Mr. *Legality*, until I thought that the Mountain that stands by his house, would have fallen upon my head: wherefore there I was forced to stop.

Good Will. That Mountain has been the death of many, and will be the death of many more: 'tis well you escaped being by it dasht in pieces.

Chr. Why, truly I do not know what had become of me there, had not *Evangelist* happily met me again as I was musing in the midst of my *dumps*: but 'twas Gods mercy that he came to me again, for else I had never come hither. But now I am come, such a one as I am, more fit indeed for death by that Mountain, than thus to stand

• *A Man may have Company when he sets out for Heaven, & yet go thither alone.*

• *Christian accuseth himself before the man at the Gate.*

talking with my Lord: But Oh, what a favour is this to me, that yet I am admitted entrance here.

Good Will. *We make no objections against any, notwithstanding all that they have done before they come hither, **they in no wise are cast out*; and therefore, good *Christian*, come a little way with me, and I will teach thee about the way thou must go. *Look before thee; dost thou see this narrow way? That is the way thou must go. It was cast up by the Patriarchs, Prophets, Christ, and his Apostles, and it is as straight as a Rule can make it: This is the way thou must go.

• Christian is comforted again.
• John 6. 37.
• Christian directed yet on his way.

Chr. But said *Christian*, Is there no turnings nor windings, by which a Stranger **may loose the way?*

• Christian afraid of losing his way.

Good Will. Yes, there are many ways *Butt* down upon this; and they are Crooked, and Wide: But *thus* thou may'st distinguish the right from the wrong, *That* only being *straight and narrow.

• Mat. 7. 14.

Then I saw in my Dream, *That *Christian* asked him further, If he could not help him off with his burden that was upon his back; For as yet he had not got rid thereof, nor could he by any means get it off without help.

• Christian weary of his Burden

He told him, As to the burden, be content to bear it, until thou comest to the place of *Deliverance; for there it will fall from thy back it self.

• There is no deliverance from the guilt,

Then *Christian* began to gird up his loins, and to address himself to his Journey. So the other told him, that by that he was gone some distance from the Gate, he would come at the house of the *Interpreter*; at whose Door he should knock; and he would shew him excellent things. Then *Christian* took his leave of his Friend, and he again bid him God speed.

and burden of sin, but by the death and blood of Christ.

Then he went on, till he came at the house of the *Interpreter, where he knocked over, and over: at last one came to the Door, and asked *Who was there?*

• Christian comes to the House of the Interpreter.

Chr. Sir, here is a Travailer, who was bid by an acquaintance of the Good-man of this House, to call here for my profit: I would therefore speak with the Master of the House: so he called for the Master of the House; who after a little time came to *Christian*, and asked him what he would have?

Chr. Sir, said *Christian*, I am a Man that am come from the City of *Destruction*, and am going to the Mount *Zion*, and I was told by the Man that stands at the Gate, at the head of this way, that if I called here, you would shew me excellent things, such as would be an help to me in my Journey.

Inter. Then said the *Interpreter*, *come in, I will shew thee that which will be profitable to thee. So he commanded his man to *light the Candle, and bid *Christian* follow him; so he had him into a private Room, and bid his Man open a Door; the which when he had done, *Christian* saw a Picture of a very grave Person hang up against the wall, and this was the fashion of it, *It had eyes lift up to Heaven, the best of Books in its hand, the Law of Truth was written upon its lips, the World was behind its back; it stood as if it pleaded with Men, and a Crown of Gold did hang over its head.*

Chr. Then said Christian, *What means this?*

Inter. The Man whose Picture this is, is one of a thousand, he can *beget Children, Travel in birth with Children, and *Nurse them himself when they are born. And whereas thou seest *him with his eyes lift up to Heaven, the best of Books in his hand, and the Law of Truth writ on his Lips: it is to shew thee, that his work is to know, and unfold dark things to sinners; even as also thou seest *him stand as if he Pleaded with Men: And whereas thou seest the World as cast behind him, and that a Crown hangs over his head; that is, to shew thee, that slighting and despising the things that are present, for the love that he hath to his Masters service, he is sure in the world that comes next to have Glory for his Reward: Now, said the *Interpreter*, I have shewed thee this Picture first, *because the Man whose Picture this is, is the only Man, whom the Lord of the Place whither thou art going, hath Authorized, to be thy Guide in all difficult places thou mayest meet with in the way: wherefore take good heed to what I have shewed thee, and bear well in thy mind what thou hast seen; lest in thy Journey, thou meet with some that pretend to lead thee right, but their way goes down to death.

Then he took him by the hand, and led him into a very large *Parlour* that was full of dust, because never swept; the which, after he had reviewed a little while, the *Interpreter* called for a man to

He is entertained.
Illumination.
Christian sees a brave Picture.
The fashion of the Picture.
1 Cor. 4. 15.
Gal. 4. 19.
1 Thes. 2. 7.
The meaning of the Picture.
Why he shewed him the Picture first.

sweep: Now when he began to sweep, the dust began so abundantly to fly about, that *Christian* had almost therewith been choaked: Then said the *Interpreter* to a *Damsel* that stood by, Bring hither Water, and sprinkle the Room; which when she had done, was swept and cleansed with pleasure.

Chr. Then said Christian, *What means this?*

In. The *Interpreter* answered; This Parlor, is the heart of a Man that was never sanctified by the sweet Grace of the Gospel: The *dust*, is his Original Sin, and inward Corruptions that have defiled the whole Man. He that began to sweep at first, is the Law; but She that brought water, and did sprinkle it, is the Gospel: Now, whereas thou sawest that so soon as the first began to sweep, the dust did so fly about, that the Room by him could not be cleansed, but that thou wast almost choaked therewith, this is to shew thee, that the Law, instead of cleansing the heart (by its working) from sin, *doth revive, put *strength into, and *increase it in the soul, even as it doth discover and forbid it, for it doth not give power to subdue. • Rom. 7. 6. • 1 Cor. 15. 56. • Rom. 5. 20.

Again, as thou sawest the *Damsel* sprinkle the Room with Water, upon which it was cleansed with pleasure: This is to shew thee, that when the Gospel comes in the sweet and precious influences thereof to the heart, then I say, even as thou sawest the Damsel lay the dust by sprinkling the Floor with Water, so is sin vanquished and subdued, and the soul made clean, through the Faith of it; and consequently *fit for the King of Glory to inhabit. • John 15. 3. Ephes. 5. 26.

I saw moreover in my Dream, *that the *Interpreter* took him by the hand, and had him into a little Room, where sat two little Children, each one in his Chair: The name of the eldest was *Passion*, and of the other, *Patience*; *Passion* seemed to be much discontent, but *Patience* was very quiet. Then *Christian* asked, What is the reason of the discontent of *Passion*? The *Interpreter* answered, The Governour of them would have him stay for his best things till the beginning of the next year; but he will have all now: *But *Patience* is willing to wait. Acts 15. 9. Rom. 16. 25, 26. John 15. 13. *He shewed him Passion & Patience. Passion will have all now. * Patience is for waiting.

Then I saw that one came to *Passion*, and brought him a Bag of Treasure, and poured it down at his feet; the which he took up, and rejoyced therein, and withall, laughed *Patience* to scorn: But I *Passion has his desire,

• *And quickly lavishes all away.*
• *The matter expounded.*

And beheld but a while, and he had *lavished all away, and had nothing left him but Rags.†

Chr. Then said Christian *to the* Interpreter, *Expound this matter more fully to me.*

Int. So he said, These two Lads are Figures; *Passion*, of the Men of *this* World; and *Patience*, of the Men of *that* which is to come: For as here thou seest, *Passion will have all now*, this year; that is to say, in *this* World; *So* are the Men of this World: they must have all their good things now, they cannot stay till next *Year*; that is, untill the next World, for their Portion of good. That Proverb, *A Bird in the hand is worth two in the Bush*, is of more Authority with them, then are all the Divine Testimonies of the good of the world to come. But as thou sawest, that he had quickly lavished all away, and had presently left him, nothing but Raggs; So will it be with all such Men at the end of this world.

• *The Worldly Man for a Bird in the hand.*

Chr. Then said Christian; *Now I see that* Patience has the best *Wisdom, and that upon many accounts.* 1. *Because he stays for the best things.* 2 *And also because he will have the glory of His, when the other hath nothing but Raggs.*

• *Patience had the best Wisdom.*

In. Nay, you may add another; to wit, The glory of the *next* world will never wear out; but *these* are suddenly gone. Therefore *Passion* had not so much reason to laugh at *Patience*, because he had his good things first, as *Patience* will have to laugh at *Passion*, *because he had his best things *last; for *first* must give place to *last*, because *last* must have his time to come, but *last* gives place to *nothing*; for there is not another to succeed: he therefore that hath his Portion *first*, must needs have a time to spend it; but he that has his Portion *last*, must have it lastingly. Therefore it is said of *Dives, In thy life thou receivedst thy good things, and likewise Lazarus evil things; but now he is comforted, and thou art tormented.*

• *Things that are first must give place, but things that are last are lasting.*
• *Luk. 16. Dives had his good things first.*

Chr. Then I perceive, 'tis not best to covet things that are now, but to wait for things to come.

Int. You say the Truth, *For the things that are seen, are Temporal; but the things that are not seen, are Eternal: But though this be so, yet since things present, and our fleshly appetite, are such near Neighbours one to another; and again, because things to come, and carnal sense, are such strangers one to another: therefore it is, that the

• *2 Cor. 4. 18. The first things are but Temporal.*

first of these so suddenly fall into *amity*, and that *distance* is so continued between the second.

Then I saw in my Dream, that the *Interpreter* took *Christian* by the hand, and led him into a place, where was a Fire burning against a Wall, and one standing by it always, casting much Water upon it to quench it: Yet did the Fire burn higher and hotter.

Then said Christian, *What means this?*

The *Interpreter* answered, This fire is the work of Grace that is wrought in the heart; he that casts Water upon it, to extinguish and put it out, is the *Devil*: but in that thou seest the fire, notwithstanding, burn higher and hotter, thou shalt also see the reason of that: So he had him about to the back side of the Wall, where he saw a Man with a Vessel of Oyl in his hand, of the which he did also continually cast, but secretly, into the fire. Then said *Christian*, *What means this?* The *Interpreter* answered, This is *Christ*, who continually with the Oyl of his Grace, maintains the work already begun in the heart; by the means of which, notwithstanding what the Devil can do, the souls of his people prove gracious still. And in that thou sawest, that the Man stood behind the Wall to maintain the fire; this is to teach thee, that it is hard for the tempted to see how this work of Grace is maintained in the soul. 2 Cor. 12. 9.

I saw also that the *Interpreter* took him again by the hand, and led him into a pleasant place, where was builded a stately Palace, beautiful to behold; at the sight of which, *Christian* was greatly delighted; he saw also upon the top thereof, certain Persons walked, who were cloathed all in gold. Then said *Christian*, May we go in thither? Then the *Interpreter* took him, and led him up toward the door of the Palace; and behold, at the door, stood a great company of men, as desirous to go in, but durst not. There also sat a Man, at a little distance from the door, at a Table-side, with a Book, and his Inkhorn before him, to take the Name of him that should enter therein: He saw also that in the doorway, stood many Men in Armour to keep it, being resolved to do to the Man that would enter, what hurt and mischief they could. Now was *Christian* somwhat in a muse; at last, when every Man started back for fear of the Armed Men; *Christian* saw a man of a very stout countenance come up to the Man that sat there to write; saying, *Set down my Name Sir*; * The valiant man. †

the which when he had done, he saw the Man draw his Sword, and put an Helmet upon his Head, and rush toward the door upon the Armed Men, who laid upon him with deadly force; but the Man, not at all discouraged, fell to cutting and hacking most fiercely; so after he had *received and given many wounds to those that at-tempted to keep him out, he cut his way through them all, and pressed forward into the Palace; at which there was a pleasant voice heard from those that were within, even of the Three†that walked upon the top of the Palace, saying,

* Acts 14. 22.

> Come in, Come in;
> Eternal Glory thou shalt win.

So he went in, and was cloathed with such Garments as they. Then *Christian* smiled, and said, I think verily I know the meaning of this.

Now, said *Christian*, let me go hence: Nay stay (said the *Inter-preter*,) till I have shewed thee a little more, and after that, thou shalt go on thy way. So he took him by the hand again, and led him into a very dark Room, where there sat a Man in an Iron *Cage.†

* Despair like an Iron Cage.

Now the Man, to look on, seemed very sad: he sat with his eyes looking down to the ground, his hands folded together; and he sighed as if he would break his heart. Then said *Christian*, *What means this?* At which the *Interpreter* bid him talk with the Man.

Chr. Then said *Christian* to the Man, *What art thou?* The Man answered, *I am what I was not once.*

Chr. What wast thou once?

Man. The *Man* said, I was once a fair *and flourishing Professor, both in mine own eyes, and also in the eyes of others: I once was, as I thought, fair for the Cœlestial City, and had then even joy at the thoughts that I should get thither.

* Luke 8. 13.

Chr. Well, but what art thou now?

Man. I am *now* a Man of Despair,†and am shut up in it, as in this Iron Cage. I cannot get out; O *now* I cannot.

Chr. But how camest thou in this condition?

Man. I left off to watch, and be sober; I laid the reins upon the neck of my lusts; I sinned against the light of the Word, and the goodness of God: I have grieved the Spirit, and he is gone; I tempted the Devil, and he is come to me; I have provoked God to anger, and he has left me; I have so hardened my heart, that I *cannot* repent.

Then said *Christian* to the *Interpreter*, But is there no hopes for such a Man as this? Ask him, said the *Interpreter*.

Chr. Then said *Christian*, *Is there no hope but you must be kept in this Iron Cage of Despair?*

Man. No, none at all.

Chr. *Why? The Son of the Blessed is very pitiful,*

Man. I have Crucified him to my self afresh, I have despised *his ·Luke 19. Person, I have despised his Righteousness, I have counted his Blood 14· an unholy thing, I have done despite *to the Spirit of Grace: There- ·Heb. 10. fore I have shut my self out of all the Promises; and there now 28, 29. remains to me nothing but threatnings, dreadful threatnings, fearful threatnings of certain Judgement and firy Indignation, which shall devour me as an Adversary.

Chr. *For what did you bring your self into this condition?*

Man. For the Lusts, Pleasures, and Profits of this World; in the injoyment of which, I did then promise my self much delight: but now even every one of those things also bite me, and gnaw me like a burning worm.

Chr. *But canst thou not now repent and turn?*

Man. God hath denied me repentance; his Word gives me no encouragement to believe; yea, himself hath shut me up in this Iron Cage: nor can all the men in the World let me out. O Eternity! Eternity! how shall I grapple with the misery that I must meet with in Eternity?

Inter. Then said the *Interpreter* to *Christian*, Let this mans misery be remembred by thee, and be an everlasting caution to thee.

Chr. Well, said *Christian*, this is fearful; God help me to watch and be sober; and to pray, that I may shun the cause of this mans misery. Sir, is it not time for me to go on my way now?

Int. Tarry till I shall shew thee one thing more, and then thou shalt go on thy way.

So he took *Christian* by the hand again, and led him into a Chamber, where there was one a rising out of Bed; and as he put on his Rayment, he shook and trembled. Then said *Christian*, Why doth this man thus tremble? The *Interpreter* then bid him tell to *Christian* the reason of his so doing: So he began, and said, This night as I was in my sleep, I Dreamed, and behold the Heavens grew exceeding black; also it thundered and lightned in most fearful wise, that it put me into an Agony. So I looked up in my Dream, and saw the Clouds rack at an unusual rate, upon which I heard a great sound of a Trumpet, and saw also a Man sit upon a Cloud, attended with the thousands of Heaven; they were all in flaming fire, also the Heavens was on a burning flame. I heard then a voice, saying, *Arise ye Dead, and come to Judgement*; and with that the Rocks rent, the Graves opened, & the Dead that were therein came forth; some of them were exceeding glad, and looked upward; and some sought to hide themselves under the Mountains: Then I saw the Man that sat upon the Cloud, open the Book; and bid the World draw near. Yet there was by reason of a Fiery flame that issued out and came from before him, a convenient distance betwixt him and them, as betwixt the Judge and the Prisoners at the Bar. I heard it also proclaimed to them that attended on the Man that sat on the Cloud, *Gather together the Tares, the Chaff, and Stubble, and cast them into the burning Lake*; and with that the Bottomless pit opened, just whereabout I stood; out of the mouth of which there came in an abundant manner Smoak, and Coals of fire, with hideous noises. It was also said to the same persons *Gather my Wheat into my Garner*. And with that I saw many catch'd up *and carried away into the Clouds, but I was left behind. I also sought to hide my self, but I could not; for the Man that sat upon the Cloud, still kept his eye upon me: my sins also came into mind, and my Conscience did accuse me on every side. Upon this I awaked from my sleep.

Chr. But what was it that made you so afraid of this sight?

Man. Why, I thought that the day of Judgement was come, and that I was not ready for it: but this frighted me most, that the Angels gathered up several, and left me behind; also the pit of Hell opened her mouth just where I stood: my Conscience too within afflicted me; and as I thought, the Judge had always his eye upon me, shewing indignation in his countenance.

Then said the *Interpreter* to *Christian*, Hast thou considered all these things?

[Marginal references, left column:]

1 Cor. 15.
1 Thess. 4.
Jude 15.
2 Thess. 1.
8.
John 5. 28.
Rev. 20. 11,
12, 13, 14.
Isa. 26. 21.
Mich. 7.
16, 17.
Psal. 5. 1,
2, 3.
Dan. 7. 10.
Mal. 3. 2,
3.
Dan. 7. 9,
10.

• Mat. 3. 2.
Ch. 13. 30.
Mal. 4. 1.

• Luke 3.
17.
• 1 Thess.
4. 16, 17.

Rom. 2. 14,
15.

Chr. Yes, and they put me in *hope* and *fear*.

Inter. Well, keep all things so in thy mind, that they may be as a *Goad* in thy sides, to prick thee forward in the way thou must go. Then *Christian* began to gird up his loins, and to address himself to his Journey. Then said the *Interpreter*, The Comforter[†] be always with thee good *Christian*, to guide thee in the way that leads to the City.

So *Christian* went on his way, saying,

> *Here I have seen things rare, and profitable;*
> *Things pleasant, dreadful, things to make me stable*
> *In what I have began to take in hand:*
> *Then let me think on them, and understand*
> *Wherefore they shewed me was, and let me be*
> *Thankful, O good Interpreter, to thee.*

Now I saw in my Dream, that the high way up which *Christian* was to go, was fenced on either side with a Wall, and that Wall is called *Salvation. Up this way therefore did burdened *Christian* run, but not without great difficulty, because of the load on his back. **• Isa. 26. 1.**

He ran thus till he came at a place somewhat ascending; and upon that place stood a *Cross*,[†]and a little below in the bottom, a Sepulcher. So I saw in my Dream, that just as *Christian* came up with the *Cross*, his burden loosed from off his Shoulders, and fell from off his back; and began to tumble; and so continued to do, till it came to the mouth of the Sepulcher, where it fell in, and I saw it no more.

Then was *Christian* glad *and lightsom, and said with a merry heart, *He hath given me rest, by his sorrow; and life, by his death.* Then he stood still a while, to look and wonder; for it was very surprizing to him, that the sight of the Cross should thus ease him of his burden. He looked therefore, and looked again, even till the springs that were in his head sent the *waters down his cheeks. Now as he stood looking and weeping, behold three shining ones came to him, and saluted him, with *Peace be to thee:* so the first said to him, *Thy sins be forgiven.* The second stript him of his Rags, and *cloathed him with change of Raiment. The third also set *a mark in his fore-head, and gave him a Roll with a Seal upon it,[†]which he bid him look on

• When God releases us of our guilt and burden, we are as those that leap for joy.
• Zech. 12. 10.
• Mark 2. 5.
• Zech. 3. 4.
• Eph. 1. 13.

as he ran, and that he should give it in at the Cœlestial Gate: so they went their way. Then *Christian* gave three leaps for joy, and went on singing.

A Christian can sing tho alone, when God doth give him the joy of his heart.

> *Thus far did I come loaden with my sin,*
> *Nor could ought ease the grief that I was in,*
> *Till I came hither: What a place is this!*
> *Must here be the beginning of my bliss?*
> *Must here the burden fall from off my back?*
> *Must here the strings that bound it to me, crack?*
> *Blest Cross! blest Sepulcher! blest rather be*
> *The Man that there was put to shame for me.*

I saw then in my Dream that he went on thus, even untill he came at a bottom, where he saw, a little out of the way, three Men fast asleep, with Fetters upon their heels. The name of the one was *Simple*, another *Sloth*, and the third *Presumption*.

* *Simple, Sloth,* and *Presumption.*

Christian then seeing them lye in this case, went to them, if peradventure he might awake them. And cried, You are like them that sleep on the top of *a Mast, for the dead Sea is under you, a Gulf that hath no bottom: Awake therefore, and come away; be willing also, and I will help you off with your Irons. He also told them, If he that goeth about like *a roaring Lion, comes by, you will certainly become a prey to his teeth. With that they lookt upon him, and began to reply in this sort: *Simple said, I see no danger;* Sloth said, *Yet a little more sleep:* and Presumption said, *Every Fatt must stand upon his own bottom, what is the answer else that I should give thee?* And so *they* lay down to sleep again, and *Christian* went on his way.

* *Prov. 23. 24.*

* *1 Pet. 5. 8.*

* *There is no perswasion will do, if God openeth not the eyes.*

Yet was he troubled to think, That men in that danger should so little esteem the kindness of him that so freely offered to help them; both by awakening of them, counselling of them, and proffering to help them off with their Irons. And as he was troubled thereabout, he espied two men come tumbling over the Wall, on the left hand of the narrow way; and they made up a pace to him. The name of the one was *Formalist*, and the name of the other *Hypocrisie*. So, as I said, they drew up unto him, who thus entered with them into discourse.

* *Christian talked with them.*

Chr. *Gentlemen, Whence came you, and whither do you go?*

Form. and *Hyp.* We were born in the Land of Vain-glory, and are going for praise to Mount *Sion*.

Chr. Why came you not in at the Gate which standeth at the beginning of the way? Know you not that it is written, *That he that cometh not in by the door, but climbeth up some other way, the same is a thief and a robber. • John 10. 1.

Form. and *Hyp.* They said, That to go to the Gate for entrance, was by all their Countrey-men counted too far about; and that therefore their usual way was to make a short cut of it, and to climb over the Wall as they had done.

Chr. But will it not be counted a Trespass, against the Lord of the City whither we are bound, thus to violate his revealed will?

Form. and *Hyp.* They told him, *That as for that, he needed not to trouble his head thereabout: for what they did they had custom for; and could produce, if need were, Testimony that would witness it, for more then a thousand years.†

• *They that come into the way, but not by the door, think that they can say something in vindication of their own Practice.*

Chr. But said Christian, *Will your Practice stand a Trial at Law?*

Form. and *Hyp.* They told him, That Custom, it being of so long a standing, as above a thousand years, would doubtless now be admitted as a thing legal, by any Impartial Judge. And besides, said they, so be we get into the way, what's matter which way we get in? if we are in, we are in:† thou art but in the way, who, as we perceive, came in at the Gate; and we are also in the way that came tumbling over the wall: Wherein now is thy condition better then ours?

Chr. I walk by the Rule of my Master, you walk by the rude working of your fancies. You are counted thieves already, by the Lord of the way; therefore I doubt you will not be found true men at the end of the way. You come in by your selves without his direction, and shall go out by your selves without his mercy.

To this they made him but little answer; only they bid him look to himself. Then I saw that they went on every man in his way, without much conference one with another; save that these two men told *Christian,* That, as to *Laws and Ordinances,* they doubted not, but they should as conscienciously do them as he. Therefore said they, We see not wherein thou differest from us, but by the Coat† that is on thy back, which was, as we tro, given thee by some of thy Neighbours, to hide the shame of thy nakedness.

• Gal. 2. 16. *Chr.* By *Laws and Ordinances, you will not be saved, since you came not in by the door. And as for this Coat that is on my back, it was given me by the Lord of the place whither I go; and that, as you say, to cover my nakedness with. And I take it as a token of his kindness to me, for I had nothing but rags before; and besides, • Christian *has got his Lord's Coat on his back, and is comforted therewith, he is comforted also with his Mark, and his Roll.* *thus I comfort my self as I go: Surely, think I, when I come to the Gate of the City, the Lord thereof will know me for good, since I have his Coat on my back; a *Coat* that he gave me freely in the day that he stript me of my rags. I have moreover a mark in my forehead, of which perhaps you have taken no notice, which one of my Lords most intimate Associates fixed there in the day that my burden fell off my shoulders. I will tell you moreover, that I had then given me a Roll sealed to comfort me by reading, as I go in the way; I was also bid to give it in at the Cœlestial Gate, in token of my certain going in after it: all which things I doubt you want; and want them, because you came not in at the Gate.

To these things they gave him no answer, only they looked upon each other, and *laughed.* Then I saw that they went on all, save that *Christian* kept before, who had no more talk but with himself, and that somtimes sighingly, and somtimes comfortably: also he would be often reading in the Roll, that one of the shining ones gave him, by which he was refreshed.

I believe then, that they all went on till they came to the foot • He comes to the hill Difficulty. of an Hill, *at the bottom of which was a Spring. There was also in the same place two other ways besides that which came straight from the Gate; one turned to the left hand, and the other to the right, at the bottom of the Hill: but the narrow way lay right up the Hill, (and the name of the going up the side of the Hill, is called • Isa. 49. 10. *Difficulty.*) *Christian* now went to the *Spring and drank thereof to refresh himself, and then began to go up the Hill; saying,

> *This Hill, though high, I covet to ascend,*
> *The difficulty will not me offend:*
> *For I perceive the way to life lies here;*
> *Come, pluck up, Heart; lets neither faint nor fear:*
> *Better, tho difficult, th' right way to go,*
> *Then wrong, though easie, where the end is wo.*

The other two also came to the foot of the Hill. But when they saw that the Hill was steep and high, and that there was two other ways to go; and supposing also that these two ways might meet again, with that up which *Christian* went, on the other side of the Hill: Therefore they were resolved to go in those ways; (now the name of one of those ways was *Danger*, and the name of the other *Destruction*) So *the one took the way which is called *Danger*, which led him into a great Wood; and the other took directly up the way to *Destruction*, which led him into a wide field full of dark Mountains, where he stumbled and fell,† and rose no more.

* *The danger of turning out of the way.*

I looked then after *Christian*, to see him go up the Hill, where I perceived he fell from running to going, and from going to clambering upon his hands and his knees, because of the steepness of the place. Now about the midway to the top of the Hill, was a pleasant *Arbour, made by the Lord of the Hill, for the refreshing of weary Travailers. Thither therefore *Christian* got, where also he sat down to rest him. Then he pull'd his Roll out of his bosom, and read therein to his comfort; he also now began afresh to take a review of the Coat or Garment that was given him as he stood by the Cross. Thus pleasing himself a while, he at last fell into a slumber, and thence into a fast sleep, which detained him in that place untill it was almost night, and in his sleep his *Roll fell out of his hand. Now as he was sleeping, there came one to him & awaked him, saying *Go to the Ant, thou sluggard, consider her ways, and be wise: and with that *Christian* suddenly started up, and sped him on his way, and went a pace till he came to the top of the Hill.

* *A award of grace.*

* *He that sleeps is a loser.*

* Prov. 6. 6.

Now when he was got up to the top of the Hill, there came two men running against him amain; the name of the one was *Timorous*, and the name of the other *Mistrust*. To whom *Christian* said, Sirs, what's the matter you run the wrong way? *Timorous* answered, That they were going to the City of *Zion*, and had got up that *difficult* place; but, said he, the further we go, the more danger we meet with, wherefore we turned, and are going back again.

* *Christian meets with Mistrust and Timorous.*

Yes, said *Mistrust*, for just before us lye a couple of Lions† in the way, whether sleeping or wakeing we know not and we could not think, if we came within reach, but they would presently pull us in pieces.

Chr. Then said *Christian*, You make me afraid, but whither shall I fly to be safe? If I go back to mine own Countrey, *That* is prepared for Fire and Brimstone; and I shall certainly perish there. If I can get to the Cœlestial City, I am sure to be in safety there. *I must venture: To go back is nothing but death, to go forward is fear of death, and life everlasting beyond it. I will yet go forward. So *Mistrust* and *Timorous* ran down the Hill; and *Christian* went on his way. But thinking again of what he heard from the men, he felt in his bosom for his Roll: that he might read therein and be comforted; but he felt, and *found it not. Then was *Christian* in great distress, and knew not what to do, for he wanted that which used to relieve him, and that which should have been his Pass into the Cœlestial City. Here therefore he began to be much *perplexed, and knew not what to do; at last he bethought himself that he had slept in the *Arbour* that is on the side of the Hill: and falling down upon his knees, he asked God forgiveness for that his foolish Fact, and then went back to look for his Roll. But all the way he went back, who can sufficiently set forth the sorrow of *Christians* heart? somtimes he sighed, somtimes he wept, and often times he chid himself, for being so foolish to fall asleep in that place which was erected only for a little refreshment from his weariness. Thus therefore he went back, carefully looking on this side, and on that, all the way as he went, if happily he might find his Roll, that had been his comfort so many times in his Journey. He went thus till he came again within sight of the *Arbour*, where he sat and slept; but that sight renewed *his sorrow the more, by bringing again, even a fresh, his evil of sleeping unto his mind. Thus therefore he now went on, bewailing his sinful sleep, saying, *O wretched Man that I am*, that I should sleep in the day time! that I should sleep in the midst of difficulty! that I should so indulge the flesh, as to use that rest for ease to my flesh, which the Lord of the Hill hath erected only for the relief of the spirits of Pilgrims! How many steps have I took in vain! (Thus it happened to *Israel* for their sin, they were sent back again by the way of the Red-Sea) and I am made to tread those steps with sorrow, which I might have trod with delight, had it not been for this sinful sleep. How far might I have been on my way by this time! I am made to

• Christian shakes off fear.

• Christian missed his Roll, wherein he used to take comfort.

• He is perplexed for his Roll.

• Christian bewails his foolish sleeping.
Rev. 2. 2.
1 Thess. 5. 7, 8.

tread those steps thrice over, which I needed not to have trod but once: Yea now also I am like to be benighted, for the day is almost spent. O that I had not slept! Now by this time he was come to the *Arbour* again, where for a while he sat down and wept, but at last (as *Christian* would have it) looking sorrowfully down under the Settle, there he *espied his Roll; the which he with trembling and haste catch'd up, and put it into his bosom; but who can tell how joyful this man was, when he had gotten his Roll again! For this Roll was the assurance of his life, and acceptance at the desired Haven. Therefore he laid it up in his bosom, gave thanks to God for directing his eye to the place where it lay, and with joy and tears betook him self again to his Journey. But Oh how nimbly now did he go up the rest of the Hill! Yet before he got up, the Sun went down upon *Christian*; and this made him again recall the vanity of his sleeping to his remembrance, and thus he again began to condole with himself: *Ah thou sinful sleep! how for thy sake am I like to be benighted in my Journey! I must walk without the Sun, darkness must cover the path of my feet, and I must hear the noise of doleful Creatures, because of my sinful sleep!* Now also he remembred the story that *Mistrust* and *Timorous* told him of, how they were frighted with the sight of the Lions. Then said *Christian* to himself again, These Beasts range in the night for their prey, and if they should meet with me in the dark, how should I shift them? how should I escape being by them torn in pieces? Thus he went on his way, but while he was thus bewayling his unhappy miscarriage, he lift up his eyes, and behold there was a very stately Palace before him, the name whereof was *Beautiful*,†and it stood just by the High-way side.

So I saw in my Dream, that he made haste and went forward, that if possible he might get Lodging there; Now before he had gone far, he entered into a very narrow passage, which was about a furlong off of the Porters Lodge, and looking very narrowly before him as he went, he espied two Lions in the way. Now, thought he, I see the dangers that *Mistrust* and *Timorous* were driven back by, (The Lions were chained, but he saw not the Chains) Then he was afraid, and thought also himself to go back after them, for he thought nothing but death was before him: But the *Porter* at the Lodge,

*Christian *findeth his Roll where* he lost it.*

' Mark 13.
14. whose name is *Watchful, perceiving that *Christian* made a halt, as
if he would go back, cried unto him saying, Is thy strength so
small? fear not the Lions, for they are Chained; and are placed there
for trial of faith where it is; and for discovery of those that have
none: keep in the midst of the Path, and no hurt shall come unto
thee.

Then I saw that he went on, trembling for fear of the Lions; but
taking good heed to the directions of the *Porter*; he heard them
roar, but they did him no harm. Then he clapt his hands, and went
on till he came and stood before the Gate where the *Porter* was.
Then said *Christian* to the *Porter*, Sir, What house is this? and may
I lodge here to night? The *Porter* answered, This House was built
by the Lord of the Hill: and he built it for the relief and security
of Pilgrims. The *Porter* also asked whence he was, and whither he
was going?

Chr. I am come from the City of *Destruction*, and am going to
Mount *Zion*; but because the Sun is now set, I desire, if I may, to
lodge here to night.

Por. What is your name?

Chr. My name is, now, *Christian*; but my name at the first was
* Gen. 9. *Graceless*: I came of the Race of *Japhet, whom God will perswade
27. to dwell in the Tents of *Shem*.

Por. But how doth it happen that you come so late, the Sun is set?

Chr. I had been here sooner, but that, wretched man that I am!
I slept in the *Arbour* that stands on the Hill side; nay, I had notwithstanding that, been here much sooner, but that in my sleep I lost
my Evidence, and came without it to the brow of the Hill; and then
feeling for it, and finding it not, I was forced with sorrow of heart,
to go back to the place where I slept my sleep, where I found it,
and now I am come.

Por. Well, I will call out one of the Virgins of this place, who will,
if she likes your talk, bring you in to the rest of the Family, according to the Rules of the House. So *Watchful* the Porter rang a Bell;
at the sound of which, came out at the door of the House, a Grave
and Beautiful Damsel, named *Discretion*, and asked why she was
called.

The *Porter* answered, This Man is in a Journey from the City of
Destruction to Mount *Zion*, but being weary, and benighted, he

asked me if he might lodge here to night; so I told him I would call for thee, who after discourse had with him, mayest do as seemeth thee good, even according to the Law of the House.

Then she asked him whence he was, and whither he was going, and he told her. She asked him also, how he got into the way and he told her; Then she asked him, What he had seen, and met with in the way, and he told her; and last, she asked his name, so he said, It is *Christian*; and I have so much the more a desire to lodge here to night, because, by what I perceive, this place was built by the Lord of the Hill, for the relief and security of Pilgrims. So she smiled, but the water stood in her eyes: And after a little pause, she said, I will call forth two or three more of the Family. So she ran to the door, and called out *Prudence, Piety* and *Charity*, who after a little more discourse with him, had him in to the Family; and many of them meeting him at the threshold of the house, said, Come in thou blessed of the Lord; this house was built by the Lord of the Hill, on purpose to entertain such Pilgrims in. Then he bowed his head, and followed them into the House. So when he was come in, and set down, they gave him somthing to drink; and consented together, that until supper was ready, some one or two of them should have some particular discourse with *Christian*, for the best improvement of time: and they appointed *Piety* and *Prudence* and *Charity* to discourse with him; and thus they began.

Piety. Come good Christian, *since we have been so loving to you, to receive you in to our House this night; let us, if perhaps we may better our selves thereby, talk with you of all things that have happened to you in your Pilgrimage.* Piety discourses him.

Chr. With a very good will, and I am glad that you are so well disposed.

Piety. What moved you at first to betake yourself to a Pilgrim's life?

Chr. I was *driven out of my Native Countrey, by a dreadful sound that was in mine ears, to wit, That unavoidable destruction did attend me, if I abode in that place where I was. * How Christian was driven out of his own Countrey.

Piety. But how did it happen that you came out of your Countrey this way?

Chr. It was as God would have it; for when I was under the fears of destruction, I did not know whither to go; but by chance there came a man, even to me, (as I was trembling and weeping) whose

• *How he got into the way to Sion.* name is *Evangelist, and he directed me to the Wicket-gate, which else I should never have found; and so set me into the way that hath led me directly to this House.

Piety. But did you not come by the House of the Interpreter?

Chr. Yes, and did see such things there, the remembrance of • *A rehearsal of what he saw in the way.* which will stick by me as long as I live; specially three *things; to wit, How Christ, in despite of Satan, maintains his work of Grace in the heart; how the Man had sinned himself quite out of hopes of Gods mercy; and also the Dream of him that thought in his sleep the day of Judgement was come.

Piety. Why? Did you hear him tell his Dream?

Chr. Yes, and a dreadful one it was, I thought. It made my heart ake as he was telling of it, but yet I am glad I heard it.

Piety. Was that all that you saw at the house of the Interpreter?

Chr. No, he took me and had me where he shewed me a stately Palace, and how the People were clad in Gold that were in it; and how there came a venturous Man, and cut his way through the armed men that stood in the door to keep him out; and how he was bid to come in, and win eternal Glory. Methought those things did ravish my heart; I could have staid at that good Mans house a twelve-month, but that I knew I had further to go.

Piety. And what saw you else in the way?

Chr. Saw! Why, I went but a little further, and I saw one, as I thought in my mind, hang bleeding upon the Tree; and the very sight of him made my burden fall off my back (for I groaned under a weary burden) but then it fell down from off me. 'Twas a strange thing to me, for I never saw such a thing before: Yea, and while I stood looking up, (for then I could not forbear looking) three shining ones came to me: one of them testified that my sins were forgiven me: another stript me of my rags, and gave me this Broidred Coat which you see; and the third set the mark which you see in my forehead, and gave me this sealed Roll (and with that he plucked it out of his bosom.)

Piety. But you saw more then this, did you not?

Chr. The things that I have told you were the best: yet some other matters I saw, as namely I saw three Men, *Simple, Sloth,* and *Presump-*

tion, lye a sleep a little out of the way as I came, with Irons upon their heels; but do you think I could awake them? I also saw *Formalist* and *Hypocrisie* come tumbling over the wall, to go, as they pretended, to *Sion*, but they were quickly lost; even as I my self did tell them, but they would not believe: but, above all, I found it *hard* work to get up this Hill, and as *hard* to come by the Lions mouths; and truly if it had not been for the good Man, the Porter that stands at the Gate, I do not know, but that after all, I might have gone back again: but now I thank God I am here, and I thank you for receiving of me.

Then *Prudence* thought good to ask him a few questions, and desired his answer to them. <small>Prudence discourses him.</small>

Pru. *Do you not think somtimes of the Countrey from whence you came?*

Chr. Yes, *but with much shame and detestation; *Truly, if I had been mindful of that Countrey from whence I came out, I might have had opportunity to have returned: but now I desire a better Countrey; that is, an* Heavenly. <small>*Christians thoughts of his Native Countrey. Heb. 11. 15, 16.</small>

Pru. *Do you not yet bear away with you some of the things that then you were conversant withal?*

Chr. Yes but greatly against my will; especially my inward and *carnal cogitations; with which all my Countrey-men, as well as my self, were delighted; but now all those things are my grief: and might I but chuse mine own things, I would *chuse never to think of those things more; but when I would be doing of that which is best, that which is worst is with me. <small>* Christian distasted with carnal cogitations. *Christians choice.</small>

Pru. *Do you not find sometimes, as if those things were vanquished, which at other times are your perplexity?*

Chr. Yes, but that is but seldom; but they are to me *Golden hours, in which such things happen to me. <small>*Christians golden hours.</small>

Pru. *Can you remember by what means you find your anoyances at times, as if they were vanquished?*

Chr. Yes, when *I think what I saw at the Cross, that will do it; and when I look upon my Broidered Coat, that will do it; also when I look into the Roll that I carry in my bosom, that will do it; and when my thoughts wax warm about whither I am going, that will do it. <small>* How Christian gets power against his corruptions.</small>

Pru. *And what is it that makes you so desirous to go to Mount* Zion?

Chr. Why, *there I hope to see him *alive,* that did hang *dead* on the Cross; and there I hope to be rid of all those things, that to this day are in me, an anoiance to me; there they say there is no *death, and there I shall dwell with such Company as I like best. For to tell you truth, I love him, because I was by him eased of my burden, and I am weary of my inward sickness; I would fain be where I shall die no more, and with the Company that shall continually cry, *Holy, Holy, Holy.

Then said *Charity* to *Christian,* *Have you a family? are you a married man?*

Chr. I have a Wife and four small Children.†

Cha. And why did you not bring them along with you?

Chr. Then *Christian* *wept, and said, Oh how willingly would I have done it, but they were all of them utterly averse to my going on Pilgrimage.

Cha. But you should have talked to them, and have endeavoured to have shewen them the danger of being behind.

Chr. So I did, and told them also what God had shewed to me of the destruction of our City; but I seemed to them as one that mocked, and they believed me not.

Cha. And did you pray to God that he would bless your counsel to them?

Chr. Yes, and that with much affection; for you must think that my Wife and poor Children were very dear unto me.

Cha. But did you tell them of your own sorrow, and fear of destruction? for I suppose that destruction was visible enough to you?

Chr. Yes, over, and over, and over. They might also *see my fears in my countenance, in my tears, and also in my trembling under the apprehension of the Judgment that did hang over our heads; but all was not sufficient to prevail with them to come with me.

Cha. But what could they say for themselves why they came not?

Chr. Why, *my Wife was afraid of losing this World; and my Children were given to the foolish delights of youth: so what by one thing, and what by another, they left me to wander in this manner alone.

Cha. But did you not with your vain life, damp all that you by words used by way of perswasion to bring them away with you?

• *Why Christian would be at Mount Zion.*
• Isa. 25. 8. Rev. 21. 4.

• *Charity discourses him.*

• *Christian's love to his Wife and Children.*

Gen. 19. 14.

• *Christian's fears of perishing might be read in his very countenance.*

• *The cause why his Wife and Children did not go with him.*

Chr. Indeed I cannot commend my life; for I am conscious to my self of many failings: therein, I know also that a man by his conversation, may soon overthrow what by argument or perswasion he doth labour to fasten upon others for their good: Yet, this I can say, I was very wary of giving them occasion, by any unseemly action, to make them averse to going on Pilgrimage. Yea, for this very thing, they would tell me I was too precise,† and that I denied my self of things (for their sakes) in which they saw no evil. Nay, I think I may say, that, if what they saw in me did hinder them, it was my great tenderness in sinning against God, or of doing any wrong to my Neighbour.

Cha. Indeed *Cain hated his Brother, because his own works were evil, and his Brothers righteous; and if thy Wife and Children have been offended with thee for this, they thereby shew themselves to be implacable to *good; and thou hast delivered thy soul from their blood.*

Now I saw in my Dream, that thus they sat talking together until supper was ready. So when they had made ready, they sat down to meat; Now the Table was furnished *with fat things, and with Wine that was well refined; and all their talk *at the Table was about the Lord of the Hill: As namely, about what he had done, and wherefore he did what he did, and why he had builded that House: and by what they said, I perceived that he had been a *great Warriour*, and had fought with and slain *him that had the power of Death, but not without great danger to himself, which made me love him the more.

For, as they said, and as I believe, (said *Christian*) he did it with the loss of much blood; but that which put Glory of Grace into all he did, was, that he did it of pure love to his Countrey. And besides, there were some of them of the Household that said, they had seen, and spoke with him since he did dye on the Cross; and they have attested, that they had it from his own lips, that he is such a lover of poor Pilgrims, that the like is not to be found from the East to the West.

They moreover gave an instance of what they affirmed, and that was, He had stript himself of his glory† that he might do this for the Poor; and that they heard him say and affirm, That he would not dwell in the Mountain of *Zion* alone. They said moreover, That he

Christian's good conversation before his Wife and Children.

* 1 John 3. 12. Christian clear of their blood if they perish.
* Ezek. 3. 19.

* What Christian had to his supper.
* Their talk at supper time.

* Heb. 2. 14, 15.

• *Christia* had made many Pilgrims *Princes, though by nature they were
makes *Beggars born, and their original had been the Dunghil.
Princes of
Beggars.　　Thus they discoursed together till late at night; and after they
• 1 Sam. 2. had committed themselves to their Lord for Protection, they betook
8. themselves to rest. The Pilgrim they laid in a large upper *Chamber,
Psal. 113.
7.
• *Christians* whose window opened towards the Sun rising; the name of the
Bed- Chamber was *Peace*, where he slept till break of day; and then he
Chamber. awoke and sang,

> *Where am I now? is this the love and care*
> *Of Jesus, for the men that Pilgrims are?*
> *Thus to provide! That I should be forgiven!*
> *And dwell already the next door to Heaven.*

So in the Morning they all got up, and after some more discourse,
they told him that he should not depart, till they had shewed him
the *Rarities* of that place. And first they had him into the Study,
• *Christian* *where they shewed him Records of the greatest Antiquity; in
had into the which, as I remember my Dream, they shewed him first the Pedi-
Study, and
what he gree of the Lord of the Hill, that he was the Son of the Ancient of
saw there. Days,†and came by an eternal Generation. Here also was more fully
Recorded the Acts that he had done, and the names of many
hundreds that he had taken into his service; and how he had placed
them in such Habitations that could neither by length of Days, nor
decaies of Nature, be dissolved.

　　Then they read to him some of the worthy Acts that some of his
servants had done: As how they had subdued Kingdoms, wrought
Righteousness, obtained Promises, stopped the mouths of Lions,
• Heb. 11. quenched the *violence of Fire, escaped the edge of the Sword; out
33, 34. of weakness were made strong, waxed valiant in fight, and turned
to flight the Armies of the *Aliens*.

　　Then they read again in another part of the Records of the House,
where it was shewed how willing their Lord was to receive into his
favour, any, even any, though they in time past had offered great
affronts to his Person and proceedings. Here also were several other
Histories of many other famous things; of all which *Christian* had a
view. As of things both Ancient and Modern; together with Prophe-

cies and Predictions of things that have their certain accomplishment, both to the dread and amazement of enemies, and the comfort and solace of Pilgrims.

The next day they took him, and had him into the *Armory;[†] where they shewed him all manner of Furniture, which their Lord had provided for Pilgrims, as Sword, Shield, Helmet, Brest plate, *All-Prayer*, and Shooes that would not wear out. And there was here enough of this, to harness out as many men for the service of their Lord, as there be Stars in the Heaven for multitude. • Christian *had into the Armory.*

They also shewed him some of the Engines with which some of his Servants had done wonderful things. *They shewed him *Moses* Rod, the Hammer and Nail with which *Jael* slew *Sisera*, the Pitchers, Trumpets, and Lamps too, with which *Gideon* put to flight the Armies of *Midian*. Then they shewed him the Oxes goad wherewith *Shamger* slew six hundred men. They shewed him also the Jaw bone with which *Sampson* did such mighty feats; they shewed him moreover the Sling and Stone with which *David* slew *Goliah* of *Gath*: and the Sword also with which their Lord will kill the Man of Sin,[†] in the day that he shall rise up to the prey. They shewed him besides many excellent things, with which *Christian* was much delighted. This done, they went to their rest again. • Christian *is made to see Ancient things.*

Then I saw in my Dream, that on the morrow he got up to go forwards, but they desired him to stay till the next day also; and then said they, we will, (if the day be clear) shew you the *Delectable Mountains;[†] which they said, would yet further add to his comfort; because they were nearer the desired Haven, then the place where at present he was. So he consented and staid. When the Morning was up, they had him to the top of the House, *and bid him look South; so he did; and behold at a great distance he saw a most pleasant Mountainous Country, beautified with Woods, Vinyards, Fruits of all sorts; Flowers also, with Springs and Fountains, very delectable to behold. Then he asked the name of the Countrey; they said it was *Immanuels[†]Land*: and it is as common, said they, as this *Hill* is to, and for all the Pilgrims. And when thou comest there, from thence, thou maist see to the Gate of the Cœlestial City, as the Shepherds that live there will make appear. • Christian *shewed the Delectable Mountains.*

• Isa. 33. 16, 17.

• Christian
sets for-
ward.
• Christian
sent away
Armed.

Now he bethought himself of setting forward,* and they were willing he should: but first, said they, let us go again into the Armory, so they did; and when he came there, they *harnessed him from head to foot, with what was of proof, lest perhaps he should meet with assaults in the way. He being therefore thus accoutred, walketh out with his friends to the Gate, and there he asked the *Porter* if he saw any Pilgrims pass by; then the *Porter* answered, Yes.

Ch. Pray did you know him?

Por. I asked his name, and he told me it was *Faithful*.

Chr. O, said *Christian*, I know him, he is my Townsman, my near Neighbour, he comes from the place where I was born: how far do you think he may be before?

Por. He is got by this time below the Hill.

• How
Christian
and the
Porter greet
at parting.

Chr. Well, *said *Christian*, good Porter the Lord be with thee, and add to all thy blessings much increase, for the kindness that thou hast shewed to me.

Then he began to go forward, but *Discretion*, *Piety*, *Charity*, and *Prudence* would accompany him down to the foot of the Hill. So they went on together, reiterating their former discourses till they came to go down the Hill. Then said *Christian*, as it was *difficult* coming up, so (so far as I can see) it is *dangerous* going down. Yes, said *Prudence*, so it is; for it is an hard matter for a man to go down into the valley of *Humiliation*, as thou art now, and to catch no slip by the way; therefore, said they, are we come out to accompany thee down the Hill. So he began to go down, but very warily, yet he caught a slip or two.

Then I saw in my Dream, that these good Companions (when *Christian* was gone down to the bottom of the Hill) gave him a loaf of Bread, a bottle of Wine, and a cluster of Raisins; and then he went on his way.

But now in this Valley of *Humiliation* poor *Christian* was hard put to it, for he had gone but a little way before he espied a foul *Fiend* coming over the field to meet him; his name is *Apollyon*.† Then did *Christian* begin to be afraid, and to cast in his mind whether to go back, or to stand his ground. But he considered again, that he had no Armour for his back, and therefore thought that to turn the back to him, might give him greater advantage with ease to pierce him

with his Darts; therefore he resolved to venture, and *stand his
ground. For thought he, had I no more in mine eye, then the saving
of my life, 'twould be the best way to stand.

*Christians resolution at the approach of Apollyon.

So he went on, and *Apollyon* met him; now the Monster†was
hidious to behold, he was cloathed with scales like a Fish (and they
are his pride) he had Wings like a Dragon, feet like a Bear, and out
of his belly came Fire and Smoak, and his mouth was as the mouth
of a Lion. When he was come up to *Christian*, he beheld him with a
disdainful countenance, and thus began to question with him.

Apol. *Whence come you, and whither are you bound?*

Chr. I come from the City of *Destruction*, *which is the place of
all evil, and am going to the City of *Zion*.

* Discourse betwixt Christian and Apollyon.

Apol. *By this I perceive thou art one of my Subjects, for all that Countrey
is mine; and I am the Prince and God of it. How is it then that thou hast
ran away from thy King? Were it not that I hope thou maiest do me more
service, I would strike thee now at one blow to the ground.*

Chr. I was born indeed in your Dominions, but your service was
hard, and your wages such as a man could not live on, *for the wages
of Sin is death; therefore when I was come to years, I did as other
considerate persons do, look out, if perhaps I might mend my self.

* Rom. 6. 23.

Apol. *There is no Prince that will thus lightly lose his Subjects; neither
will I as yet lose thee. But since thou complainest of thy service and wages,
*be content to go back; what our Countrey will afford, I do here promise to
give thee.*

* Apollyons flattery.

Chr. But I have let my self to another, even to the King of Princes,
and how can I with fairness go back with thee?

Apol. *Thou hast done in this, according to the Proverb, *changed a bad
for a worse: but it is ordinary for those that have professed themselves his
Servants, after a while to give him the slip; and return again to me: do thou
so too, and all shall be well.*

* Apollyon undervalues Christs service.

Chr. I have given him my faith, and sworn my Allegiance to him;
how then can I go back from this, and not be hanged as a Traitor?

Apol. *Thou didest the same to me, *and yet I am willing to pass by all, if
now thou wilt yet turn again, and go back.*

* Apollyon pretends to be merciful.

Chr. What I promised thee was in my none-age; and besides, I
count that the Prince under whose Banner now I stand, is able to
absolve me; yea, and to pardon also what I did as to my compliance

with thee: and besides, (O thou destroying *Apollyon*) to speak truth, I like his Service, his Wages, his Servants, his Government, his Company, and Countrey better then thine: and therefore leave off to perswade me further, I am his Servant, and I will follow him.

Apol. *Consider again when thou art in cool blood, what thou art like to meet with in the way that thou goest. Thou knowest that for the most part, his Servants come to an ill end, because they are transgressors against me, and my ways: How many of them have been put to shameful deaths! and besides, thou countest his service better then mine, whereas he never came yet from the place where he is, to deliver any that served him out of our hands: but as for me, how many times, as all the World very well knows, have I delivered, either by power or fraud, those that have faithfully served me, from him and his, though taken by them; and so I will deliver thee.*

Chr. His forbearing at present to deliver them, is on purpose to try their love, whether they will cleave to him to the end: and as for the ill end thou sayest they come to, that is most glorious in their account: For, for present deliverance, they do not much expect it; for they stay for their Glory, and then they shall have it, when their Prince comes in his, and the Glory of the Angels.

Apol. *Thou hast already been unfaithful in thy service to him, and how dost thou think to receive wages of him?*

Chr. Wherein, O *Apollyon*, have I been unfaithful to him;

Apol. *Thou didst faint at first setting out, when thou wast almost choked in the Gulf of Dispond. Thou didst attempt wrong ways to be rid of thy burden, whereas thou shouldest have stayed till thy Prince had taken it off. Thou didst sinfully sleep, and loose thy choice thing: thou wast also almost perswaded to go back, at the sight of the Lions; and when thou talkest of thy Journey, and of what thou hast heard, and seen, thou art inwardly desirous of vainglory in all that thou sayest or doest.*

Chr. All this is true, and much more, which thou hast left out; but the Prince whom I serve and honour, is merciful, and ready to forgive: but besides, these infirmities possessed me in thy Countrey, for there I suckt them in, and I have groaned under them, been sorry for them, and have obtained Pardon of my Prince.

Apol. *Then* *Apollyon* *broke out into a grievous rage, saying, I am an enemy to this Prince: I hate his Person, his Laws, and People: I am come out on purpose to withstand thee.*

Apollyon pleads the grievous ends of Christians, to diswade Christian from persisting in his way.

Apollyon pleads Christian's infirmities against him.

Apollyon in a rage falls upon Christian.

Chr. *Apollyon*, beware what you do, for I am in the Kings Highway, the way of Holiness, therefore take heed to your self.

Apol. Then *Apollyon* strodled quite over the whole breadth of the way, and said, I am void of fear in this matter, prepare thy self to dye, for I swear by my Infernal Den, that thou shalt go no further, here will I spill thy soul: and with that he threw a flaming Dart at his brest; but *Christian* had a Shield in his hand, with which he caught it, and so prevented the danger of that. Then did *Christian* draw, for he saw 'twas time to bestir him; and *Apollyon* as fast made at him, throwing Darts as thick as hail; by the which, notwithstanding all that *Christian* could do to avoid it, *Apollyon* wounded him in his *head*, his *hand* and *foot;* this made *Christian* give a little back: *Apollyon* therefore followed his work amain, and *Christian* again took courage, and resisted as manfully as he could. This sore Combat lasted for above half a day, even till *Christian* was almost quite spent. For you must know, that *Christian*, by reason of his wounds, must needs grow weaker and weaker. • *Christian wounded in his understanding, faith and conversation.*

Then *Apollyon* espying his opportunity, began to gather up close to *Christian*, and wrestling with him, gave him a dreadful fall; and with that *Christian's* Sword flew out of his hand. Then said *Apollyon*, I am sure of thee now; and with that, he had almost prest him to death; so that *Christian* began to despair of life. But as God would have it, while *Apollyon* was fetching of his last blow, thereby to make a full end of this good Man, *Christian* nimbly reached out his hand for his Sword, and caught it, saying, *Rejoyce not against me, O mine Enemy! when I fall, I shall arise*; and with that, gave him a deadly thrust, which made him give back, as one that had received his mortal wound: *Christian* perceiving that, made at him again, saying *Nay, in all these things we are more then Conquerours, through him that loved us.* And with that, *Apollyon* spread forth his Dragons wings, and sped him away; that *Christian* saw him no more. *Apollyon casteth down to the ground Christian. Christian's victory over Apollyon.* • Mic. 7. 8. • Rom. 8. 37. James 4. 7.

In this Combat no man can imagine, unless he had seen and heard as I did, what yelling, and hideous roaring *Apollyon* made; all the time of the fight, he spake like a Dragon: and on the other side, what sighs and groans brast from *Christians* heart. I never saw him all the while give so much as one pleasant look, till he perceived he had wounded *Apollyon* with his two-edg'd Sword, then indeed he *A brief relation of the Combat by the spectator.*

did smile, and look upward: but twas the dreadfullest sight that ever I saw.

Christian gives God thanks for deliverance. So when the Battel was over, *Christian* said, I will here give thanks to him that hath delivered me out of the mouth of the Lion; to him that did help me against *Apollyon*: and so he did, saying,

> *Great* Beelzebub, *the Captain of this Fiend,*
> *Design'd my ruin; therefore to this end*
> *He sent him harnest out, and he with rage*
> *That hellish was, did fiercely me Ingage:*
> *But blessed* Michael *helped me, and I*
> *By dint of Sword did quickly make him flye;*
> *Therefore to him let me give lasting praise,*
> *And thank and bless his holy name always.*

Then there came to him an hand with some of the leaves of the Tree of Life, the which *Christian* took, and applyed to the wounds that he had received in the Battel, and was healed immediately. He also sat down in that place to eat Bread, and to drink of the Bottle that was given him a little before; so being refreshed, he addressed himself to his Journey, with his *Sword drawn in his hand; for he said, I know not but some other enemy may be at hand. But he met with no other affront from *Apollyon*, quite through this Valley.

• Christian goes on his Journey with his Sword drawn in his hand.

Now at the end of this Valley, was another, called the Valley of the *Shadow of Death*, and *Christian* must needs go through it, because the way to the Cœlestial City lay through the midst of it. Now this Valley is a very solitary place: The Prophet *Jeremiah thus describes it, *A Wilderness, a Land of desarts, and of Pits, a Land of drought, and of the shadow of death, a Land that no Man* (but a Christian) *passeth through, and where no man dwelt.*

• Jer. 2. 6.

Now here *Christian* was worse put to it then in his fight with *Apollyon*, as by the sequel you shall see.

I saw then in my Dream, that when *Christian* was got to the Borders of the Shadow of Death, there met him two Men, *Children of them that brought up an *evil report of the good Land, making haste to go back: to whom *Christian* spake as follows.

• The children of the Spies go back.
• Numb. 13.

Chr. *Whither are you going?*

Men. They said, Back, back; and would have you to do so too, if either life or peace is prized by you.

Chr. Why? what's the matter? said Christian.

Men. Matter! said they; we were going that way as you are going, and went as far as we durst; and indeed we were almost past coming back, for had we gone a little further, we had not been here to bring the news to thee.

Chr. But what have you met with? said Christian.

Men. Why we were almost in the Valley of the Shadow of Death, but that by good hap we looked before us, and saw the danger before we came to it. Psal. 44. 19. Psal. 107. 19.

Chr. But what have you seen? said Christian.

Men. Seen! Why the Valley it self, which is as dark as pitch; we also saw there the Hobgoblins, Satyrs,†and Dragons of the Pit: we heard also in that Valley a continual howling and yelling, as of a People under unutterable misery, who there sat bound in affliction and Irons: and over that Valley hangs the discouraging *Clouds of confusion, death also doth always spread his wings over it: in a word, it is every whit dreadful, being utterly without Order. *Job 3.5. ch. 10.22.

Ch. Then said Christian, *I perceive not yet, by what you have said, but that *this is my way to the desired Haven.* *Jer. 2.6.

Men. Be it thy way, we will not chuse it for ours; so they parted, and *Christian* went on his way, but still with his Sword drawn in his hand, for fear lest he should be assaulted.

I saw then in my Dream, so far as this Valley reached, there was on the right hand a very deep Ditch; that Ditch is it into which the blind have led the blind†in all Ages, and have both there miserably perished. Again, behold on the left hand, there was a very dangerous Quagg, into which, if even a good Man falls, he can find no bottom for his foot to stand on: Into that Quagg *King* David *once did fall,*† and had no doubt therein been smothered, had not He that is able, pluckt him out. Psal. 69. 14.

The path-way was here also exceeding narrow, and therefore good *Christian* was the more put to it; for when he sought in the dark to shun the ditch on the one hand, he was ready to tip over into the mire on the other; also when he sought to escape the mire, without great carefulness he would be ready to fall into the ditch.

Thus he went on, and I heard him here sigh bitterly: for, besides the dangers mentioned above, the path way was here so dark, that oft times when he lift up his foot to set forward, he knew not where, or upon what he should set it next.

About the midst of this Valley, I perceived the mouth of Hell to be, and it stood also hard by the way side: Now thought *Christian*, what shall I do? And ever and anon the flame and smoke would come out in such abundance, with sparks and hideous noises, (things that cared not for *Christians* Sword, as did *Apollyon* before) that he was forced to put up his Sword, and betake himself to another

• Ephes. 6.
18.
• Psal. 116.
4.

weapon called *All-prayer: so he cried in my hearing, **O Lord I beseech thee deliver my Soul*. Thus he went on a great while, yet still the flames would be reaching towards him: also he heard doleful voices, and rushings too and fro, so that sometimes he thought he should be torn in pieces, or trodden down like mire in the Streets. This frightful sight was seen, and these dreadful noises were heard

Christian
put to a
stand, but
for a while.

by him for several miles together: and coming to a place, where he thought he heard a company of *Fiends* coming forward to meet him, he stopt; and began to muse what he had best to do. Somtimes he had half a thought to go back. Then again he thought he might be half way through the Valley; he remembred also how he had already vanquished many a danger: and that the danger of going back might be much more then for to go forward; so he resolved to go on. Yet the *Fiends* seemed to come nearer and nearer, but when they were come even almost at him, he cried out with a most vehement voice, *I will walk in the strength of the Lord God*; so they gave back, and came no further.

One thing I would not let slip, I took notice that now poor *Christian* was so confounded, that he did not know his own voice: and thus I perceived it: Just when he was come over against the mouth of the burning Pit, one of the wicked ones got behind him,

• Christian
made be-
lieve that
he spake
blasphemies,
when 'twas
Satan that
suggested
them into
his mind.

and stept up softly to him, and whisperingly suggested many grievous blasphemies to him, which he *verily thought had proceeded from his own mind. This put *Christian* more to it than any thing that he met with before, even to think that he should now blaspheme him that he loved so much before; yet, could he have helped it, he would not have done it: but he had not the discretion neither to stop his ears, nor to know from whence those blasphemies came.

When *Christian* had travelled in this disconsolate condition some considerable time, he thought he heard the voice of a man, as going before him, saying, *Though I walk through the valley of the shaddow of death, I will fear none ill, for thou art with me.* Psalm 23. 4.

Then was he glad, and that for these reasons:

First, because he gathered from thence, that some who feared God were in this Valley as well as himself.

Secondly, For that he perceived, God was with them, though in that dark and dismal state; and why not, thought he, with me? though by reason of the impediment that attends this place, I cannot perceive it. Job 9. 10.

Thirdly, For that he hoped (could he over-take them) to have company by and by. So he went on, and called to him that was before, but he knew not what to answer; for that he also thought himself to be alone: And by and by, the day broke; then said *Christian, *He hath turned the shadow of death into the morning.* * Amos 5. 8.

Now morning being come, he looked back, not of desire to return, but to see, by the light of the day, what hazards he had gone through in the dark. So he saw more perfectly the Ditch that was on the one hand, and the Quag that was on the other; also how narrow the way was which lay betwixt them both; also now he saw the Hobgoblins, and Satyrs, and Dragons of the Pit, but all afar off; for after break of day, they came not nigh; yet they were discovered to him, according to that which is written, *He discovereth deep things out of darkness, and bringeth out to light the shadow of death.* *Christian glad at break of day.*

* Job. 12. 22.

Now was *Christian* much affected with his deliverance from all the dangers of his solitary way, which dangers, though he feared them more before, yet he saw them more clearly now, because the light of the day made them conspicuous to him; and about this time the Sun was rising, and this was another mercy to *Christian:* for you must note, that tho the first part of the Valley of the Shadow of Death was dangerous, *yet this second part which he was yet to go, was, if possible, far more dangerous: for from the place where he now stood, even to the end of the Valley, the way was all along set so full of Snares, Traps, Gins, and Nets here, and so full of Pits, *The second part of this Valley very dangerous.*

Pitfalls, deep holes, and shelvings down there, that had it now been dark, as it was when he came the first part of the way, had he had a thousand souls, they had in reason been cast away; but, as I said, just now the Sun was rising. Then said he *His candle shineth on my head, and by his light I go through darkness.*

* Job 29. 3.

In this light therefore he came to the end of the Valley. Now I saw in my Dream, that at the end of this Valley lay blood, bones, ashes, and mangled bodies of men, even of Pilgrims that had gone this way formerly: And while I was musing what should be the reason, I espied a little before me a Cave, where two Giants, *Pope* and *Pagan*, dwelt in old time, by whose Power and Tyranny the Men whose bones, blood, ashes, &c. lay there, were cruelly put to death. But by this place *Christian* went without much danger, whereat I somewhat wondered; but I have learnt since, that *Pagan* has been dead many a day; and as for the other, though he be yet alive, he is by reason of age, and also of the many shrewd brushes that he met with in his younger dayes, grown so crazy and stiff in his joynts, that he can now do little more then sit in his Caves mouth, grinning at Pilgrims as they go by, and biting his nails, because he cannot come at them.

So I saw that *Christian* went on his way, yet at the sight of the *old Man* that sat in the mouth of the *Cave*, he could not tell what to think, specially because he spake to him, though he could not go after him; saying, *You will never mend, till more of you be burned:* but he held his peace, and set a good face on't, and so went by, and catcht no hurt. Then sang *Christian,*

> *O world of wonders! (I can say no less)*
> *That I should be preserv'd in that distress*
> *That I have met with here! O blessed bee*
> *That hand that from it hath delivered me!*
> *Dangers in darkness, Devils, Hell, and Sin,*
> *Did compass me, while I this Vale was in:*
> *Yea, Snares, and Pits, and Traps, and Nets did lie*
> *My path about, that worthless·silly I*
> *Might have been catch't, intangled, and cast down:*
> *But since I live, let JESUS wear the Crown.*

Now as *Christian* went on his way, he came to a little ascent, which was cast up on purpose, that Pilgrims might see before them: up there therefore *Christian* went, and looking forward, he saw *Faithful* before him, upon his Journey. Then said *Christian* aloud, Ho, ho, So-ho; stay, and I will be your Companion. At that *Faithful* looked behind him, to whom *Christian* cried again, Stay, stay, till I come up to you: but *Faithful* answered, *No*, I am upon my life, and the Avenger of Blood is behind me. At this *Christian* was somwhat moved, and putting to all his strength, he quickly got up with *Faithful*, and did also over-run him, so the *last was first*. Then did *Christian* vain-gloriously smile, because he had gotten the start of his Brother: but not taking good heed to his feet, he suddenly stumbled and fell, and could not rise again, until *Faithful* came up to help him. *Christian overtakes Faithful. Christians fall makes Faithful*

Then I saw in my Dream, they went very lovingly on together; and had sweet discourse of all things that had happened to them in their Pilgrimage: and thus *Christian* began. *and he go lovingly together.*

Chr. *My honoured and well beloved Brother* Faithful, *I am glad that I have overtaken you; and that God has so tempered our spirits, that we can walk as Companions in this so pleasant a path.*

Faith. I had thought dear friend, to have had your company quite from our Town, but you did get the start of me; wherefore I was forced to come thus much of the way alone.

Chr. *How long did you stay in the City of* Destruction, *before you set out after me on your Pilgrimage?*

Faith. Till I could stay no longer; for there was great talk presently after you was gone out, that our City would in short time with Fire from Heaven be burned down to the ground.

Chr. *What? Did your Neighbours talk so?*

Faith. Yes, 'twas for a while in every bodies mouth. *Their talk about the Countrey*

Chr. *What, and did no more of them but you come out to escape the danger?* *from whence they came.*

Faith. Though there was, as I said, a great talk thereabout, yet I do not think they did firmly believe it. For in the heat of the discourse, I heard some of them deridingly speak of you, and of your desperate Journey, (for so they called this your Pilgrimage); but I did believe, and do still, that the end of our City will be with Fire and Brimstone from above: and therefore I have made mine escape.

Chr. Did you hear no talk of Neighbour Pliable?

Faith. Yes, *Christian*, I heard that he followed you till he came at the Slow of *Dispond*; where, as some said, he fell in; but he would not be known to have so done: but I am sure he was soundly bedabled with that kind of dirt.

Chr. And what said the Neighbours to him?

How Plyable was accounted of when he got home.

Faith. He hath since his going back been had greatly in derision, and that among all sorts of People: some do mock and despise him, and scarce will any set him on work. He is now seven times worse then if he had never gone out of the City.

Chr. But why should they be so set against him, since they also despise the way that he forsook?

Faith. Oh, they say, Hang him; he is a Turn-Coat, he was not true to his profession: I think God has stired up even his enemies to hiss at him, and make him a Proverb, because he hath forsaken the way.

Prov. 15. 10.

Chr. Had you no talk with him before you came out?

Faith. I met him once in the Streets, but he leered away on the other side, as one ashamed of what he had done; so I spake not to him.

Chr. Well, at my first setting out, I had hopes of that Man; but now I fear he will perish in the overthrow of the City, *for it is happened to him according to the true Proverb, The Dog is turned to his Vomit again, and the Sow that was Washed to her wallowing in the mire.

* 2 Pet. 2. 22.
The Dog and Sow.

Faith. They are my fears of him too: But who can hinder that which will be?

Chr. Well Neighbour *Faithful* said *Christian*, let us leave him, and talk of things that more immediately concern our selves. *Tell me now, what you have met with in the way as you came; for I know you have met with some things, or else it may be writ for a wonder.*

Faith. I escaped the Slow that I perceive you fell into, and got up to the Gate without that danger; only I met with one whose name was *Wanton*, that had like to have done me a mischief.

Faithfull assaulted by Wanton.
* Gen. 39. 11, 12, 13.

Chr. 'Twas well you escaped her Net; *Joseph was hard put to it by her, and he escaped her as you did, but it had like to have cost him his life. But what did she do to you?*

Faith. You cannot think (but that you know somthing) what a flattering tongue she had: she lay at me hard to turn aside with her, promising me all manner of content.

Chr. Nay, she did not promise you the content of a good conscience.

Faith. You know what I mean, all carnal and fleshly content.

Chr. Thank God you have escaped her: The *abhorred of the Lord shall fall into her Ditch. • Prov. 22. 14.

Faith. Nay, I know not whether I did wholly escape her, or no.

Chr. Why, I tro you did not consent to her desires?

Faith. No, not to defile my self; for I remembered an old writing that I had seen, which saith, *Her steps take hold of Hell.* So I shut mine eyes, because I would not be bewitched with her looks: then she railed on me, and I went my way. Prov. 5. 5. Job. 31. 1.

Chr. Did you meet with no other assault as you came?

Faith. When I came to the foot of the Hill called *Difficulty*, I met with a very aged Man, who asked me, *What I was, and whither bound?* I told him that I was a Pilgrim, going to the Cœlestial City: Then said the old Man, *Thou lookest like an honest fellow; Wilt thou be content to dwell with me, for the wages that I shall give thee?* Then I asked him his name, and where he dwelt? He said he name was *Adam the first,* and I dwell in the Town of *Deceit. I asked him then, What was his work? and what the wages that he would give? He told me, That his work was *many delights*; and his wages, that I should be his Heir at last. I further asked him, What House he kept, and what other Servants he had? so he told me, *That his House was maintained with all the dainties in the world, and that his Servants were those of his own begetting.* Then I asked how many children he had, He said, that he had but three Daughters, *The *lust of the flesh, the lust of the eyes, and the pride of life,* and that I should marry them all, if I would. Then I asked, how long time he would have me live with him? And he told me, *As long as he lived himself.* *He is assaulted by* Adam *the first.* • Eph. 4. 22. • 1 Joh. 2. 16.

Chr. Well, and what conclusion came the Old Man, *and you to, at last?*

Faith. Why, at first I found my self somewhat inclinable to go with the Man, for I thought he spake very fair; but looking in his forehead as I talked with him, I saw there written, *Put off the old Man with his deeds.*

Chr. And how then?

Faith. Then it came burning hot into my mind, whatever he said, and however he flattered, when he got me home to his House, he would sell me for a Slave. So I bid him forbear to talk, for I would not come near the door of his House. Then he reviled me, and told me, that he would send such a one after me, that should make my way bitter to my soul: So I turned to go away from him: but just as I turned my self to go thence, I felt him take hold of my flesh, and give me such a deadly twitch back, that I thought he had pull'd part of me after himself: This made me cry, *O wretched Man!* So I went on my way up the Hill.

• Rom. 7. 24.

Now when I had got about half way up, I looked behind me, and saw one coming after me, swift as the wind; so he overtook me just about the place where the Settle stands.

Chr. *Just there, said* Christian, *did I sit down to rest me; but being overcome with sleep, I there lost this Roll out of my bosom.*

Faith. But good Brother hear me out: So soon as the Man overtook me, he was but a word and a blow: for down he knockt me, and laid me for dead. But when I was a little come to my self again, I asked him wherefore he served me so? he said, Because of my secret inclining to *Adam the first*; and with that, he strook me another deadly blow on the brest, and beat me down backward; so I lay at his foot as dead as before. So when I came to my self again, I cried him mercy; but he said, I know not how to show mercy, and with that knockt me down again. He had doubtless made an end of me, but that one came by, and bid him forbear.

Chr. Who was that, that bid him forbear?

Faith. I did not know him at first, but as he went by, I perceived the holes in his hands, and his side; then I concluded that he was our Lord. So I went up the Hill.

• The temper of Moses.

Chr. That Man that overtook you, was Moses;* he spareth none, neither knoweth he how to shew mercy to those that transgress his Law.*

Faith. I know it very well, it was not the first time that he has met with me. 'Twas he that came to me when I dwelt securely at home, and that told me, He would burn my house over my head, if I staid there.

Chr. But did not you see the house that stood there on the top of that Hill on the side of which Moses *met you?*

Faith. Yes, and the Lions too, before I came at it, but for the Lions, I think they were a sleep, for it was about Noon; and because I had so much of the day before me, I passed by the Porter, and came down the Hill.

Chr. He told me indeed that he saw you go by, but I wish you had called at the House; for they would have shewed you so many Rarities, that you would scarce have forgot them to the day of your death. But pray tell me, did you meet no body in the Valley of Humility?

Faith. Yes, I met with one *Discontent,* who would willingly have perswaded me to go back again with him: his reason was, for that the Valley was altogether without *Honour;* he told me moreover, That there to go, was the way to disobey all my Friends, as Pride, Arrogancy, Self-conceit, worldly Glory, with others, who he knew, as he said, would be very much offended, if I made such a Fool of my self, as to wade through this Valley. Faithful *assaulted by* Discontent.

Chr. Well, and how did you answer him?

Faith. I told him, that although all these that he named might claim kindred of me, and that rightly, (for indeed they were my Relations, *according to the flesh*) yet since I became a Pilgrim, they have disowned me, as I also have rejected them; and therefore they were to me now no more then if they had never been of my Linage; I told him moreover, That as to this Valley, he had quite mis-represented the thing: *for before Honour is Humility, and a haughty spirit before a fall.* Therefore said I, I had rather go through this Valley to the Honour that was so accounted by the wisest, then chuse that which he esteemed most worth our affections. Faithful's *answer to* Discontent.

Chr. Met you with nothing else in that Valley?

Faith. Yes, I met with *Shame,* But of all the Men that I met with in my Pilgrimage, he, I think, bears the wrong name: the other would be said nay, after a little argumentation (and some what else) but this bold faced *Shame* would never have done. *He is assaulted with* Shame.

Chr. Why, what did he say to you?

Faith. What! why he objected against Religion it self; he said it was a pitiful, low, sneaking business for a man to mind Religion; he said that a tender conscience was an unmanly thing, and that for Man to watch over his words and ways, so as to tye up himself from that hectoring liberty, that the brave spirits of the times accustom

themselves unto, would make him the Ridicule of the times. He
objected also, that but few of the Mighty, Rich, or Wise, were ever
of my opinion; nor any of them neither, before they were perswaded
to be Fools, and to be of a voluntary fondness, to venture the loss
of all, *for no body else knows what.* He moreover objected *the base and
low estate and condition of those that were chiefly the Pilgrims;
also their ignorance of the times in which they lived, and want of
understanding in all natural Science. Yea, he did hold me to it at
that rate also, about a great many more things then here I relate;
as, that it was a *shame* to sit whining and mourning under a Sermon,
and a *shame* to come sighing and groaning home. That it was a *shame*
to ask my Neighbour forgiveness for petty faults, or to make resti-
tution where I had taken from any: He said also that Religion made
a man grow strange to the great, because of a few vices (which he
called by finer names) and made him own and respect the base,
because of the same Religious fraternity. And is not this, said he,
a *shame?*

 Chr. *And what did you say to him?*

 Faith. Say! I could not tell what to say at the first. Yea, he put
me so to it, that my blood came up in my face, even this *Shame*
fetch'd it up, and had almost beat me quite off. But at last I began
to consider, *That that which is highly esteemed among Men, is had in
abomination with God.* And I thought again, this *Shame* tells me what
men are, but it tells me nothing what God, or the Word of God is.
And I thought moreover, That at the day of doom, we shall not be
doomed to death or life, according to the hectoring spirits of the
world; but according to the Wisdom and Law of the Highest.
Therefore thought I, what God says, is best, though all the men in
the world are against it. Seeing then, that God prefers his Religion,
seeing God prefers a tender Conscience, seeing they that make
themselves Fools for the Kingdom of Heaven, are wisest; and that
the poor man that loveth Christ, is richer then the greatest man
in the world that hates him; *Shame* depart, thou art an enemy to
my Salvation: shall I entertain thee against my Soveraign Lord?
How then shall I look him in the face at his coming? Should I now
be *ashamed* of his ways and Servants, how can I expect the blessing?
But indeed this *Shame* was a bold Villain; I could scarce shake him

1 Cor. 1.
26. ch. 3.
18.
Phi. 3. 7, 8.
• John 7.
48.

• Luke 16.
15.

Mar. 8. 38.

out of my company; yea, he would be haunting of me, and continually whispering me in the ear, with some one or other of the infirmities that attend Religion: but at last I told him, 'Twas but in vain to attempt further in this business; for those things that he disdained, in those did I see most glory: And so at last I got past this *importunate* one: And when I had shaken him off, then I began to sing.

> *The tryals that those men do meet withal*
> *That are obedient to the Heavenly call,*
> *Are manifold and suited to the flesh,*
> *And come, and come, and come again afresh;*
> *That now, or somtime else, we by them may*
> *Be taken, overcome, and cast away.*
> *O let the Pilgrims, let the Pilgrims then,*
> *Be vigilant, and quit themselves like Men.*

Chr. I am glad, my Brother, that thou didst withstand this *Villain* so bravely; for of all, as thou sayst, I think he has the wrong name: for he is so bold as to follow us in the Streets, and to attempt to put us to shame *before all men*; that is, to make us ashamed of that which is good: but if he was not himself audacious, he would never attempt to do as he does, but let us still resist him: for notwithstanding all his Bravadoes, he promoteth the Fool, and none else. The Wise shall Inherit Glory, said *Solomon*, but shame shall Prov. 3. 35. be the promotion of Fools.

Faith. I think we must cry to him for help against shame, that would have us be valiant for Truth upon the Earth.

Chr. You say true. But did you meet no body else in that Valley?

Faith. No not I, for I had Sun-shine all the rest of the way, through that, and also through the Valley of the Shadow of Death.

Chr. 'Twas well for you, I am sure it fared far otherwise with me. I had for a long season, as soon almost as I entred into that Valley, a dreadful Combat with that foul Fiend *Apollyon*: Yea, I thought verily he would have killed me; especially when he got me down, and crusht me under him, as if he would have crusht me to pieces. For as he threw me, my Sword flew out of my hand; nay he told me, *He was sure of me*: but *I cried to God, and he heard me, and delivered me out of all my troubles.* Then I entred into the Valley of the Shadow

of Death, and had no light for almost half the way through it. I
thought I should a been killed there, over, and over: but at last,
day brake, and the Sun rose, and I went through that which was
behind with far more ease and quiet.

Moreover, I saw in my Dream, that as they went on, *Faithful*, as
he chanced to look on one side, saw a Man whose name is *Talkative*,
walking at a distance besides them, (for in this place there was room
enough for them all to walk). *He was a tall Man, and somthing more
comely at a distance then at hand.* To this Man, *Faithful* addressed himself in this manner.

Talkative described.

Faith. *Friend, Whither away? Are you going to the Heavenly Countrey?*

Talk. I am going to that same place.

Faith. *That is well: Then I hope we may have your good company.*

Talk. With a very good will, will I be your companion.

Faithful and Talkative enter discourse.

Faith. *Come on then, and let us go together, and let us spend our time in
discoursing of things that are profitable.*

Talk. To talk of things that are good, to me is very acceptable,
with you or with any other; and I am glad that I have met with
those that incline to so good a work. For to speak the truth, there
are but few that care thus to spend their time (as they are in their
travels) but chuse much rather to be speaking of things to no profit,
and this hath been a trouble to me.

Talkatives dislike of bad discourse.

Faith. *That is indeed a thing to be lamented; for what things so worthy
of the use of the tongue and mouth of men on Earth, as are the things of the
God of Heaven?*

Talk. I like you wonderful well, for your saying is full of conviction; and I will add, What thing so pleasant, and what so profitable,
as to talk of the things of God?

What things so pleasant? (that is, if a man hath any delight in
things that are wonderful) for instance: If a man doth delight to
talk of the History or the Mystery of things; or if a man doth love
to talk of Miracles, Wonders, or Signs, where shall he find things
Recorded so delightful, and so sweetly penned, as in the holy
Scripture?

Faith. *That's true: but to be profited by such things in our talk, should
be that which we design.*

Talk. That is it that I said; for to talk of such things is most profitable, for by so doing, a Man may get knowledge of many things;[†]as of the vanity of earthly things, and the benefit of things above: (thus in general) but more particularly, by this a man may learn the necessity of the New-birth, the insufficiency of our works, the need of Christs righteousness, *&c*. Besides, by this a man may learn by *talk*, what it is to repent, to believe, to pray, to suffer, or the like: by this also a Man may learn what are the great promises & consolations of the Gospel, to his own comfort. Further, by this a Man may learn to refute false opinions, to vindicate the truth, and also to instruct the ignorant.

> Talkatives fine discourse.

Faith. *All this is true, and glad am I to hear these things from you.*

Talk. Alas! the want of this is the cause that so few understand the need of faith, and the necessity of a work of Grace in their Soul, in order to eternal life: but ignorantly live in the works of the Law, by which a man can by no means obtain the Kingdom of Heaven.

Faith. *But by your leave, Heavenly knowledge of these, is the gift of God; no man attaineth to them by humane industry, or only by the talk of them.*

Talk. All this I know very well. For a man can receive nothing except it be given him from Heaven; all is of Grace, not of works: I could give you an hundred Scriptures for the confirmation of this.

> O brave Talkative.

Faith. *Well then, said* Faithful; *what is that one thing, that we shall at this time found our discourse upon?*

Talk. What you will: I will talk of things heavenly, or things earthly; things Moral, or things Evangelical; things Sacred, or things Prophane; things past, or things to come; things forraign, or things at home; things more Essential, or things Circumstantial: provided that all be done to our profit.

> O brave Talkative.

Faith. Now did *Faithful* begin to wonder; *and stepping to* Christian, *(for he walked all this while by himself) he said to him, (but softly) What a brave Companion have we got! Surely this man will make a very excellent Pilgrim.*

> Faithful beguiled by Talkative.

Chr. At this *Christian* modestly smiled, and said, This man with whom you are so taken, will beguile with this tongue of his, twenty of them that know him not.

> Christian makes a discovery of Talkative, telling Faithful who he was.

Faith. *Do you know him then?*

Chr. Know him! Yes, better then he knows himself.

Faith. *Pray what is he?*

Chr. His name is *Talkative*, he dwelleth in our Town; I wonder that you should be a stranger to him, only I consider that our Town is large.

Faith. *Whose Son is he? And whereabout doth he dwell?*

Chr. He is the Son of one *Saywell*, he dwelt in *Prating-row*; and he is known of all that are acquainted with him, by the name of *Talkative* in *Prating-row*, and notwithstanding his fine tongue, he is but a sorry fellow.

Faith. *Well, he seems to be a very pretty man.*

Chr. That is, to them that have not through acquaintance with him, for he is best abroad, near home he is ugly enough: your saying, That he is a *pretty man*, brings to my mind what I have observed in the work of the Painter, whose Pictures shews best at a distance; but very near, more unpleasing.

Faith. *But I am ready to think you do but jest, because you smiled.*

Chr. God-forbid that I should *jest*, (though I smiled) in this matter, or that I should accuse any falsely; I will give you a further discovery of him: This man is for any company, and for any *talk*; as he *talketh now* with you, so will he *talk* when he is on the *Ale-bench*: And the more drink he hath in his crown, the more of these things he hath in his mouth: Religion hath no place in his heart, or house, or conversation; all he hath lieth in his *tongue*, and his Religion is to make a noise *therewith*.

Faith. *Say you so! Then I am in this man greatly deceived.*

Chr. Deceived? you may be sure of it. Remember the Proverb, *They say and do not: but the Kingdom of God is not in word, but in power.* He *talketh* of Prayer, of Repentance, of Faith, and of the New-birth: but he knows but only to *talk* of them. I have been in his Family, and have observed him both at home and abroad; and I know what I say of him is the truth. His house is as empty of Religion, *as the white of an Egg is of savour.* There is there, neither Prayer, nor sign of Repentance for sin: Yea, the bruit in his kind serves God far better than he. He is the very stain, reproach, and shame of Religion to all that know him; it can hardly have a good word in all that end of the Town where he dwells, through him. Thus say the common People that know him, *A Saint abroad, and a Devil at home*: His poor Family finds it so, he is such a *churl*, such a railer at,

Mat. 23.
1 Cor. 4. 20.
Talkative talks, but does not.

His house is empty of Religion.

He is a stain to Religion, Rom. 2. 24, 25.

The Proverb that goes of him.

and so unreasonable with his Servants, that they neither know how to do for, or speak to him. Men that have any dealings with him, say, 'tis better to deal with a *Turk* then with him, for fairer dealing they shall have at their hands. This *Talkative*, if it be possible, will go beyond them, defraud, beguile, and over-reach them. Besides, he brings up his Sons to follow his steps; and if he findeth in any of them *a foolish timorousness*, (for so he calls the first appearance of a tender conscience) he calls them fools and blockheads; and by no means will imploy them in much, or speak to their commendations before others. For my part I am of opinion, that he has, by his wicked life, caused many to stumble and fall; and will be, if God prevent not, the ruine of many more. *Men shun to deal with him.*

Faith. *Well, my Brother, I am bound to believe you; not only because you say you know him, but also because like a Christian, you make your reports of men. For I cannot think that you speak these things of ill will, but because it is even so as you say.*

Chr. Had I known him no more than you, I might perhaps have thought of him as at the first you did: Yea, had he received this report, at *their* hands only, that are enemies to Religion, I should have thought it had been a slander (A Lot that often falls from bad mens mouths upon good mens names and professions): But all these things, yea, and a great many more as bad, of my own knowledge I can prove him guilty of. Besides, good men are ashamed of him, they can neither call him *Brother* nor *Friend*: the very naming of him among them, makes them blush, if they know him.

Faith. *Well, I see that Saying, and Doing are two things, and hereafter I shall better observe this distinction.*

Chr. They are two things indeed, and are as diverse as are the Soul and the Body: For as the Body without the Soul, is but a dead Carkass; so, *Saying*, if it be alone, is but a dead Carkass also. The Soul of Religion is the practick part: *Pure Religion and undefiled, before God and the Father, is this, To visit the Fatherless and Widows in their affliction, and to keep himself unspotted from the World.* This *Talkative* is not aware of, he thinks that *hearing* and *saying* will make a good Christian and thus he deceiveth his own Soul. Hearing is but as the sowing of the Seed; talking is not sufficient to prove that *The Carkass of Religion.* *James 1. 27. see ver. 22, 23, 24, 25, 26.*

fruit is indeed in the heart and life; and let us assure our selves, that at the day of Doom, men shall be judged according to their fruits. It will not be said then, *Did you believe?* but, Were you *Doers*, or *Talkers* only? and accordingly shall they be judged. The end of the world is compared to our Harvest, and you know men at Harvest regard nothing but Fruit. Not that any thing can be accepted that is not of Faith: But I speak this to shew you how insignificant the profession of *Talkative* will be at that day.

See Mat. 13. and ch. 25.

Faith. This brings to my mind that of Moses, *by which he describeth the beast that is clean. He is such an one that parteth the Hoof, and cheweth the Cud: Not that parteth the Hoof only, or that cheweth the Cud only. The Hare cheweth the Cud, but yet is unclean, because he parteth not the Hoof. And this truly resembleth* Talkative; *he cheweth the Cud, he seeketh knowledge, he cheweth upon the Word, but he divideth not the Hoof, he parteth not with the way of sinners; but as the Hare he retaineth the foot of a Dog, or Bear, and therefore he is unclean.*

Lev. 11. Deut. 14.

Faithful convinced of the badness of Talkative.

Chr. You have spoken, for ought I know, the true Gospel sense of those Texts; and I will add an other thing. *Paul* calleth some men, yea, and those great Talkers too, *sounding Brass, and Tinckling Cymbals*; that is, as he Expounds them in another place, *Things without life, giving sound.* Things without life, that is, without the true Faith and Grace of the Gospel; and consequently, things that shall never be placed in the Kingdom of Heaven among those that are the Children of life: Though their *sound* by their *talk*, be as if it were the *Tongue*, or voice of an Angel.

1 Cor. 13. 1, 2, 3, ch. 14. 7.

Talkative, like to things that sound without life.

Faith. Well, I was not so fond of his company at first, but I am as sick of it now. What shall we do to be rid of him?

Chr. Take my advice, and do as I bid you, and you shall find that he will soon be sick of your Company too, except God shall touch his heart and turn it.

Faith. What would you have me to do?

Chr. Why, go to him, and enter into some serious discourse about *the power of Religion*: And ask him plainly (when he has approved of it, for that he will) whether this thing be set up in his Heart, House, or Conversation.

Faith. Then *Faithful* stept forward again, and said to *Talkative: Come, what chear? how is it now?*

Talk. Thank you, well. I thought we should have had a great deal of *Talk* by this time.

Faith. *Well, if you will, we will fall to it now; and since you left it with me to state the question, let it be this: How doth the saving Grace of God discover it self, when it is in the heart of man?*

Talk. I perceive then that our talk must be *about the power of things;* Well, 'tis a very good question, and I shall be willing to answer you. And take my answer in brief thus. First, *Where the Grace of God is in the heart, it causeth* there *a great out-cry against sin.* Secondly——

Faith. *Nay hold, let us consider of one at once: I think you should rather say, It showes it self by inclining the Soul to abhor its sin.*

Talk. Why, what difference is there between crying out against, and abhoring of sin?

Faith. *Oh! a great deal; a man may cry out against sin, of policy; but he cannot abhor it, but by vertue of a godly antipathy against it: I have heard many cry out against sin in the Pulpit, who yet can abide it well enough in the heart, and house, and conversation.* Josephs *Mistris cried out with a loud voice, as if she had been very holy; but she would willingly, notwithstanding that, have committed uncleanness with him. Some cry out against sin, even as the Mother cries out against her Child in her lap, when she calleth it slut, and naughty Girl, and then falls to hugging and kissing it.*

Talk. You lie at the catch, I perceive.†

Faith. *No not I, I am only for setting things right. But what is the second thing whereby you would prove a discovery of a work of grace in the heart?*

Talk. Great knowledge of Gospel Mysteries.

Faith. *This sign should have been first, but first or last, it is also false; Knowledge, great knowledge may be obtained in the mysteries of the Gospel, and yet no work of grace in the Soul. Yea, if a man have all knowledge, he may yet be nothing, and so consequently be no child of God. When Christ said, Do you know all these things? And the Disciples had answered, Yes: He addeth, Blessed are ye if ye do them.†He doth not lay the blessing in the knowing of them, but in the doing of them. For there is a knowledge that is not attended with doing: He that knoweth his Masters will and doth it not. A man may know like an Angel, and yet be no Christian: therefore your sign is not true. Indeed to know, is a thing that pleaseth Talkers and Boasters; but to do, is that which pleaseth God. Not that the heart can be good without knowledge; for without that the heart is naught:*

Marginal notes:

Talkatives false discovery of a work of grace.

To cry out against sin, no sign of Grace.

Gen. 39. 15.

Great knowledge no sign of grace. 1 Cor. 13.

Knowledge and knowledge. There is therefore knowledge, and knowledge. *Knowledge that resteth in the bare speculation of things, and knowledge that is accompanied with the grace of faith and love, which puts a man upon doing even the will of God from the heart: the first of these will serve the Talker, but without the other the* *True* *true Christian is not content.* Give me understanding, and I shall keep *Knowledge attended with en-deavours.* thy Law, yea, I shall observe it with my whole heart, *Psal.* 119. 34.

Talk. You lie at the catch again, this is not for edification.

Faith. *Well, if you please propound another sign how this work of grace discovereth it self where it is.*

Talk. Not I, for I see we shall not agree.

Faith. *Well, if you will not, will you give me leave to do it?*

Talk. You may use your Liberty.

One good sign of grace. Faith. *A work of grace in the soul discovereth it self, either to him that hath it, or to standers by.*

John 16. 8.
Rom. 7. 24.
John 16. 9.
Mark 16. 16.
Psal. 38. 18.
Jer. 31. 19.
Gal. 2. 16.
Acts 4. 12.
Matth. 5. 6.
Rev. 21. 6.

To him that hath it, thus. It gives him conviction of sin, especially of the defilement of his nature, and the sin of unbelief, (for the sake of which he is sure to be damned, if he findeth not mercy at Gods hand by faith in Jesus Christ.) This sight and sense of things worketh in him sorrow and shame for sin; he findeth moreover revealed in him the Saviour of the World, and the absolute necessity of closing with him, for life, at the which he findeth hungrings and thirstings after him, to which hungrings, &c. the promise is made. Now according to the strength or weakness of his Faith in his Saviour, so is his joy and peace, so is his love to holiness, so are his desires to know him more, and also to serve him in this World. But though I say it discovereth it self thus unto him; yet it is but seldom that he is able to conclude that this is a work of Grace, because his corruptions now, and his abused reason, makes his mind to mis-judge in this matter; therefore in him that hath this work, there is required a very sound Judgement, before he can with steddiness conclude that this is a work of Grace.

Rom. 10. 10.
Phil. 1. 27.
Matth. 5. 9.
John 24.
Psal. 50. 20.
Job 42. 5, 6.
Ezek. 29. 43.

To others it is thus discovered.

1. *By an experimental confession of his Faith in Christ.* 2. *By a life answerable to that confession, to wit, a life of holiness; heart-holiness, family-holiness (if he hath a Family) and by Conversation-holiness in the world: which in the general teacheth him, inwardly to abhor his sin, and himself for that in secret, to suppress it in his Family, and to promote holiness, in the World; not by talk only, as an Hypocrite or* Talkative *person may do: but*

by a practical Subjection in Faith, and Love, to the power of the word: And now Sir, as to this brief description of the work of Grace, and also the discovery of it, if you have ought to object, object: if not, then give me leave to propound to you a second question.

Talk. Nay, my part is not now to object, but to hear, let me therefore have your second question. *Another good sign of grace.*

Faith. It is this, *Do you experience the first part of this description of it? and doth your life and conversation testifie the same? or standeth your Religion in Word, or in Tongue, and not in Deed and Truth: pray, if you incline to answer me in this, say no more then you know the God above will say Amen to; and also, nothing but what your Conscience can justifie you in. For, not he that commendeth himself is approved, but whom the Lord commendeth.[†] Besides, to say I am thus, and thus, when my Conversation, and all my Neighbours tell me, I lye, is great wickedness.*

Talk. Then *Talkative* at first began to blush, but recovering himself, thus he replyed, You come now to Experience, to Conscience, and God: and to appeals to him for justification of what is spoken: This kind of discourse I did not expect, nor am I disposed to give an answer to such questions, because, I count not my self bound thereto, unless you take upon you to be a *Catechizer*; and, though you should so do, yet I may refuse to make you my Judge: But I pray will you tell me, why you ask me such questions? *Talkative not pleased with Faithfuls question.*

Faith. Because I saw you forward to talk, and because I knew not that you had ought else but notion. Besides, to tell you all the truth, I have heard of you, that you are a Man whose Religion lies in talk, and that your conversation gives this your Mouth-profession the lye. They say You are a spot among Christians, and that Religion fareth the worse for your ungodly conversation, and some already have stumbled at your wicked ways, and that more are in danger of being destroyed thereby; your Religion, and an Ale-house, and Covetousness, and uncleanness, and swearing, and lying, and vain Company-keeping, &c. will stand together. The Proverb is true of you, which is said of a Whore; to wit, That she is a shame to all Women; so you are a shame to all Professors. *The reasons why Faithful put to him that question. Faithfuls plain dealing to Talkative.*

Talk. Since you are ready to take up reports, and to judge so rashly as you do; I cannot but conclude you are some peevish, or melancholly man not fit to be discoursed with, and so adieu. *Talkative flings away from Faithful.*

Chr. Then came up *Christian* and said to his Brother, I told you how it would happen, your words and his lusts could not agree; he had rather leave your company, then reform his life: but he is *A good* gone as I said, let him go; the loss is no mans but his own, he has *riddance.* saved us the trouble of going from him: for he continuing, as I suppose he will do, as he is, he would have been but a blot in our Company: besides, the Apostle says, *From such withdraw thy self.*[†]

Faith. *But I am glad we had this little discourse with him, it may happen that he will think of it again; however, I have dealt plainly with him; and so am clear of his blood, if he perisheth.*

Chr. You did well to talk so plainly to him as you did; there is but little of this faithful dealing with men now a days, and that makes Religion to stink in the nostrills of many, as it doth: for they are these *Talkative* Fools, whose Religion is only in word, and are debauched and vain in their Conversation, that (being so much admitted into the Fellowship of the Godly) do stumble the World, blemish Christianity, and grieve the Sincere. I wish that all Men would deal with such, as you have done, then should they either be made more conformable to Religion, or the company of Saints would be too hot for them. Then did Faithful say,

> How Talkative *at first lifts up his Plumes!*
> How bravely *doth he speak! how he presumes*
> To drive down *all before him! but so soon*
> As Faithful *talks of* Heartwork, *like the Moon*
> That's past the full, *into the wain he goes;*
> And so will all, *but he that* Heartwork *knows.*

Thus they went on talking of what they had seen by the way; and so made that way easie, which would otherwise, no doubt, have been tedious to them: for now they went through a Wilderness.

Now when they were got almost quite out of this Wilderness, *Faithful* chanced to cast his eye back, and espied one coming after them, and he knew him. Oh! said *Faithful* to his Brother, who comes yonder? Then *Christian* looked, and said, It is my good friend *Evangelist.* Ai, and my good friend too, said *Faithful*; for 'twas he that set *Evangelist* me the way to the Gate. Now was *Evangelist* come up unto them, *overtakes* and thus saluted them.
them again.

Evan. Peace be with you, dearly beloved, and, peace be to your helpers.[†]

Chr. *Welcome, welcome, my good* Evangelist, *the sight of thy countenance brings to my remembrance, thy ancient kindness, and unwearied laboring for my eternal good.* **They are glad at the sight of him.**

Faith. *And, a thousand times welcome, said good* Faithful; *Thy company, O sweet* Evangelist, *how desirable is it to us, poor Pilgrims!*

Evan. Then, said *Evangelist,* How hath it fared with you, my friends, since the time of our last parting? *what* have you met with, and *how* have you behaved your selves?

Chr. *Then* Christian, *and* Faithful *told him of all things that had happened to them in the way; and* how, *and with* what *difficulty they had arrived to that place.*

Evang. Right glad am I, said *Evangelist*; not that you met with trials, but that you have been victors; and for that you have (notwithstanding many weaknesses,) continued in the way to this very day. **His exhortation to them.**

I say, right glad am I of this thing,[†] and that for mine own sake and yours; I have sowed, and you have reaped, and the day is coming, when both he that sowed, and they that reaped shall rejoyce together; that is, if you hold out: for, in due time ye shall reap, if you faint not. The Crown is before you, and it is an incorruptible one; so run that you may obtain it. Some there be that set out for this Crown, and after they have gone far for it, another comes in, and takes it from them; hold fast therefore that you have, let no man take your Crown; you are not yet out of the gun-shot of the Devil: you have not resisted unto blood, striving against sin: let the Kingdom be always before you, and believe stedfastly concerning things that are invisible. Let nothing that is on this side the other world get within you; and above all, look well to your own hearts; and to the lusts thereof; for they are deceitful above all things, and desperately wicked; set your faces like a flint, you have all power in Heaven and Earth on your side. **John 4. 36. Gal. 6. 9. 1 Cor. 9. 24, 25, 26, 27. Rev. 3. 11.**

Chr. Then **Christian thanked him for his exhortation, but told him withal, that they would have him speak farther to them for their help, the rest of the way; and the rather, for that they well knew that he was a Prophet, and could tell them of things that might happen unto them; and also* *** They do thank him for his exhortation.**

how they might resist and overcome them. To which request Faithful *also* consented. So Evangelist *began as followeth.*

• He pre-
dicteth
what
troubles
they shall
meet with
in Vanity
Fair, and
encourageth
them to
stedfastness.

*Evan.** My Sons, you have heard in the words of the truth of the Gospel, that you must through many tribulations enter into the Kingdom of Heaven. And again, that in every City, bonds and afflictions abide in you; and therefore you cannot expect that you should go long on your Pilgrimage without them, in some sort or other. You have found something of the truth of these testimonies upon you already, and more will immediately follow: for now, as you see, you are almost out of this Wilderness, and therefore you will soon come into a Town that you will by and by see before you: and in that Town you will be hardly beset with enemies, who will strain hard but they will kill you: and be you sure that one or both of you must seal the testimony which you hold, with blood: but be you faithful unto death, and the King will give you a Crown of life.

• He whose
lot it will be
there to
suffer, will
have the
better of his
brother.

*He that shall die there, although his death will be unnatural, and his pain perhaps great, he will yet have the better of his fellow; not only because he will be arrived at the Cœlestial City soonest, but because he will escape many miseries that the other will meet with in the rest of his Journey. But when you are come to the Town, and shall find fulfilled what I have here related, then remember your friend and quit your selves like men; and commit the keeping of your souls to your God, as unto a faithful Creator.

Then I saw in my Dream, that when they were got out of the Wilderness, they presently saw a Town before them, and the name of that Town is *Vanity*; and at the Town there is a *Fair* kept called *Vanity-Fair*:†It is kept all the year long, it beareth the name of

Vanity-Fair, because the Town where tis kept, *is lighter then* Vanity; and also, because all that is there sold, or that cometh thither, is *Vanity.* As is the saying of the wise, *All that cometh is vanity.*

This Fair is no new erected business, but a thing of Ancient standing; I will shew you the original of it.

Almost five thousand years agone,†there were Pilgrims walking to the Cœlestial City, as these two honest persons are; and *Beelzebub, Apollyon,* and *Legion,*†with their Companions, perceiving by the path that the Pilgrims made, that their way to the City lay through *this Town* of *Vanity,* they contrived here to set up a Fair; a Fair wherein should be sold of *all sorts of Vanity*, and that it should last all the year long. Therefore at *this Fair* are all such Merchandize sold, as Houses, Lands, Trades, Places, Honours, Preferments, Titles, Countreys, Kingdoms, Lusts, Pleasures, and Delights of all sorts, as Whores, Bauds, Wives, Husbands, Children, Masters, Servants, Lives, Blood, Bodies, Souls, Silver, Gold, Pearls, Precious Stones, and what not. *The Merchandise of this Fair.*

And moreover, at this Fair there is at all times to be seen Jugglings, Cheats, Games, Plays, Fools, Apes, Knaves, and Rogues, and that of all sorts.

Here are to be seen too, and that for nothing, Thefts, Murders, Adultries, False-swearers, and that of a blood-red colour.

And as in other Fairs of less moment, there are the several Rows and Streets under their proper names, where such and such Wares are vended, so here likewise, you have the proper Places, Rows, Streets, (*viz.* Countreys, and Kingdoms) where the Wares of this Fair are soonest to be found: Here is the *Britain* Row, the *French* Row, the *Italian* Row, the *Spanish* Row, the *German* Row, where several sorts of Vanities are to be sold. But as in other *fairs*, some one Commodity is as the chief of all the *fair*, so the Ware of *Rome* and her Merchandize is greatly promoted in *this fair*: Only our *English* Nation, with some others, have taken a dislike thereat. *The Streets of this fair.*

Now, as I said, the way to the Cœlestial City lyes just thorow *this Town*, where this lusty Fair is kept; and he that will go to the City, and yet not go thorow this Town, must needs *go out of the World*. The Prince of Princes himself, when here, went through *this Town* to his own Countrey, and that upon a *Fair-day* too: Yea, and as I think it was *Beelzebub*, the chief Lord of this *Fair*, that invited him to buy of his *Vanities*; yea, would have made him Lord of the *Fair*, would he but have done him Reverence as he went thorow the *Town*. Yea, because he was such a person of Honour, *Beelzebub* had him from *Street* to *Street*, and shewed him all the Kingdoms of the World in a little time, that he might, if possible alure that Blessed One, to *cheapen* and *buy* some of his *Vanities*. But he had no mind to the Merchandize, and therefore left the *Town*; without laying out so much as one Farthing upon these *Vanities*. This *Fair* therefore is an Ancient thing, of long standing, and a very great *Fair*. *1 Cor. 5. 10. Christ went through this Fair. Matth. 4. 8. Luk. 4. 5, 6, 7. Christ bought nothing in this Fair.*

The Pil-
grims enter
the Fair. Now these Pilgrims, as I said, must needs go thorow this *Fair*:
Well, so they did; but behold, even as they entred into the *Fair*, all
the people in the *Fair* were moved, and the Town it self as it were
The Fair in in a Hubbub about them; and that for several reasons: For,
a hubbub
about them. First, The Pilgrims were cloathed with such kind of Raiment, as
The first was diverse from the Raiment of any that traded in that *fair*. The
cause of the
hubbub. people therefore of the *fair* made a great gazing upon them: Some
said they were Fools, some they were Bedlams, and some they were
Outlandish-men.

1 Cor. 2. 7, Secondly, And as they wondred at their Apparel, so they did
8.
The second likewise at their Speech; for few could understand what they said;
cause of the
hubbub. they naturally spoke the Language of *Canaan*; But they that kept
the *fair*, were the men of this World: So that from one end of the
fair to the other, they seemed *Barbarians*†each to the other.

Third cause
of the
hubbub. Thirdly, But that which did not a little amuse the Merchandizers,
was, that these Pilgrims set very light by all their Wares, they
cared not so much as to look upon them: and if they called upon
them to buy, they would put their fingers in their ears, and cry,
Psal. 119. *Turn away mine eyes from beholding vanity*; and look upwards, signi-
Phil. 3. 19, fying that their Trade and Traffick was in Heaven.

20.
Fourth One chanced mockingly, beholding the carriages of the men, to
cause of the
hubbub. say unto them, What will ye buy? but they, looking gravely upon
Prov. 23. him, said, *We buy the Truth*. At that, there was an occasion taken
23.
They are to despise the men the more; some mocking, some taunting, some
mocked. speaking reproachfully, and some calling upon others to smite them.
The fair in At last things came to an hubbub, and great stir in the *fair*; inso-
a hubbub. much that all order was confounded. Now was word presently
brought to the *great one* of the *fair*, who quickly came down, and
They are deputed some of his most trusty friends to take these men into
examined. examination, about whom the *fair* was almost overturned. So the
men were brought to examination; and they that sat upon them,
asked them whence they came, whither they went, and what they
* *They tell* did there in such an unusual Garb? *The men told them, that they
who they
are and were Pilgrims and Strangers in the world, and that they were going
whence to their own Countrey, which was the Heavenly *Jerusalem*; and that
they came. they had given none occasion to the men of the Town, nor yet to
Heb. 11. the Merchandizers, thus to abuse them, and to let them in their
13, 14, 15,
16.

Journey, except it was, for that, when one asked them what they
would buy, they said, they would *buy the Truth*. But they that were *They are*
appointed to examine them, did not believe them to be any other *not believed.*
then Bedlams and Mad, or else such as came to put all things into
a confusion in the *fair*. Therefore they took them, and beat them,
and besmeared them with dirt, and then put them into the Cage, *They are*
that they might be made a Spectacle to all the men of the *fair*. *put in the*
There therefore they lay for some time, and were made the objects *Cage.*
of any mans sport, or malice, or revenge, the great one of the *fair* *Their*
laughing still at all that befel them. But the men being patient, and *behaviour*
not rendering railing for railing, but contrarywise blessing, and *in the Cage.*
giving good words for bad, and kindness for injuries done: Some men *The men of*
in the *fair* that were more observing, and less prejudiced then the *the Fair do*
rest, began to check and blame the baser sort for their continual *fall out*
abuses done by them to the men: They therefore in angry manner *among them-*
let fly at them again, counting them as bad as the men in the Cage, *selves about*
and telling them that they seemed confederates, and should be made *these two*
partakers of their misfortunes. The other replied, That for ought *men.*
they could see, the men were quiet, and sober, and intended no body
any harm; and that there were many that Traded in their *fair*, that
were more worthy to be put into the Cage, yea, and Pillory too,
then were the men that they had abused. Thus, after divers words
had passed on both sides, (the men behaving themselves all the
while very wisely, and soberly before them) they fell to some Blows,
among themselves, and did harm one to another. Then were these *They are*
two poor men brought before their Examiners again, and there *made the*
charged as being guilty of the late Hubbub that had been in the *Authors of*
fair. So they beat them pitifully, and hanged Irons upon them, and *turbance.*
led them in Chaines up and down the *fair*, for an example and a *They are*
terror to others, lest any should further speak in their behalf, or *fair in*
joyn themselves unto them. But *Christian and Faithful* behaved them- *Chaines, for*
selves yet more wisely, and received the ignominy and shame that *others.*
was cast upon them, with so much meekness and patience, that it *Some of the*
won to their side (though but few in comparison of the rest) several *fair won to*
of the men in the *fair*. This put the other party yet into a greater *them.*
rage, insomuch that they concluded the death of these two men. *Their ad-*
Wherefore they threatned that neither the Cage, nor Irons, should *versaries*
resolve to
kill them.

serve their turn, but that they should die, for the abuse they had done, and for deluding the men of the *fair*.

Then were they remanded to the Cage again, until further order should be taken with them. So they put them in, and made their feet fast in the Stocks.

Here also they called again to mind what they had heard from their faithful friend *Evangelist*, and was the more confirmed in their way and sufferings, by what he told them would happen to them. They also now comforted each other, that whose lot it was to suffer, even he should have the best on't; therefore each man secretly wished that he might have that preferment: but committing themselves to the All-wise dispose of him that ruleth all things, with much content they abode in the condition in which they were, until they should be otherwise disposed of.

Then a convenient time being appointed, they brought them forth to their Tryal in order to their Condemnation. When the time was come, they were brought before their Enemies and arraigned; the Judges name was Lord *Hategood*. Their Indictment was one and the same in substance, though somewhat varying in form; the Contents whereof was this:

That they were enemies to, and disturbers of their Trade; that they had made Commotions and Divisions in the Town, and had won a party to their own most dangerous Opinions, in contempt of the Law of their Prince.

Then *Faithful* began to answer, That he had only set himself against that which had set it self against him that is higher then the highest. And, said he, As for disturbance, I make none, being my self a man of Peace; the Party that were won to us, were won, by beholding our Truth and Innocence, and they are only turned from the worse to the better. And as to the King you talk of; since he is *Beelzebub*, the Enemy of our Lord, I defie him and all his Angels.

Then Proclamation was made, that they that had ought to say for their Lord the King against the Prisoner at the Bar, should forthwith appear, and give in their evidence. So there came in three Witnesses, to wit, *Envy*, *Superstition*, *and Pickthank*. They was then asked, If they knew the Prisoner at the Bar? and what they had to say for their Lord the King against him?

Then stood forth *Envy, and said to this effect; My Lord, I have known this man a long time, and will attest upon my Oath before this honourable Bench, That he is— • *Envy begins.*

Judge. Hold, give him his Oath: So they sware him. Then he said, My Lord, this man, notwithstanding his plausible name, is one of the vilest men in our Countrey; He neither regardeth Prince nor People, Law nor Custom; but doth all that he can to possess all men with certain of his disloyal notions, which he in the general calls Principles of Faith and Holiness. And in particular, I heard him once my self affirm, *That Christianity, and the Customs of our Town of* Vanity, *were Diametrically opposite, and could not be reconciled.* By which saying, my Lord, he doth at once, not only condemn all our laudable doings, but us in the doing of them.

Judg. Then did the Judge say to him, Hast thou any more to say?

Envy. My Lord, I could say much more, only I would not be tedious to the Court. Yet if need be, when the other Gentlemen have given in their Evidence, rather then any thing shall be wanting that will dispatch him, I will enlarge my Testimony against him. So he was bid stand by. Then they called *Superstition*, and bid him look upon the Prisoner; they also asked, What he could say for their Lord the King against him? Then they sware him, so he began.

Super.* My Lord, I have no great acquaintance with this man, nor do I desire to have further knowledge of him; However this I know, that he is a very pestilent fellow, from some discourse that the other day I had with him in this *Town*; for then talking with him, I heard him say, That our Religion was naught, and such by which a man could by no means please God: which sayings of his, my Lord, your Lordship very well knows, what necessarily thence will follow, *to wit*, That we still do worship in vain, are yet in our Sins, and finally shall be damned; and this is that which I have to say. • *Super-stition follows.*

Then was *Pickthank* sworn, and bid say what he knew, in behalf of their Lord the King against the Prisoner at the Bar.

Pick. My Lord, and you Gentlemen all, This fellow I have known of a long time, and have heard him speak things that ought not to be spoke. For he hath railed on our noble Prince *Beelzebub*, and hath spoke contemptibly of his honourable Friends, whose names are the Lord *Old man*, the Lord *Carnal delight*, the Lord *Luxurious*, the Lord *Pickthanks Testimony.* *Sins are all Lords and Great ones.*

Desire of Vain-glory, my old Lord *Lechery*, Sir *Having Greedy*, with all the rest of our Nobility; and he hath said moreover, that if all men were of his mind, if possible, there is not one of these Noble-men should have any longer a being in this Town. Besides, he hath not been afraid to rail on you, my Lord, who are now appointed to be his Judge, calling you an ungodly villain, with many other such like vilifying terms, with which he hath bespattered most of the Gentry of our Town. When this *Pickthank* had told his tale, the Judge directed his speech to the Prisoner at the Bar, saying, Thou Runagate, Heretick, and Traitor, hast thou heard what these honest Gentlemen have witnessed against thee.

Faith. *May I speak a few words in my own defence?*

Judg. Sirrah, Sirrah, thou deservest to live no longer, but to be slain immediately upon the place; yet that all men may see our gentleness towards thee, let us hear what thou hast to say.

Faith. 1. I say then in answer to what Mr. *Envy* hath spoken, I never said ought but this, *That what Rule, or Laws, or Custom, or People, were flat against the Word of God, are diametrically opposite to Christianity.* If I have said a miss in this, convince me of my errour, and I am ready here before you to make my recantation.

Faithfuls defence of himself.

2. As to the second, to wit, Mr. *Superstition,* and his charge against me, I said only this, *That in the worship of God there is required a divine Faith; but there can be no divine Faith, without a divine Revelation of the will of God: therefore whatever is thrust into the worship of God, that is not agreeable to divine Revelation, cannot be done but by an humane Faith, which Faith will not profit to Eternal Life.*

3. As to what Mr. *Pickthank* hath said, I say, (avoiding terms, as that I am said to rail, and the like) That the Prince of this Town, with all the Rablement his Attendants, by this Gentleman named, are more fit for a being in Hell, then in this Town and Countrey; *and so the Lord have mercy upon me.*

The Judge his speech to the Jury.

Then the Judge called to the Jury (who all this while stood by, to hear and observe;) Gentlemen of the Jury, you see this man about whom so great an uproar hath been made in this Town: you have also heard what these worthy Gentlemen have witnessed against him; also you have heard his reply and confession: It lieth now in your brests to hang him, or save his life. But yet I think meet to instruct you into our Law.

There was an Act made in the days of *Pharaoh* the Great, Servant **Exod. 1.** to our Prince, That lest those of a contrary Religion should multiply and grow too strong for him, their Males should be thrown into the River. There was also an Act made in the days of *Nebuchadnezzar* **Dan. 3.** the Great, another of his Servants, That whoever would not fall down and worship his golden Image, should be thrown into a fiery Furnace. There was also an Act made in the days of *Darius*, That **Dan. 6.** who so, for some time, called upon any God but his, should be cast into the Lions Den. Now the substance of these Laws this Rebel has broken, not only in thought, (which is not to be born) but also in word and deed; which must therefore needs be intolerable.

For that of *Pharaoh*, his Law was made upon a supposition, to prevent mischief, no Crime being yet apparent; but here is a Crime apparent. For the second and third, you see he disputeth against our Religion; and for the Treason he hath confessed, he deserveth to die the death.

Then went the Jury out, *whose names were Mr. *Blind-man*, • *The Jury* Mr. *No-good*, Mr. *Malice*, Mr. *Love-lust*, Mr. *Live-loose*, Mr. *Heady*, *and their* Mr. *High-mind*, Mr. *Enmity*, Mr. *Lyar*, Mr. *Cruelty*, Mr. *Hate-light*, *names.* and Mr. *Implacable*, who every one gave in his private Verdict against him among themselves, and afterwards unanimously concluded to bring him in guilty before the Judge. And first Mr. *Blind-* *Every ones* man, the foreman, said, *I see clearly that this man is an Heretick.* Then *private* said Mr. *No-good*, *Away with such a fellow from the Earth. Ay*, said *verdict.* Mr. *Malice*, *for I hate the very looks of him.* Then said Mr. *Love-lust*, *I could never indure him. Nor I*, said Mr. *Live-loose*, *for he would always be condemning my way. Hang him, hang him*, said Mr. *Heady. A sorry Scrub*, said Mr. *High-mind. My heart riseth against him*, said Mr. *Enmity. He is a Rogue*, said Mr. *Lyar. Hanging is too good for him*, said Mr. *Cruelty. Lets dispatch him out of the way*, said Mr. *Hate-light.* Then said Mr. *Implacable, Might I have all the World given me, I could not be reconciled to him, therefore let us forthwith bring him in guilty of death:** • *They con-* And so they did, therefore he was presently Condemned, To be had *clude to* from the place where he was, to the place from whence he came, and *bring him in* there to be put to the most cruel death that could be invented. *guilty of* *death.*

The Cruel death of Faithful. They therefore brought him out, to do with him according to their Law; and first they Scourged him, then they Buffetted him, then they Lanced his flesh with Knives; after that they Stoned him with Stones, then prickt him with their Swords, and last of all they burned him to Ashes at the Stake. Thus came *Faithful* to his end.

• A Chariot and Horses wait to take away Faithful. *Now, I saw that there stood behind the multitude, a Chariot†and a couple of Horses, waiting for *Faithful*, who (so soon as his adversaries had dispatched him) was taken up into it, and straightway was carried up through the Clouds, with sound of Trumpet, the *Christian is still alive.* nearest way to the Cœlestial Gate. But as for *Christian*, he had some respit, and was remanded back to prison; so he there remained for a space: But he that over-rules all things, having the power of their rage in his own hand, so wrought it about, that *Christian* for that time escaped them, and went his way.

And as he went he Sang.

• The Song that Christian made of Faithful after his death.

> *Well* Faithful, *thou hast faithfully profest*
> *Unto thy Lord: with him thou shalt be blest;*
> *When* Faithless *ones, with all their vain delights,*
> *Are crying out under their hellish plights.*
> *Sing*, Faithful, *sing; and let thy name survive;*
> *For though they kill'd thee, thou art yet alive.*

Now I saw in my Dream, that *Christian* went not forth alone, for *Christian has another Companion.* there was one whose name was *Hopeful*, (being made so by the beholding of *Christian* and *Faithful* in their words and behaviour, in their sufferings at the *fair*) who joyned himself unto him, and entring into a brotherly covenant, told him that he would be his Companion. Thus one died to make Testimony to the Truth, and another rises out of his Ashes to be a Companion with *Christian*. *There is more of the men of the fair will follow.* This *Hopeful* also told *Christian*, that there were many more of the men in the *fair* that would take their time and follow after.

They over- take By-ends. So I saw that quickly after they were got out of the *fair*, they overtook one that was going before them, whose name was *By-ends*;† so they said to him, What Countrey-man, Sir? and how far go you this way? He told them, That he came from the Town of *Fair-speech*, and he was going to the Cœlestial City, (but told them not his name.)

From **Fair-speech, said* Christian; *is there any that be good live there?* • Prov. 26.
25.

By-ends. Yes, said *By-ends,* I hope.

Chr. *Pray Sir, what may I call you?* said *Christian.*

By-ends. I am a Stranger to you, and you to me; if you be going By-ends
this way, I shall be glad of your Company; if not, I must be content. *loth to tell his name.*

Chr. *This Town of* Fair-speech *said* Christian, *I have heard of it, and, as I remember, they say its a Wealthy place.*

By-ends. Yes, I will assure you that it is, and I have very many Rich Kindred there.

Chr. *Pray who are your Kindred there, if a man may be so bold;*

By-ends. Almost the whole Town; and in particular, my Lord *Turn-about,* my Lord *Time-server,* my Lord *Fair-speech,* (from whose Ancestors that Town first took its name:) Also Mr. *Smooth-man,* Mr. *Facing-bothways,* Mr. *Any-thing,* and the Parson of our Parish, Mr. *Two-tongues,* was my Mothers own Brother by Father's side: And to tell you the Truth, I am become a Gentleman of good Quality; yet my Great Grand-father was but a Water-man, looking one way, and Rowing another: and I got most of my estate by the same occupation.

Chr. *Are you a Married man?*

By-ends. Yes, and my Wife is a very Virtuous woman, the Daughter *The wife* of a Virtuous woman: She was my Lady *Fainings* Daughter, therefore *and Kindred of* she came of a very Honourable Family, and is arrived to such a pitch *By-ends.* of Breeding, that she knows how to carry it to all, even to Prince and Peasant. 'Tis true, we somewhat differ in religion from those of *Where* the stricter sort,†yet but in two small points: First, we never strive *By-ends differs from* against Wind and Tide. Secondly, we are always most zealous *others in* when Religion goes in his Silver Slippers; we love much to walk *Religion.* with him in the Street, if the Sun shines, and the people applaud it.

Then *Christian* stept a little a toside to his fellow *Hopeful,* saying, It runs in my mind that this is one *By-ends,* of *Fair-speech,* and if it be he, we have as very a Knave in our company, as dwelleth in all these parts. Then said *Hopeful, Ask him; methinks he should not be ashamed of his name.* So *Christian* came up with him again; and said, Sir, you talk as if you knew something more then all the world doth, and if I take not my mark amiss, I deem I have half a guess of you: Is not your name Mr. *By-ends* of Fair-speech?

By-ends. That is not my name, but indeed it is a Nick-name that is given me by some that cannot abide me, and I must be content to bear it as a reproach, as other good men have born theirs before me.

How
By-ends
got his
name. Chr. *But did you never give an occasion to men to call you by this name?*

By-ends. Never, never! The worst that ever I did to give them an occasion to give me this name, was, that I had alwayes the luck to jump in my Judgement with the present way of the times, whatever it was, and my chance was to get thereby; but if things are thus cast upon me, let me count them a blessing, but let not the malicious load me therefore with reproach.

Chr. *I thought indeed that you was the man that I had heard of, and to tell you what I think, I fear this name belongs to you more properly then you are willing we should think it doth.*

He desires
to keep
Company
with
Christian. *By-ends*. Well, if you will thus imagine, I cannot help it. You shall find me a fair Company-keeper, if you will still admit me your associate.

Chr. *If you will go with us, you must go against Wind and Tide, the which, I perceive, is against your opinion: You must also own Religion in his Rags, as well as when in his Silver Slippers, and stand by him too, when bound in Irons, as well as when he walketh the Streets with applause.*

By-ends. You must not impose, nor Lord it over my Faith; leave me to my liberty, and let me go with you.

Chr. *Not a step further, unless you will do in what I propound, as we.*

Then said *By-ends*, I shall never desert my old Principles, since they are harmless and profitable. If I may not go with you, I must do as I did before you overtook me, even go by my self, until some overtake me that will be glad of my company.

Now I saw in my dream, that *Christian* and *Hopeful*, forsook him, and kept their distance before him, but one of them looking back, saw three men following Mr. *By-ends*, and behold, as they came up with him, he made them a very low *Conje*, and they also gave him a *Complement*. The mens names were Mr. *Hold-the-World*, Mr. *Mony-love*, and Mr. *Save-all*;[†] men that Mr. *By-ends*, had formerly bin acquainted with; for in their minority they were Schoolfellows, and were taught by one Mr. *Gripe-man*, a Schoolmaster in *Love-gain*, which is a market town in the County of *Coveting* in the North.

This Schoolmaster taught them the art of getting, either by violence, cousenage, flattery, lying or by putting on a guise of Religion, and these four Gentlemen had attained much of the art of their Master, so that they could each of them have kept such a School themselves.

Well when they had, as I said, thus saluted each other, Mr. *Monylove* said to Mr. *By-ends*, Who are they upon the Road before us? for *Christian* and *Hopeful* were yet within view.

By-ends. They are a couple of far countrey-men, that after *their mode*, are going on Pilgrimage.

Mony-love. Alas, why did they not stay that we might have had their good company, for *they*, and *we*, and *you* Sir, I hope, are all going on Pilgrimage.

By-ends. We are so indeed, but the men before us, are so ridged, and love so much their own notions, and do also so lightly esteem the opinions of others; that let a man be never so godly, yet if he jumps not with them in all things, they thrust him quite out of their company.

Mr. *Save-all*. That's bad; But we read of some, *that are righteous over-much*,† and such mens ridgedness prevails with them to judge and condemn all but themselves. But I pray what and how many, were the things wherein you differed?

By-ends. Why they after their head-strong manner, conclude that it is duty to rush on their Journy *all* weathers, and I am for waiting for *Wind* and *Tide*. They are for hazzarding all for God, at a clap, and I am for taking *all* advantages to secure my life and estate. They are for holding *their notions*, though all other men are against them, but I am for Religion in what, and so far as the times, and my safety will bear it. They are for Religion, when in rags, and contempt, but I am for him when he walks in his golden slipers in the Sunshine, and with applause.

Mr. *Hold-the-world*. Ai, and hold you there still, good Mr. *By-ends*, for, for my part, I can count him but a fool, that having the liberty to keep what he has, shall be so unwise as to lose it. Let us be wise *as Serpents*,† 'tis best to make hay when the Sun shines; you see how the Bee lieth still all winter and bestirs her then only when she can have profit with pleasure. God sends sometimes Rain, and sometimes Sunshine;† if they be such fools to go through the first, yet let us be content to take fair weather along with us. For my part I like that Religion best, that will stand with the security of Gods good blessings unto us; for who can imagin that is ruled by his reason, since God has bestowed upon us the good things of this life, but that he would have us keep them for his sake? *Abraham* and *Solomon* grew rich in Religion. And *Job* saies,† that a good man *shall lay up gold as dust*. He must not be such as the men before us, if they be as you have discribed them.

Mr. *Save-all*. I think that we are all agreed in this matter, and therefore there needs no more words about it.

Mr. *Mony-love*. No, there needs no more words about this matter indeed, for he that believes neither Scripture nor reason (and you see we have both on our side) neither knows his own liberty, nor seeks his own safety.

Mr. *By-ends*. My Brethren, we are, as you see, going all on Pilgrimage, and for our better diversion from things that are bad, give me leave to propound unto you this question.

Suppose a man; a Minister, or a Tradesman, &c. should have an advantage lie before him to get the good blessings of this life. Yet so, as that he can by no means come by them, except, in appearance at least, he becomes extraordinary Zealous† in some points of Religion, that he medled not with before, may he not use this means to attain his end, and yet be a right honest man?

Mr. *Mony-love*. I see the bottom of your question, and with these Gentlemens good leave, I will endeavour to shape you an answer. And first to speak to your question, as it concerns a *Minister* himself. *Suppose a Minister, a worthy man, possessed but of a very small benefice, and has in his eye a greater, more fat, and plump by far; he has also now an opportunity of getting of it; yet so as by being more studious, by preaching more frequently, and zealously, and because the temper of the people requires it, by altering of some of his principles, for my part I see no reason but a man may do this (provided he has a call.) Ai, and more a great deal besides, and yet be an honest man.* For why,

1. His desire of a greater benefice is lawful (this cannot be contradicted) since 'tis set before him by providence; so then, he may get it if he can, *making no question for conscience sake.*

2. Besides, his desire after that benefice, makes him more studious, a more zealous preacher, *&c.* and so makes him a better man.[†] Yea, makes him better improve his parts, which is according to the mind of God.

3. Now as for his complying with the temper of his people, by disserting, to serve them, some of his principles, this argueth, 1. That he is of a self-denying temper. 2. Of a sweet and winning deportment. 3. And so more fit for the Ministerial function.

4. I conclude then, that a Minister that changes a *small* for a *great*, should not for so doing, be judged as covetous, but rather, since he is improved in his parts and industry thereby, be counted as one that pursues his call, and the opportunity put into his hand to do good.

And now to the second part of the question which concerns the *Tradesman* you mentioned: suppose such an one to have but a poor imploy in the world, but by becoming Religious, he may mend his market, perhaps get a rich wife, or more and far better customers to his shop. For my part I see no reason but that this may be lawfully done. For why,

1. To become religious is a vertue, by what means soever a man becomes so.

2. Nor is it unlawful to get a rich wife, or more custome to my shop.

3. Besides the man that gets these by becoming religious, gets that which is good, of them that are good, by becoming good himself; so then here is a good wife, and good customers, and good gaine, and all these by becoming religious, which is good. Therefore to become religious to get all these is a good and profitable design.

This answer, thus made by this Mr. *Mony-love*, to Mr. *By-ends'* question, was highly applauded by them all; wherefore they concluded upon the whole, that it was most wholsome and advantagious. And because, as they thought, no man was able to contradict it, and because *Christian* and *Hopeful* was yet within call; they joyfully agreed to assault them with the question as soon as they overtook them, and the rather because they had opposed Mr. *By-ends* before. So they called after them, and they stopt, and stood still till they came up to them, but they concluded as they went, that not *By-ends*, but old Mr. *Hold-the-world* should propound the question to them, because, as they supposed, their answer to him would be without the remainder of that heat that was kindled betwixt Mr. *By-ends* and them, at their parting a little before.

So they came up to each other and after a short salutation, Mr. *Hold-the-world* propounded the question to *Christian* and his fellow, and bid them to answer it if they could.

Chr. Then said *Christian*, Even a babe in Religion may answer ten thousand such questions. For if it be unlawful to follow Christ for loaves, as it is, *Joh.* 6. how much more abominable is it to make of him and religion a stalking horse to get and enjoy the world. Nor do we find any other than Heathens, Hypocrites, Devils and Witches that are of this opinion.

1. *Heathens*, for when *Hamor* and *Shechem* had a mind to the Daughter and Cattle of *Jacob*, and saw that there was no waies for them to come at them, but by becoming circumcised, they say to their companions; If every male of us be circumcised, as they are circumcised, shall not their Cattle, and their substance, and every beast of theirs be ours? Their Daughters and their Cattle were that which they sought to obtain, and their Religion the stalking horse they made use of to come at them. Read the whole story, *Gen.* 34. 20, 21, 22, 23.

2. The Hypocritical Pharisees were also of this Religion, long prayers were their pretence, but to get widdows houses were their intent, and greater damnation was from God their Judgment, *Luke* 20. 46, 47.

3. *Judas* the Devil was also of this Religion, he was religious for the bag,†that he might be possessed of what was therein, but he was lost, cast away, and the very Son of perdition.

4. *Simon* the witch was of this Religion too, for he would have had the Holy Ghost, that he might have got money therewith, and his sentence from *Peters* mouth was according, *Act.* 8. 19, 20, 21, 22.

5. Neither will it out of my mind, but that that man that takes up Religion for the world, will throw away Religion for the world; for so surely as *Judas* designed the world in becoming religious: so surely did he also sell Religion, and his Master for the same. To answer the question therefore affirmatively, as I perceive you have

done, and to accept of as authentick such answer, is both Heathenish, Hypocritical and Devilish, and your reward will be according to your works. Then they stood stareing one upon another, but had not wherewith to answer *Christian*. *Hopeful* also approved of the soundness of *Christians* answer, so there was a great silence among them. Mr. *By-ends* and his company also staggered, and kept behind, that *Christian* and *Hopeful* might outgo them. Then said *Christian* to his fellow, if these men cannot stand before the sentence of men, what will they do with the sentence of God? & if they are mute when dealt with by vessels of clay, what will they do when they shall be rebuked by the flames of a devouring fire?

Then *Christian* and *Hopeful* outwent them again, and went till they came at a delicate Plain, called *Ease*, where they went with much content; but that Plain was but *narrow*, so they were quickly got over it. Now at the further side of that Plain, was a little Hill called *Lucre*, and in that Hill a *Silver-Mine*, which some of them that had formerly gone that way, because of the rarity of it, had turned aside to see; but going too near the brink of the pit, the ground being deceitful under them, broke, and they were slain; some also had been maimed there, and could not to their dying day be their own men again.

The ease that Pilgrims have is but little in this life.

Lucre Hill a dangerous Hill.

Then I saw in my Dream, that a little off the Road, over against the *Silver-Mine*, stood *Demas,†(Gentleman-like) to call to Passengers to come and see: who said to *Christian* and his fellow; *Ho, turn aside hither, and I will shew you a thing.

* Demas at the Hill Lucre.

* He calls to Christian and Hopeful to come to him.

Chr. *What thing so deserving as to turn us out of the way?*

Dem. Here is a Silver-*Mine*, and some digging in it for Treasure; if you will come, with a little paines you may richly provide for your selves.

Hopef. Then said Hopeful, *Let us go see.*

Chr. Not I, said *Christian*; I have heard of this place before now, and how many have there been slain; and besides, that Treasure is a snare to those that seek it, for it hindreth them in their Pilgrimage. Then *Christian* called to *Demas*, saying, *Is not the place dangerous? hath it not hindred many in their Pilgrimage?*

Hopeful tempted to go, but Christian holds him back.

Hos. 4. 18.

Dem. Not very dangerous, except to those that are careless: but withal, he *blushed* as he spake.

Chr. Then said *Christian* to *Hopeful*, Let us not stir a step, but still keep on our way.

Hope. I will warrant you, when By-ends *comes up, if he hath the same invitation as we, he will turn in thither to see.*

Chr. No doubt thereof, for his principles lead him that way, and a hundred to one but he dies there.

Christian
roundeth up
Demas.

Dem. Then *Demas* called again, saying, But will you not come over and see?

Chr. Then *Christian* roundly answered, saying, *Demas*, Thou art an Enemy to the right ways of the Lord of this way, and hast been already condemned for thine own turning aside, by one of his Majesties Judges;†and why seekest thou to bring us into the like condemnation? Besides, if we at all turn aside, our Lord the King will certainly hear thereof; and will there put us to shame, where we would stand with boldness before him.

2 Tim. 4.
10.

Demas cried again, That he also was one of their fraternity; and that if they would tarry a little, he also himself would walk with them.

Chr. Then said *Christian*, What is thy name? is it not the same by the which I have called thee?

De. Yes, my name is *Demas*, I am the Son of *Abraham*.

2 Kings 5.
20.
Mat. 26.
14, 15.
chap. 27. 1,
2, 3, 4, 5, 6.

Chr. I know you, *Gehazi* was your Great Grandfather, and *Judas* your Father, and you have trod their steps. It is but a devilish prank that thou usest: Thy Father was hanged for a Traitor, and thou deservest no better reward. Assure thy self, that when we come to the King, we will do him word of this thy behaviour. Thus they went their way.

By-ends
goes over
to Demas.

By this time *By-ends* and his companions was come again within sight, and they at the first beck went over to *Demas*. Now whether they fell into the Pit, by looking over the brink thereof, or whether they went down to dig, or whether they was smothered in the bottom, by the damps that commonly arise, of these things I am not certain: But this I observed, that they never was seen again in the way.

Then Sang Christian,

> By-ends, *and* Silver-Demas, *both agree;*
> *One calls, the other runs, that he may be*
> *A sharer in his Lucre: so these two*
> *Take up in this World, and no further go.*

Now I saw, that just on the other side of this Plain, the Pilgrims *They see* came to a place where stood an old *Monument*, hard by the High- *a strange* way-side, at the sight of which they were both concerned, because *monument.* of the strangeness of the form therof; for it seemed to them as if it had been a *Woman* transformed into the shape of a Pillar: here there- fore they stood looking, and looking upon it, but could not for a time tell what they should make thereof. At last *Hopeful* espied written above upon the head thereof, a Writing in an unusual hand; but he being no Scholar, called to *Christian* (for he was learned) to see if he could pick out the meaning: so he came, and after a little laying of Letters together, he found the same to be this, *Remember Lot's Wife*. So he read it to his fellow; after which, they both con- cluded, that that was the *Pillar of Salt into which *Lot's Wife* was • Gen. 19. turned for her looking back with a *covetous heart*, when she was 26. going from *Sodom* for safety. Which sudden and amazing sight, gave them occasion of this discourse.

Chr. Ah my Brother, this is a seasonable sight, it came opportunely to us after the invitation which *Demas* gave us to come over to view the Hill *Lucre*: and had we gone over as he desired us, and as thou wast inclining to do (my Brother) we had, for ought I know, been made our selves a spectacle for those that shall come after to behold.

Hope. I am sorry that I Was so foolish, and am made to wonder that I am not now as *Lot's* Wife; for wherein was the difference 'twixt her sin and mine? she only looked back, and I had a desire to go see; let Grace be adored, and let me be ashamed, that ever such a thing should be in mine heart.

Chr. Let us take notice of what we see here, for our help for time to come: *This* woman escaped one Judgment; for she fell not by the destruction of *Sodom*, yet she was destroyed by another; as we see, she is turned into a Pillar of Salt.

Hope. True, and she may be to us both *Caution*, and *Example*; *Caution* that we should shun her sin, or a sign of what judgment will overtake such as shall not be prevented by this caution: So *Korah*, *Dathan*, and *Abiram*, with the two hundred and fifty men, that perished in their sin, did also become *a sign, or example to others • Numb. to beware; but above all, I muse at one thing, to wit, how *Demas* 26. 9, 10. and his fellows can stand so confidently yonder to look for that treasure, which this Woman, but for looking behind her, after (for we read not that she stept one foot out of the way) was turned into a pillar of Salt; specially since the Judgment which overtook her, did make her an example, within sight of where they are: for they cannot chuse but see her, did they but lift up their eyes.

Chr. It is a thing to be wondered at, and it argueth that their heart is grown desperate in the case; and I cannot tell who to compare them to so fitly, as to them that pick Pockets in the presence of the Judge, or that will cut purses under the Gallows. It is said of • Gen. 13. the men of *Sodom, That they were sinners* exceedingly, because they 13. were sinners *before the Lord*; that is, in his eyesight; and notwithstanding the kindnesses that he had shewed them, for the Land of • Vers. 10. *Sodom*, was now, like the *Garden of *Eden heretofore*. This therefore provoked him the more to jealousie, and made their plague as hot as the fire of the Lord out of Heaven could make it. And it is most rationally to be concluded, that such, even such as these are, that shall sin in the sight, yea, and that too in despite of such examples that are set continually before them, to caution them to the contrary, must be partakers of severest Judgments.

Hope. Doubtless thou hast said the truth, but what a mercy is it, that neither thou, but especially I, am not made, my self, this example: this ministreth occasion to us to thank God, to fear before him, and always to remember *Lot*'s Wife.

A River. I saw then that they went on their way to a pleasant River,
Psal. 65. 9. which *David the King* called the *River of God*; but, *John, The River*
Rev. 22. *of the water of life.* Now their way lay just upon the bank of the River:
Ezek. 47. here therefore *Christian* and his Companion walked with great delight; they drank also of the water of the River, which was pleasant and enlivening to their weary Spirits: besides, on the banks
Trees by of this River, on either side, were *green Trees*, that bore all manner
the River. of Fruit; and the leaves of the Trees were good for Medicine; with
The Fruit *and leaves* the Fruit of these Trees they were also much delighted; and the
of the trees. leaves they eat to prevent Surfeits, and other Diseases that are
A Meadow *in which* incident to those that heat their blood by Travels. On either side
they lie of the River was also a Meadow, curiously beautified with Lilies;
down to *sleep.* And it was green all the year long. In this Meadow they lay down
Psal. 23. 2. and slept, for here they might *lie down safely.* When they awoke, they
Isa. 14. 30.

gathered again of the Fruit of the Trees, and drank again of the Water of the River: and then lay down again to sleep. Thus they did several days and nights. Then they sang,

> *Behold ye how these Christal streams do glide*
> *(To comfort Pilgrims) by the High-way side;*
> *The Meadows green, besides their fragrant smell,*
> *Yield dainties for them: And he that can tell*
> *What pleasant Fruit, yea Leaves, these Trees do yield,*
> *Will soon sell all, that he may buy this Field.*

So when they were disposed to go on (for they were not, as yet, at their Journeys end) they eat and drank, and departed.

Now I beheld in my Dream, that they had not journied far, but the River and the way, for a time, parted. At which they were not a little sorry, yet they durst not go out of the way. Now the way from the River was rough, and their feet tender by reason of their Travels; *So the soul of the Pilgrims was much discouraged, because of the way.* Wherefore still as they went on, they wished for better way. Now a little before them, there was on the left hand of the Road, a *Meadow*, and a Stile to go over into it, and that *Meadow* is called *By-Path-Meadow*. Then said *Christian* to his fellow. If this Meadow lieth along by our way side, lets go over into it. Then he went to the Stile to see, and behold a Path lay along by the way on the other side of the fence. 'Tis according to my wish, said *Christian*, here is the easiest going; come good *Hopeful*, and let us go over.

Hope. *But how if this Path should lead us out of the way?*

Chr. That's not like, said the other; look, doth it not go along by the way side? So *Hopeful*, being perswaded by his fellow, went after him over the Stile. When they were gone over, and were got into the Path, they found it very easie for their feet; and withal, they looking before them, espied a Man walking as they did, (and his name was. *Vain-confidence*) so they called after him, and asked him whither that way led? he said, To the Cœlestial Gate.* Look, said *Christian*, did not I tell you so? by this you may see we are right: so they followed, and he went before them. But behold the night came on, and it grew very dark; so that they that were behind, lost the sight of him that went before.

Numb. 21. 4.

By-Path-Meadow. One temptation does make way for another.

Strong Christians may lead weak ones out of the way.

* *See what it is too suddenly to fall in with strangers.*

He therefore that went before (*Vain-confidence* by name) not seeing
the way before him, fell into a deep Pit, which was on purpose there
made by the Prince of those grounds, to catch *vain-glorious* fools
withall; and was dashed in pieces with his fall.

Isa. 9. 16.
A Pit to catch the vain-glorious in.

Now *Christian* and his fellow heard him fall. So they called, to
know the matter, but there was none to answer, only they heard
a groaning. Then said *Hopeful*, Where are we now? Then was his
fellow silent, as mistrusting that he had led him out of the way.
And now it began to rain, and thunder, and lighten in a very dreadful manner, and the water rose amain.

Reasoning between Christian and Hopeful.

Then *Hopeful* groaned in himself, saying, *Oh that I had kept on my
way!*

Chr. Who could have thought that this path should have led us
out of the way?

*Hope. I was afraid on't at very first, and therefore gave you that gentle
caution. I would have spoke plainer, but that you are older then I.*

Chr. Good Brother be not offended, I am sorry I have brought
thee out of the way, and that I have put thee into such eminent
danger; pray my Brother forgive me, I did not do it of an evil intent.

Christians repentance for leading of his Brother out of the way.

*Hope. Be comforted my Brother, for I forgive thee; and believe too, that
this shall be for our good.*

Chr. I am glad I have with me a merciful Brother: but we must
not stand thus, let's try to go back again.

Hope. But good Brother let me go before.

Chr. No, if you please let me go first; that if there be any danger,
I may be first therein, because by my means we are both gone out
of the way.

*Hope. No, said Hopeful, you shall not go first, for your mind being
troubled, may lead you out of the way again.* Then for their encouragement, they heard the voice of one, saying, *Let thine Heart be towards
the High-way, even the way that thou wentest, turn again.* But by this
time the Waters were greatly risen, by reason of which, the way of
going back was very dangerous. (Then I thought that it is easier
going out of the way when we are in, then going in, when we are
out.) Yet they adventured to go back; but it was so dark, and the
flood was so high, that in their going back, they had like to have
been drowned nine or ten times.

Jer. 31. 21.
They are in danger of drowning as they go back.

Neither could they, with all the skill they had, get again to the Stile that night. Wherefore, at last, lighting under a little shelter, they sat down there till the day brake; but being weary, they fell asleep. Now there was not far from the place where they lay, a Castle, called *Doubting-Castle*, the owner whereof was *Giant Despair*,† and it was in his grounds they now were sleeping; wherefore he getting up in the morning early, and walking up and down in his Fields, caught *Christian* and *Hopeful* asleep in his grounds. Then with a *grim* and *surly* voice he bid them awake, and asked them whence they were? and what they did in his grounds? They told him, they were Pilgrims, and that they had lost their way. Then said the Giant, You have this night trespassed on me, by trampling in, and lying on my grounds, and therefore you must go along with me. So they were forced to go, because he was stronger then they. They also had but little to say, for they knew themselves in a fault. The Giant therefore drove them before him, and put them into his Castle, into a very dark Dungeon, nasty and stinking to the spirit of these two men: Here then they lay, from *Wednesday* morning till *Saturday* night, without one bit of bread, or drop of drink, or any light, or any to ask how they did. They were therefore here in evil case, and were far from friends and acquaintance. Now in this place, *Christian* had double sorrow, because 'twas through his unadvised haste that they were brought into this distress.

They sleep in the grounds of Giant Despair.

He finds them in his ground, and carries them to Doubting Castle.

The Grievousness of their Imprisonment.

Psal. 88. 18.

Now *Giant Despair* had a Wife, and her name was *Diffidence*: so when he was gone to bed, he told his Wife what he had done, to wit, that he had taken a couple of Prisoners, and cast them into his *Dungeon*, for trespassing on his grounds. Then he asked her also what he had best to do further to them. So she asked him what they were, whence they came, and whither they were bound; and he told her; Then she counselled him, that when he arose in the morning, he should beat them without any mercy: So when he arose, he getteth him a grievous Crab-tree Cudgel, and goes down into the *Dungeon* to them; and there, first falls to rateing of them as if they were dogs, although they gave him never a word of distaste; then he falls upon them, and beats them fearfully, in such sort, that they were not able to help themselves, or to turn them upon the floor. This done, he withdraws and leaves them, there to condole their

On Thursday Giant Despair beats his Prisoners.

misery, and to mourn under their distress: so all that day they spent the time in nothing but sighs and bitter lamentations. The next night she talking with her Husband about them further, and understanding that they were yet alive, did advise him to counsel them, to make away themselves: So when morning was come, he goes to them in a surly manner, as before, and perceiving them to be very sore with the stripes that he had given them the day before; he told them, that since they were never like to come out of that place, their only way would be, forthwith to make *an end of themselves, either with Knife, Halter or Poison: For why, said he, should you chuse life, seeing it is attended with so much bitterness. But they desired him to let them go; with that he looked ugly upon them, and rushing to them, had doubtless made an end of them himself, but that he fell into one of his *fits; (for he sometimes in Sun-shine weather fell into fits) and lost (for a time) the use of his hand: wherefore he withdrew, and left them, (as before) to consider what to do. Then did the Prisoners consult between themselves, whether 'twas best to take his counsel or no: and thus they began to discourse.

Chr. Brother, said *Christian*,* what shall we do? the life that we now live is miserable: for my part, I know not whether is best, to live thus, or to die out of hand? *My soul chuseth strangling rather than life*; and the Grave is more easie for me than this Dungeon: Shall we be ruled by the Giant?

Hope. *Indeed our present condition is dreadful, and death would be far more welcome to me than thus for ever to abide: but yet let us consider, the Lord of the Country to which we are going, hath said, Thou shalt do no murther, no not to another man's person; much more then are we forbidden to take his counsel to kill our selves. Besides, he that kills another, can but commit murder upon his body; but for one to kill himself, is to kill body and soul at once. And moreover, my Brother, thou talkest of ease in the Grave; but hast thou forgotten the Hell whither, for certain, the murderers go? for no murderer hath eternal life, &c. And, let us consider again, that all the Law is not in the hand of Giant Despair: Others, so far as I can understand, have been taken by him, as well as we; and yet have escaped out of his hand: Who knows, but that God that made the world, may cause that Giant Despair may die; or that, at some time or other he may forget to lock us in;*

• On Friday Giant Despair counsels them to kill themselves.

• The Giant sometimes has fits.

• Christian crushed.

• Job 7. 15.

• Hopeful comforts him.

or, but he may in short time have another of his fits before us, and may lose the use of his limbs; and if ever that should come to pass again, for my part, I am resolved to pluck up the heart of a man, and to try my utmost to get from under his hand. I was a fool that I did not try to do it before, but however, my Brother, let's be patient, and endure a while; the time may come that may give us a happy release: but let us not be our own murderers. With these words, Hopeful, at present did moderate the mind of his Brother; so they continued together (in the dark) that day, in their sad and doleful condition.

Well, towards evening the Giant goes down into the Dungeon again, to see if his Prisoners had taken his counsel; but when he came there, he found them alive, and truly, alive was all: for now, what for want of Bread and Water, and by reason of the Wounds they received when he beat them, they could do little but breath: But, I say, he found them alive; at which he fell into a grievous rage, and told them, that seeing they had disobeyed his counsel, it should be worse with them, than if they had never been born.

At this they trembled greatly, and I think that *Christian* fell into a Swound; but coming a little to himself again, they renewed their discourse about the *Giants* counsel; and whether yet they had best to take it or no. *Now *Christian* again seemed to be for doing it, but Hopeful made his second reply as followeth.

* Christian still dejected.

Hope. *My Brother, said he, remembrest thou not how valiant thou hast been heretofore; Apollyon could not crush thee, nor could all that thou didst hear, or see, or feel in the Valley of the Shadow of Death; what hardship, terror, and amazement hast thou already gone through, and art thou now nothing but fear? Thou seest that I am in the Dungeon with thee, a far weaker man by nature than thou art: Also this Giant has wounded me as well as thee; and hath also cut off the Bread and Water from my mouth; and with thee I mourn without the light: but let's exercise a little more patience. Remember how thou playedst the man at Vanity-Fair, and wast neither afraid of the Chain nor Cage; nor yet of bloody Death: wherefore let us (at least to avoid the shame, that becomes not a Christian to be found in) bear up with patience as well as we can.

* Hopeful comforts him again, by calling former things to remembrance.

Now night being come again, and the *Giant* and his Wife being in bed, she asked him concerning the Prisoners, and if they had taken his counsel: To which he replied, They are sturdy Rogues,

they chuse rather to bear all hardship, than to make away themselves. Then said she, Take them into the Castle-yard to morrow, and shew them the *Bones* and *Skulls* of those that thou hast already dispatch'd; and make them believe, e're a week comes to an end, thou also wilt tear them in pieces as thou hast done their fellows before them.

So when the morning was come, the *Giant* goes to them again, and takes them into the Castle-yard, and shews them, as his Wife had bidden him. *These, said he, were Pilgrims as you are, once, and they trespassed in my grounds, as you have done; and when I thought fit, I tore them in pieces; and so within ten days I will do you. Go get you down to your Den again; and with that he beat them all the way thither: they lay therefore all day on *Saturday* in a lamentable case, as before. Now when night was come, and when Mrs. *Diffidence*, and her Husband, the *Giant*, were got to bed, they began to renew their discourse of their Prisoners: and withal, the old *Giant* wondered, that he could neither by his blows, nor counsel, bring them to an end. And with that his Wife replied, I fear, said she, that they live in hope that some will come to relieve them, or that they have pick-locks about them; by the means of which they hope to escape. And, sayest thou so, my dear, said the *Giant*, I will therefore search them in the morning.

Well, on *Saturday* about midnight they began to *pray*, and continued in Prayer till almost break of day.

Now a little before it was day, good *Christian*, as one half amazed, brake out in this passionate speech, *What a fool, quoth he, am I, thus to lie in a stinking Dungeon, when I may as well walk at liberty?* I have a Key in my bosom, called *Promise*,†that will, (I am perswaded) open any Lock in *Doubting-Castle*. Then said *Hopeful*, That's good news; good Brother pluck it out of thy bosom, and try: Then *Christian* pulled it out of his bosom, and began to try at the Dungeon door, whose bolt (as he turned the Key) gave back, and the door flew open with ease, and *Christian* and *Hopeful* both came out. Then he went to the outward door, that leads into the *Castle yard*, and with his Key opened the door also. After he went to the *Iron* Gate, for that must be opened too, but that Lock went *damnable* hard,†yet the Key did open it; then they thrust open the Gate to make their

• On Saturday the Giant threatned, that shortly he would pull them in pieces.

A Key in Christians bosom, called Promise, opens any Lock in Doubting Castle.

escape with speed; but that Gate, as it opened, made such a creaking, that it waked *Giant Despair*, who hastily rising to pursue his Prisoners, felt his Limbs to fail, for his fits took him again, so that he could by no means go after them. Then they went on, and came to the Kings high way again, and so were safe, because they were out of his Jurisdiction.

Now when they were gone over the Stile, they began to contrive with themselves what they should do at that Stile, to prevent those that should come after, from falling into the hands of *Giant Despair*. So they consented to erect there a *Pillar, and to engrave upon the side thereof; *Over this Stile is the way to* Doubting-Castle, *which is kept by* Giant Despair, *who despiseth the King of the Cœlestial Countrey, and seeks to destroy his holy Pilgrims.* Many therefore that followed after, read what was written, and escaped the danger. This done, they sang as follows.

> *Out of the way we went, and then we found*
> *What 'twas to tread upon forbidden ground:*
> *And let them that come after have a care,*
> *Lest heedlessness makes them, as we, to fare:*
> *Lest they, for trespassing, his prisoners are,*
> *Whose Castle's* Doubting, *and whose name's* Despair.

They went then, till they came to the Delectable Mountains, which Mountains belong to the Lord of that Hill of which we have spoken before; so they went up to the Mountains, to behold the Gardens, and Orchards, the Vineyards, and Fountains of water, where also they drank, and washed themselves, and did freely eat of the Vineyards. Now there was on the tops of these Mountains, Shepherds feeding their flocks, and they stood by the high-way side. The Pilgrims therefore went to them, and leaning upon their staves, (as is common with weary Pilgrims, when they stand to talk with any by the way,) they asked, **Whose Delectable Mountains are these? and whose be the sheep that feed upon them?*

Shep. These Mountains are *Immanuels Land*, and they are within sight of his City, and the sheep also are his, and he laid down his life for them.

Chr. Is this the way to the Cœlestial City?

A Pillar erected by Christian and his fellow.

The Delectable Mountains.

They are refreshed in the mountains.

**Talk with the Shepherds.*

Joh. 10. 11.

Shep. You are just in your way.

Chr. How far is it thither?

Shep. Too far for any, but those that *shall* get thither indeed.

Chr. Is the way safe, or dangerous?

Hos. 14. 9.   *Shep.* Safe for those for whom it is to be safe, *but transgressors shall fall therein.*

Chr. Is there in this place any relief for Pilgrims that are weary and faint in the way?

Heb. 13. 1, 2.   *Shep.* The Lord of these Mountains hath given us a charge, *Not to be forgetful to entertain strangers*: Therefore the good of the place is before you.

I saw also in my Dream, that when the *Shepherds* perceived that they were way-fairing men, they also put questions to them, (to which they made answer as in other places) as, Whence came you? and, How got you into the way? and, By what means have you so persevered therein? For but few of them that begin to come hither, do shew their face on these Mountains. But when the Shepherds heard their answers, being pleased therewith, they looked very

• The Shepherds welcome them.   lovingly upon them; and said, *Welcome to the Delectable Mountains.* The Shepherds, I say, whose names were, *Knowledge, Experience,*

The names of the Shepherds.   *Watchful,* and *Sincere,* took them by the hand, and had them to their Tents, and made them partake of that which was ready at present. They said moreover, We would that you should stay here a while, to acquaint with us, and yet more to solace your selves with the good of these Delectable Mountains. They then told them, That they were content to stay; and so they went to their rest that night, because it was very late.

Then I saw in my Dream, that in the morning, the Shepherds called up *Christian* and *Hopeful* to walk with them upon the Mountains: So they went forth with them, and walked a while, having a pleasant prospect on every side. Then said the Shepherds one to

• They are sure wonders.   another, shall we shew these Pilgrims some *wonders? So when they had concluded to do it, they had them first to the top of an Hill,

The Mountain of Errour.   called *Errour,* which was very steep on the furthest side, and bid them look down to the bottom. So *Christian* and *Hopeful* lookt down, and saw at the bottom several men, dashed all to pieces by a fall that they had from the top. Then said *Christian,* What meaneth this? The

Shepherds answered; Have you not heard of them that were made to err, by hearkening to *Hymeneus, and Philetus, as concerning the faith of the Resurrection of the Body? They answered, Yes. Then said the Shepherds, Those that you see lie dashed in pieces at the bottom of this Mountain, *are they*: and they have continued to this day unburied (as you see) for an example to others to take heed how they clamber too high, or how they come too near the brink of this Mountain. *2 Tim. 2. 17, 18.

Then I saw that they had them to the top of another Mountain, and the name of that is *Caution*; and bid them look a far off: Which when they did, they perceived, as they thought, several men walking up and down among the Tombs that were there. And they perceived that the men were blind, because they stumbled sometimes upon the Tombs, and because they could not get out from among them. Then said *Christian, What means this*? • *Mount Caution.*

The Shepherds then answered, Did you not see a little below these Mountains a *Stile* that led into a Meadow on the left hand of this way? They answered, Yes. Then said the Shepherds, From that Stile there goes a path that leads directly to *Doubting-Castle*, which is kept by *Giant Despair*; and these men (pointing to them among the Tombs) came once on Pilgrimage, as you do now, even till they came to that same *Stile*. And because the right way was rough in that place, they chose to go out of it into that Meadow, and there were taken by Giant *Despair*, and cast into *Doubting-Castle*; where, after they had a while been kept in the Dungeon, he at last did put out their eyes, and led them among those Tombs, where he has left them to wander to this very day, that the saying of the wise Man might be fulfilled, *He that wandereth out of the way of understanding, shall remain in the Congregation of the dead.* Then *Christian* and *Hopeful* looked one upon another, with tears gushing out; but yet said nothing to the Shepherds. Prov. 21. 16.

Then I saw in my Dream, that the Shepherds had them to another place, in a bottom, where was a door in the side of an Hill; and they opened the door, and bid them look in. They looked in therefore, and saw that within it was very dark, and smoaky; they also thought that they heard there a lumbring noise as of fire, and a cry of some tormented, and that they smelt the scent of Brimstone. Then said

Christian, what means this? The Shepherds told them, saying, This is
A by-way a By-way to Hell, a way that Hypocrites go in at; namely, such as
to Hell. sell their Birthright, with *Esau*: such as sell their Master, with *Judas*:
such as blaspheme the Gospel, with *Alexander*; and that lie, and
dissemble, with *Ananias* and *Saphira* his wife.†

Hope. Then said *Hopeful* to the Shepherds, *I perceive that these had
on them, even every one, a shew of Pilgrimage as we have now; had they
not?*

Shep. Yes, and held it a long time too.

*Hope. How far might they go on Pilgrimage in their day, since they
notwithstanding were thus miserably cast away?*

Shep. Some further, and some not so far as these Mountains.

Then said the Pilgrims one to another, *We had need cry to the
Strong for strength.*

Shep. Ay, and you will have need to use it when you have it,
too.

By this time the Pilgrims had a desire to go forwards, and the
Shepherds a desire they should; so they walked together towards
the end of the Mountains. Then said the Shepherds one to another,
Let us here shew to the Pilgrims the Gates of the Cœlestial City,
• *The* if they have skill to look through our *Perspective Glass. The
Shepherds* Pilgrims then lovingly accepted the motion: So they had them to
Perspec- the top of an high Hill called *Clear*, and gave them their Glass to
tive-Glass. look. Then they essayed to look, but the remembrance of that last
• *The Hill* thing that the Shepheards had shewed them, made their hands
Clear. shake; by means of which impediment they could not look steddily
through the Glass; yet they thought they saw something like the
The fruit Gate, and also some of the Glory of the place. Then they went
of slavish away and sang.
fear.

> *Thus by the* Shepherds, *Secrets are reveal'd,*
> *Which from all other men are kept conceal'd:*
> *Come to the* Shepherds *then, if you would see*
> *Things deep, things hid, and that mysterious be.*

• *A two* When they were about to depart, one of the Shepherds gave them
fold a *note of the way.* Another of them, *bid them *beware of the flatterer.*
Caution.

The third *bid them take heed that they sleep not upon the Inchanted Ground.* And the fourth, *bid them God speed.* So I awoke from my Dream.†

And I slept, and Dreamed again, and saw the same two Pilgrims going down the Mountains along the High-way towards the City. Now a little below these Mountains, on the left hand, lieth the Countrey of *Conceit*; from which Countrey there comes into the way in which the Pilgrims walked, a little crooked Lane. Here therefore they met with a very brisk Lad, that came out of that Countrey; and his name was *Ignorance.*†So *Christian* asked him, *From what parts he came? and whither he was going?* *The Countrey of Conceit, out of which came Ignorance.*

Ign. Sir, I was born in the Countrey that lieth off there, a little on the left hand; and I am going to the Cœlestial City. *Christian and Ignorance hath some talk.*

Chr. But how do you think to get in at the Gate, for you may find some difficulty there?

Ign. As other good People do, said he.

Chr. But what have you to shew at that Gate, that may cause that the Gate should be opened unto you?

Ign. I know my Lords will, and I have been a good Liver, I pay every man his own; I Pray, Fast, pay Tithes, and give Alms, and have left my Countrey, for whither I am going.

Chr. But thou camest not in at the Wicket-gate, that is, at the head of this way: thou camest in hither through that same crooked Lane, and therefore I fear, however thou mayest think of thy self, when the reckoning day shall come, thou wilt have laid to thy charge, that thou art a Theif and a Robber, instead of getting admitance into the City.

Ignor. Gentlemen, ye be utter strangers to me, I know you not, be content to follow the Religion of your Countrey, and I will follow the Religion of mine. I hope all will be well. And as for the Gate that you talk of, all the world knows that that is a great way off of our Countrey. I cannot think that any man in all our parts doth so much as know the way to it; nor need they matter whether they do or no, since we have, as you see, a fine, pleasant, green Lane, that comes down from our Countrey the next way into it. *He saith to every one, that he is a fool.*

When *Christian* saw that the man was wise in his own conceit; he said to *Hopeful*, whisperingly. *There is more hopes of a fool then of him.* And said moreover, *When he that is a fool walketh by the way, his wisdom faileth him, and he saith to every one that he is a fool.* What, shall *Pr. 26. 12. Eccl. 10. 3. How to carry it to a fool.*

we talk further with him? or out-go him at present? and so leave
him to think of what he hath heard already? and then stop again
for him afterwards, and see if by degrees we can do any good of
him? Then said *Hopeful*,

> *Let Ignorance a little while now muse*
> *On what is said, and let him not refuse*
> *Good Counsel to imbrace, lest he remain*
> *Still ignorant of what's the chiefest gain.*
> *God saith, Those that no understanding have,*
> *(Although he made them) them he will not save.*

Hope. He further added, It is not good, I think to say all to him
at once, let us pass him by, if you will, and talk to him anon, *even
as he is able to bear it.*

So they both went on, and *Ignorance* he came after. Now when
they had passed him a little way, they entered into a very dark
Lane, where they met a man whom seven Devils had bound with
seven strong Cords, and were carrying of him back *to the door* that
they saw in the side of the Hill. Now good *Christian* began to
tremble, and so did *Hopeful* his Companion: Yet as the Devils led
away the man, *Christian* looked to see if he knew him, and he thought
it might be one *Turn-away* that dwelt in the *Town* of *Apostacy*. But
he did not perfectly see his face, for he did hang his head like a Thief
that is found: But being gone past, *Hopeful* looked after him, and
espied on his back a Paper with this Inscription, *Wanton Professor*,
and damnable Apostate. Then said *Christian* to his Fellow, Now I call
to remembrance that which was told me of a thing that happened
to a good man hereabout. The name of the man was *Little-Faith*, but
a good man, and he dwelt in the Town of *Sincere.* The thing was this;
at the entering in of this passage there comes down from *Broad-way-gate*, a Lane, called *Dead-mans Lane*; so called, because of the Murders
that are commonly done there. And this *Little-Faith* going on
Pilgrimage, as we do now, chanced to sit down there and slept.
Now there happened at that time, to come down that *Lane* from
Broad-way-gate, three Sturdy Rogues; and their names were *Faint-heart*, *Mistrust*, and *Guilt*, (three Brothers) and they espying *Little-faith* where he was, came galloping up with speed: Now the good

Mat. 12.
45.
Prov. 5. 22.

*The destruction of
one Turn-away.*

*Christian
telleth his
Companion
a story of
Little-Faith.*
Broad-way-gate.
*Deadmans
Lane.*

man was just awaked from his sleep, and was getting up to go on his Journey. So they came all up to him, and with threatning Language bid him *stand*. At this *Little-Faith* look'd as white as a clout, and had neither power to *fight*, nor *flie*. Then said *Faint-heart*, Deliver thy Purse; but he making no haste to do it, (for he was loth to lose his Money) *Mistrust* ran up to him, and thrusting his hand into his Pocket, pull'd out thence a bag of Silver. Then he cried out, Thieves, thieves. With that *Guilt* with a great Club that was in his hand, strook *Little-Faith* on the head, and with that blow fell'd him flat to the ground, where he lay bleeding as one that would bleed to death. All this while the Thieves stood by. But at last, they hearing that some were upon the Road, and fearing lest it should be one *Great-grace* that dwells in the City of *Good-confidence*, they betook themselves to their heels, and left this good man to shift for himself. Now after a while, *Little-faith* came to himself, and getting up, made shift to scrabble on his way. This was the story.

Little-Faith robbed by Faint-heart, Mistrust, and Guilt.

They got away his Silver, and knockt him down.

Hope. *But did they take from him all that ever he had?*

Chr. No: the place where his Jewels† were, they never ransackt, so those he kept still; but as I was told, the good man was much afflicted for his loss. For the Thieves got most of his spending Money. That which they got not (as I said) were Jewels, also he had a little odd Money left, but *scarce* enough to bring him to his Journeys end; nay, (if I was not mis-informed) he was forced to beg as he went, to keep himself alive, (for his Jewels he might not sell.) But beg, and do what he could, *he went* (as we say) *with many a hungry belly* the most part of the rest of the way.

Little-faith lost not his best things.

1 Pet. 4. 18.

Little-faith forced to beg to his Journeys end.

Hope. *But is it not a wonder they got not from him his Certificate, by which he was to receive his admittance at the Cœlestial gate?*

Chr. 'Tis a wonder, but they got not that; though they mist it not through any good cunning of his, for he being dismayed with their coming upon him, had neither power nor skill to hide any thing; so 'twas more by good Providence then by his Indeavour, that they mist of *that good thing*.

He kept not his best things by his own cunning. 2 Tim. 1. 14.

Hope. *But it must needs be a comfort to him, that they got not this Jewel from him.*

Chr. It might have been great comfort to him, had he used it as

2 Pet. 1. 19.

he should; but they that told me the story, said, That he made but little use of it all the rest of the way; and that because of the dismay that he had in their taking away his Money: indeed he forgot it a great part of the rest of the Journey; and besides, when at any time, it came into his mind, and he began to be comforted therewith, then would fresh thoughts of his loss come again upon him, and those thoughts would swallow up all.

Hope. *Alas poor Man! this could not but be a great grief unto him.*

He is pitied by both. Chr. Grief! Ay, a grief indeed! would it not a been so to any of us, had we been used as he, to be Robbed and wounded too, and that in a strange place, as he was? 'Tis a wonder he did not die with grief, poor heart! I was told, that he scattered almost all the rest of the way with nothing but doleful and bitter complaints. Telling also to all that over-took him, or that he over-took in the way as he went, where he was Robbed, and how; who they were that did it, and what he lost; how he was wounded, and that he hardly escaped with life.

Hope. *But 'tis a wonder that his necessities did not put him upon selling, or pawning some of his Jewels, that he might have wherewith to relieve himself in his Journey.*

Christian snibbeth his fellow for unadvised speaking. Chr. Thou talkest like one, upon whose head is the Shell to this very day: For what should he *pawn* them? or to whom should he sell them? In all that Countrey where he was Robbed his Jewels were not accounted of, nor did he want that relief which could from thence be administred to him; besides, had his Jewels been missing at the Gate of the Cœlestial City, he had (and that he knew well enough) been excluded from an Inheritance there; and that would have been worse to him then the appearance, and villany of ten thousand Thieves.

Heb. 12. 16. Hope. *Why art thou so tart my Brother? Esau sold his Birth-right, and that for a mess of Pottage; and that Birth-right was his greatest Jewel: and if he, why might not Little-Faith do so too?*

A discourse about Esau and Little-Faith. Chr. Esau did sell his Birth-right indeed, and so do many besides; and by so doing, exclude themselves from the chief blessing, as also that *Caytiff* did. But you must put a difference betwixt *Esau* and *Little-Faith*, and also betwixt their Estates. *Esau's* Birth-right was

Typical, but *Little-Faith's* Jewels were not so. *Esau's* belly was his
God, but *Little-Faith's* belly was not so. *Esau's* want lay in his fleshly
appetite, *Little-Faith's* did not so. Besides, *Esau* could see no further
then to the fulfilling of his lusts; *For I am at the point to dye,* said he,
and what good will this Birth-right do me? But *Little-Faith,* though it
was his lot to have but a *little faith,* was by his *little faith* kept from
such extravagancies; and made to *see* and *prize* his Jewels more, then
to sell them, as *Esau* did his Birth-right. You read not any where
that *Esau* had *Faith,* no not so much as a *little*: Therefore no marvel,
if where the flesh only bears sway (as it will in that man where *no*
Faith is to resist) if he sells his *Birth-right,* and his Soul and all, and
that to the Devil of Hell; for it is with such, as it is with the Ass,
Who in her occasions cannot be turned away. When their minds are set
upon their Lusts, they will have what ever they cost. But *Little-
Faith* was of another temper, his mind was on things Divine;
his livelyhood was upon things that were Spiritual, and from above;
Therefore to what end should he that is of such a temper sell his
Jewels, (had there been any that would have bought them) to fill
his mind with empty things? Will a man give a penny to fill his
belly with Hay? or can you perswade the *Turtle-dove* to live upon
Carrion, like the *Crow?* Though *faithless* ones can for carnal Lusts,
pawn, or morgage, or sell what they have, and themselves out
right to boot; yet they that have *faith, saving faith,* though but
a *little* of it, cannot do so. Here therefore, my Brother, is thy
mistake.

*Hope. I acknowledge it; but yet your severe reflection had almost made
me angry.*

Chr. Why, I did but compare thee to some of the Birds that
are of the brisker sort, who will run to and fro in untrodden
paths with the shell upon their heads: but pass by that, and
consider the matter under debate, and all shall be well betwixt
thee and me.

Hope. But Christian, *These three fellows, I am perswaded in my heart,
are but a company of Cowards: would they have run else, think you, as
they did, at the noise of one that was coming on the road? Why did
not* Little-faith *pluck up a greater heart? He might, methinks, have stood
one brush with them, and have yielded when there had been no remedy.*

Margin notes: Esau was ruled by his lusts. Gen. 25. 32. Esau never had Faith. Jer. 2. 24. Little-Faith could not live upon Esaus Pottage. A comparison between the Turtle-dove and the Crow. Hopeful swaggers.

Chr. That they are Cowards, many have said, but few have found it so in the time of Trial. As for *a great heart*, *Little-faith* had none; and I perceive by thee, my Brother, hadst thou been the Man concerned, thou art but for a brush, and then to yield. And verily, since this is the height of thy Stomach, now they are at a distance from us, should they appear to thee, as they did to him, they might put thee to second thoughts.

No great heart for God, where there is but little faith. We have more courage when out, then when we are in.

But consider again, they are but Journey-men Thieves, they serve under the King of the Bottomless pit; who, if need be, will come in to their aid himself, and his voice is *as the roaring of a Lion.* I my self have been Ingaged†as this *Little-faith* was, and I found it a terrible thing. These three Villains set upon me, and I beginning like a *Christian* to resist, they gave but a call and in came their Master: I would, as the saying is, have given my life for a penny; but that, as God would have it, I was cloathed with Armour of proof. Ay, and yet, though I was so harnessed, I found it hard work to quit my self like a man; no man can tell what in that Combat attends us, but he that hath been in the Battle himself.

1 Pet. 5. 8. Christian tells his own experience in this case.

Hope. Well, but they ran, you see, when they did but suppose that one Great-Grace *was in the way.*

Chr. True, they often fled, both they and their Master, when *Great-grace* hath but appeared; and no marvel, for he is *the Kings Champion:* But I tro, you will put some difference between *Little-faith* and the *Kings Champion*; all the Kings Subjects are not his Champions: nor can they, when tried, do such feats of War as he. Is it meet to think that a little child should handle *Goliah* as *David* did?† or that there should be the strength of an *Ox* in a *Wren*? Some are strong, some are weak, some have *great* faith, some have *little*: this man was one of the weak, and therefore he went to the walls.†

The Kings Champion.

Hope. I would it had been Great-Grace *for their sakes.*

Chr. If it had been he, he might have had his hands full: For I must tell you, that though *Great-Grace* is excellent good at his Weapons, and has, and can, so long as he keeps them at Swords point, do well enough with them: yet if they get within him, even *Faint-heart*, *Mistrust*, or the other, it shall go hard but they will throw up his heels. And when a man is down, you know, what can he do?

Who so looks well upon *Great-graces* face, shall see those Scars and

Cuts there that shall easily give demonstration of what I say. Yea once I heard he should say, (and that when he was in the Combat) *We despaired even of life:* How did these sturdy Rogues and their Fellows make *David* groan, mourn, and roar? Yea *Heman* and *Hezekiah* too, though Champions in their day, were forced to bestir them, when by these assaulted; and yet, that notwithstanding, they had their Coats soundly brushed by them. *Peter* upon a time would go try what he could do; but, though some do say of him that he is the Prince of the Apostles, they handled him so, that they made him at last afraid of a sorry Girle.

Besides, their King is at their Whistle, he is never out of hearing; and if at any time they be put to the worst, he, if possible, comes in to help them: And, of him it is said, *The Sword of him that layeth* {Job 41. 26.} *at him cannot hold: the Spear, the Dart, nor the Habergeon; he esteemeth Iron as Straw, and Brass as rotten Wood. The Arrow cannot make him flie,* {Leviathans} *Sling-stones are turned with him into stubble, Darts are counted as stubble,* {sturdiness.} *he laugheth at the shaking of a Spear.* What can a man do in this case? 'Tis true, if a man could at every turn have *Jobs* Horse, and had skill and courage to ride him, he might do notable things. For his {The excel-} *neck is clothed with Thunder, he will not be afraid as the Grashoper, the* {lent mettle} *glory of his Nostrils is terrible, he paweth in the Valley, rejoyceth in his* {that is in} {Job's horse.} *strength, and goeth out to meet the armed men. He mocketh at fear, and is not affrighted, neither turneth back from the Sword. The Quiver rattleth against him, the glittering Spear, and the shield. He swalloweth the ground with fierceness and rage, neither believeth he that it is the sound of the Trumpet. He saith among the Trumpets, Ha, ha; and he smelleth the Battel a far off, the thundering of the Captains, and the shoutings.* {Job 39. 19.}

But for such footmen as thee and I are, let us never desire to meet with an enemy, nor vaunt as if we could do better, when we hear of others that they have been foiled, nor be tickled at the thoughts of our own manhood, for such commonly come by the worst when tried. Witness *Peter*, of whom I made mention before. He would swagger, Ay he would: He would, as his vain mind prompted him to say, do better, and stand more for his Master, then all men: But who so foiled, and run down with these *Villains* as he?

When therefore we hear that such Robberies are done on the

Kings High-way, two things become us to do: first to go out Harnessed, and to be sure *to take a Shield with us:* For it was for want of that, that he that laid so lustily at *Leviathan*†could not make him yield. For indeed, if that be wanting, he fears us not at all. There-fore he that had skill, hath said, *Above all take the Shield of Faith, wherewith ye shall be able to quench all the fiery darts of the wicked.*

'Tis good also that we desire of the King a Convoy, yea that he will go with us himself. This made *David* rejoyce when in the Valley of the shaddows of death; and *Moses* was rather for dying where he stood, then to go one step without his God. O my Brother, if he will but go along with us, what need we be afraid of ten thousands that shall set themselves against us, but without him, *the proud helpers fall under the slain.*

I for my part have been in the fray before now, and though (through the goodness of him that is best) I am as you see alive: yet I cannot boast of my manhood. Glad shall I be, if I meet with no more such brunts, though I fear we are not got beyond all danger. However, since the Lion and the Bear hath not as yet, devoured me, I hope God will also deliver us from the next uncir-cumcised *Philistine.* Then Sang *Christian.*

> *Poor* Little-Faith! *Hast been among the Thieves!*
> *Wast robb'd! Remember this, Who so believes*
> *And gets more faith, shall then a Victor be*
> *Over ten thousand, else scarce over three.*

So they went on, and *Ignorance* followed. They went then till they came at a place where they saw a *way* put it self into their *way,* and seemed withal, to lie as straight as the way which they should go; and here they knew not which of the two to take, for both seemed straight before them, therefore here they stood still to consider. And as they were thinking about the way, behold, a man black of flesh,†but covered with a very light Robe, came to them, and asked them, why they stood there? They answered, They were going to the Cœlestial City, but knew not which of these ways to take. Follow me, said the man, it is thither that I am going. So they followed him in the way that but now came into the road, which

Marginal notes (left column):

Ephes. 6. 16.

'Tis good to have a Convoy.
Exod. 33. 15.
Psal. 3. 5, 6, 7, 8.
Psal. 27. 1, 2, 3.
Isa. 10. 4.

A way and a way.

by degrees turned, and turned them so from the City that they *Christian*
desired to go to, that in little time their faces were turned away *and his*
fellow
from it; yet they followed him. But by and by, before they were *deluded.*
aware, he led them both within the compass of a Net, in which they *They are*
were both so entangled, that they knew not what to do; and with *taken in*
a Net.
that, *the white robe fell off the black mans back*: then they saw where
they were. Wherefore there they lay crying sometime, for they
could not get themselves out.

Chr. Then said *Christian* to his fellow, Now do I see my self in an *They be-*
errour. Did not the Shepherds bid us beware of the flatterers? As is *wail their*
conditions.
the saying of the Wise man, so we have found it this day: *A man* *Prov. 29.*
that flattereth his Neighbour, spreadeth a Net for his feet.

Hope. They also gave us a note of directions about the way, for
our more sure finding thereof: but therein we have also forgotten
to read, and have not kept our selves from the Paths of the destroyer.
Here *David* was wiser then wee; for saith he, *Concerning the works of*
men, by the word of thy lips, I have kept me from the paths of the destroyer. *Psal.* 17. 4.
Thus they lay bewailing themselves in the Net. At last they espied
a shining One coming towards them, with a whip of small cord in *A shining*
his hand. When he was come to the place where they were, he asked *one comes to*
them with
them whence they came? and what they did there? They told him, *a whip in*
That they were poor Pilgrims, going to *Sion*, but were led out of *his hand.*
their way by a black man, cloathed in white; who bid us, said they,
follow him; for he was going thither too. Then said he with the
Whip, it is *Flatterer*, a false Apostle, that hath transformed himself *Prov.* 29. 5.
into an Angel of Light. So he rent the Net and let the men out. *Dan.* 11.
32.
Then said he to them, Follow me, that I may set you in your way 2 *Cor.* 11.
again; so he led them back to the way, which they had left to follow 13, 14.
the *Flatterer*. Then he asked them, saying, Where did you lie the *They are*
last night? They said, with the Shepherds upon the delectable *examined,*
and con-
Mountains. He asked them then, If they had not of them Shepherds *victed of*
a note of direction for the way? They answered; Yes. But did you, said *forgetful-*
ness.
he, when you was at a stand, pluck out and read your note? They
answered, No. He asked them why? They said they forgot. He *Deceivers*
asked moreover, If the Shepherds did not bid them beware of the *fine spoken.*
Rom. 16.
Flatterer? They answered, Yes: But we did not imagine, said they, 18.
that this fine-spoken man had been he.

Deut. 25. Then I saw in my Dream, that he commanded them to *lie down*;
2 Chron. 6. which when they did, he chastized them sore, to teach them the
26, 27. good way wherein they should walk; and as he chastized them, he
Rev. 3. 19. said, *As many as I love, I rebuke and chasten; be zealous therefore, and*
They are *repent.* This done, he bids them go on their way, and take good heed
whipt, and
sent on their to the other directions of the Shepherds. So they thanked him for
way. all his kindness, and went softly along the right way, Singing.

> *Come hither, you that walk along the way;*
> *See how the Pilgrims fare, that go a stray!*
> *They catched are in an intangling Net,*
> *'Cause they good Counsel lightly did forget:*
> *'Tis true, they rescu'd were, but yet you see*
> *They're scourg'd to boot: Let this your caution be.*

Now after a while, they perceived afar off, one comeing softly and
alone all along the High-way to meet them. Then said *Christian* to
his fellow, Yonder is a man with his back toward *Sion*, and he is
coming to meet us.

Hope. I see him, let us take heed to our selves now, lest he should
prove a *Flatterer* also. So he drew nearer and nearer, and at last came
The up unto them. His name was *Atheist*, and he asked them whither
Atheist
meets them. they were going.

Chr. We are going to the Mount Sion.

He Laughs Then *Atheist* fell into a very great Laughter.
at them.
Chr. What is the meaning of your Laughter?

Atheist. I laugh to see what ignorant persons you are, to take upon
you so tedious a Journey; and yet are like to have nothing but your
travel for your paines.

They *Chr. Why man? Do you think we shall not be received?*
reason
together. *Atheist.* Received! There is no such place as you Dream of, in all
this World.

Chr. But there is in the World to come.

Atheist. When I was at home in mine own Countrey, I heard as
you now affirm, and from that hearing went out to see, and have
Jer. 22. 12. been seeking this City this twenty years: But find no more of it,
Eccl. 10. then I did the first day I set out.
15.
Chr. We have both heard and believe that there is such a place to be found.

Atheist. Had not I, when at home, believed, I had not come thus far to seek: But finding none, (and yet I should, had there been such a place to be found, for I have gone to seek it further then you) I am going back again, and will seek to refresh my self with the things that I then cast away, for hopes of that, which I now see, is not.

Chr. Then said *Christian* to *Hopeful* his Fellow, *Is it true which this man hath said?*

Hope. Take heed, he is one of the *Flatterers*; remember what it hath cost us once already for our harkning to such kind of Fellows. What! no Mount *Sion?* Did we not see from the Delectable Mountains the Gate of the City? Also, are we not now to walk by Faith? *Let us go on, said *Hopeful*, lest the man with the Whip overtakes us again.

You should have taught me that Lesson, which I will round you in the ears withal; *Cease, my Son, to hear the Instruction that causeth to err from the words of knowledge.* I say my Brother, cease to hear him, and let us believe to the saving of the Soul.

Chr. My Brother, I did not put the question to thee, for that I doubted of the Truth of our belief my self: But to prove thee, and to fetch from thee a fruit of the honesty of thy heart. As for this man, I know that he is blinded by the god of this World: Let thee and I go on knowing that we have belief of the Truth, and no lie is of the Truth.

Hope. Now do I rejoyce in hope of the glory of God: So they turned away from the man, and he, Laughing at them, went his way.

I saw then in my Dream, that they went till they came into a certain Countrey, whose Air naturally tended to make one drowsie, if he came a stranger into it. And here *Hopeful* began to be very dull and heavy of sleep, wherefore he said unto *Christian*, I do now begin to grow so drowsie, that I can scarcely hold up mine eyes; let us lie down here and take one Nap.

Chr. By no means, said the other, *lest sleeping, we never awake more.*

Hope. Why my Brother? sleep is sweet to the Labouring man; we may be refreshed if we take a Nap.

Chr. Do you not remember, that one of the Shepherds bid us beware of the Inchanted ground? He meant by that, that we should beware of sleeping; wherefore let us not sleep as do others, but let us watch and be sober.

The Atheist takes up his content in this World.

Christian proveth his Brother.

Hopeful's gracious answer.
2 Cor. 5. 7.

* *A remembrance of former chastisements is an help against present temptations.*
Prov. 19.

Heb. 10. 27. 39. *A fruit of an honest heart.*
1 Joh. 2. 21.

They are come to the inchanted ground.

Hopeful begins to be drowsie.

Christian keeps him awake.

1 Thes. 5. 6.

He is thankful.
Eccl. 4. 9.

Hope. I acknowledge my self in a fault, and had I been here alone, I had by sleeping run the danger of death. I see it is true that the wise man saith, *Two are better then one.* Hitherto hath thy Company been my mercy; *and thou shalt have a good reward for thy labour.*

To prevent drowsiness, they fall to good discourse.
Good discourse prevents drowsiness.
or
The Dreamers note.

Chr. Now then, said Christian, *to prevent drowsiness in this place, let us fall into good discourse.*

Hope. With all my heart, said the other.

Chr. Where shall we begin?

Hope. Where God began with us.†But do you begin if you please.

> *When Saints do sleepy grow, let them come hither,*
> *And hear how these two Pilgrims talk together:*
> *Yea, let them learn of them, in any wise*
> *Thus to keep ope their drowsie slumbring eyes.*
> *Saints fellowship, if it be manag'd well,*
> *Keeps them awake, and that in spite of hell.*

• *They begin at the beginning of their conversion.*

Chr. Then *Christian* began and said, *I will ask you a question. How* **came you to think at first of doing as you do now?*

Hope. Do you mean, How came I at first to look after the good of my soul?

Chr. Yes, that is my meaning.

Hope. I continued a great while in the delight of those things which were seen, and sold at our *fair;* things which, as I believe now, would have (had I continued in them still) drownded me in perdition and destruction.

Chr. What things were they?

• *Hopeful's life before conversion.*

Hope. All the Treasures and Riches of the World. *Also I delighted much in Rioting, Revelling, Drinking, Swearing, Lying, Uncleanness, Sabbath-breaking, and what not, that tended to destroy the Soul. But I found at last, by hearing and considering of things that are Divine, which indeed I heard of you, as also of beloved *Faithful,* that was put to death for his Faith and good-living in *Vanity-fair. That the end of these things is death.* And that for these things sake the wrath of God cometh upon the children of disobedience.

Rom. 6. 21, 22, 23.
Ephes. 5. 6.

• *Hopeful at first shuts his eyes against the light.*

Chr. And did you presently fall under the power of this conviction?

Hope. No,* I was not willing presently to know the evil of sin, nor the damnation that follows upon the commission of it, but endeavoured, when my mind at first began to be shaken with the word, to shut mine eyes against the light thereof.

Chr. But what was the cause of your carrying of it thus to the first workings of Gods blessed Spirit upon you?

Hope. *The causes were, 1. I was ignorant that this was the work of God upon me. I never thought that by awaknings for sin, God at first begins the conversion of a sinner. 2. Sin was yet very sweet to my flesh, and I was loth to leave it. 3. I could not tell how to part with mine old Companions, their presence and actions were so desirable unto me. 4. The hours in which convictions were upon me, were such troublesome and such heart-affrighting hours that I could not bear, no not so much as the remembrance of them upon my heart. • *Reasons of his resisting of light.*

Chr. Then as it seems, sometimes you got rid of your trouble.

Hope. Yes verily, but it would come into my mind again; and then I should be as bad, nay worse then I was before.

Chr. Why, what was it that brought your sins to mind again?

Hope. Many things, As,

1. *If I did but meet a good man in the Streets; or,
2. If I have heard any read in the Bible; or,
3. If mine Head did begin to Ake; or,
4. If I were told that some of my Neighbours were sick; or,
5. If I heard the Bell Toull for some that were dead; or,
6. If I thought of dying my self; or,
7. If I heard that suddain death happened to others.
8. But especially, when I thought of my self, that I must quickly come to Judgement.

• *When he had lost his sense of sin, what brought it again.*

Chr. And could you at any time with ease get off the guilt of sin when by any of these wayes it came upon you?

Hope. No, not latterly, for then they got faster hold of my Conscience. And then, if I did but think of going back to sin (though my mind was turned against it) it would be double torment to me.

Chr. And how did you do then?

Hope. I thought I must endeavour to mend my life, for else thought I, I am sure to be damned. *When he could no longer shake off his guilt by sinful courses, then he endeavours to mend.*

Chr. And did you indeavour to mend?

Hope. Yes, and fled from, not only my sins, but sinful Company too; and betook me to Religious Duties, as Praying, Reading, weeping for Sin, speaking Truth to my Neighbours, &c. These things I did, with many others, too much here to relate.

Chr. *And did you think your self well then?*

Then he thought himself well.

Hope. Yes, for a while; but at the last my trouble came tumbling upon me again, and that over the neck of all my Reformations.

Chr. *How came that about, since you was now Reformed?*

Reformation at last could not help, and why.
Isa. 64. 6.
Gal. 2. 16.
Luke 17. 10.

Hope. There were several things brought it upon me, especially such sayings as these; *All our righteousnesses are as filthy rags, By the works of the Law no man shall be justified. When you have done all things, say, We are unprofitable:* with many more the like: From whence I began to reason with my self thus: If *all* my righteousnesses are filthy rags, if by the deeds of the Law, *no* man can be justified; And, if when we have done *all*, we are yet unprofitable: Then 'tis but a folly to think of heaven by the Law. I further thought thus: *If a man runs an 100*l. into the Shop-keepers debt, and after that shall pay for all that he shall fetch, yet his old debt stands still in the Book un-crossed; for the which the Shop-keeper may sue him, and cast him into Prison till he shall pay the debt.

His being a debtor by the Law troubled him.

Chr. *Well, and how did you apply this to your self?*

Hope. Why, I thought thus with my self; I have by my sins run a great way into Gods Book, and that my now reforming will not pay off that score; therefore I should think still under all my present amendments, But how shall I be freed from that damnation that I have brought my self in danger of by my former transgressions?

Chr. *A very good application: but pray go on.*

Hope. Another thing that hath troubled me, even since my late amendments, is, that if I look narrowly into the best of what I do now, I still see sin, new sin, mixing it self with the best of that I do. So that now I am forced to conclude, that notwithstanding my former fond conceits of my self and duties, I have committed sin enough in one duty to send me to Hell, though my former life had been faultless.

His espying bad things in his best duties, troubled him.

Chr. *And what did you do then?*

Hope. Do! I could not tell what to do, till I brake my mind to *Faithful;* for he and I were well acquainted: And he told me, That unless I could obtain the righteousness of a man that never had sinned, neither mine own, nor all the righteousness of the World could save me.

This made him break his mind to Faithful, who told him the way to be saved.

Chr. *And did you think he spake true?*

Hope. Had he told me so when I was pleased and satisfied with mine own amendments, I had called him Fool for his pains: but now, since I see my own infirmity, and the sin that cleaves to my best performance, I have been forced to be of his opinion.

Chr. *But did you think, when at first he suggested it to you, that there was such a man to be found, of whom it might justly be said, That he never committed sin?*

Hope. I must confess the words at first sounded strangely, but after a little more talk and company with him, I had full conviction about it.

At which he started at present.

Chr. *And did you ask him what man this was, and how you must be justified by him?*

Hope. Yes, and he told me it was the Lord Jesus, that dwelleth on the right hand of the most High: *And thus, said he, you must be justified by him, even by trusting to what he hath done by himself in the days of his flesh, and suffered when he did hang on the Tree. I asked him further, How that mans righteousness could be of that efficacy, to justifie another before God? And he told me, He was the mighty God, and did what he did, and died the death also, not for himself, but for me; to whom his doings, and the worthiness of them should be imputed, if I believed on him.

Heb. 10.
Rom. 4.
Col. 1.
1 Pet. 1.
* *A more particular discovery of the way to be saved.*

Chr. *And what did you do then?*

Hope. I made my objections against my believing, for that I thought he was not willing to save me.

He doubts of acceptation.

Chr. *And what said Faithful to you then?*

Hope. He bid me go to him and see: Then I said, It was presumption: but he said, No: for I was invited to come. *Then he gave me a Book of *Jesus* his inditing, to incourage me the more freely to come: And he said concerning that Book, That every jot and tittle thereof stood firmer then Heaven and earth. Then I asked him, What I must do when I came? and he told me, I must intreat upon my knees with all my heart and soul, the Father to reveal him to me. Then I asked him further, How I must make my supplication to him? And he said, Go, and thou shalt find him upon a mercy-seat, where he sits all the year long, to give pardon and forgiveness to them that come. I told him that I knew not what to say when I

Mat. 11.
28.
* *He is better instructed.*
Mat. 24.
35.
Psal. 95. 6.
Dan. 6. 10.
Jer. 29. 12,
13.
Exod. 25. 22.
Lev. 16. 9.
Numb. 7.
8, 9.
Heb. 4. 16.

• *He is bid to pray.* came: *and he bid me say to this effect, *God be merciful to me a sinner, and make me to know and believe in Jesus Christ; for I see that if his righteousness had not been, or I have not faith in that righteousness, I am utterly cast away: Lord, I have heard that thou art a merciful God, and hast ordained that thy Son Jesus Christ should be the Saviour of the world; and moreover, that thou art willing to bestow him upon such a poor sinner as I am, (and I am a sinner indeed) Lord take therefore this opportunity, and magnifie thy grace in the Salvation of my soul, through thy Son Jesus Christ. Amen.*

Chr. *And did you do as you were bidden?*

Hope. *Yes, over, and over, and over.*

Chr. *And did the Father reveal his Son to you?*

He prays. Hope. Not at the first, nor second, nor third, nor fourth, nor fifth; no, nor at the sixth time neither.

Chr. *What did you do then?*

Hope. What! why I could not tell what to do.

Chr. *Had you not thoughts of leaving off praying?*

• *He thought to leave off praying.* Hope. *Yes, an hundred times, twice told.

Chr. *And what was the reason you did not?*

• *He durst not leave off praying, and why.* Hope. *I believed that that was true which had been told me, *to wit*, That without the righteousness of this Christ, all the World could not save me: And therefore thought I with my self, If I leave off, I die; and I can but die at the Throne of Grace. And withall, this came into my mind, *If it tarry, wait for it, because it will surely come, and will not tarry.* So I continued Praying untill the Father shewed me his Son.

Habb. 2. 3.

Chr. *And how was he revealed unto you?*

Hope. I did *not* see him with my bodily eyes, but with the eyes of mine understanding; and thus it was. One day I was very sad, I think sader then at any one time in my life; and this sadness was through a fresh sight of the greatness and vileness of my sins: And as I was then looking for nothing but *Hell*, and the everlasting damnation of my Soul, suddenly, as I thought, I saw the Lord Jesus look down from Heaven upon me, and saying, *Believe on the Lord Jesus Christ, and thou shalt be saved.*

Ephes. 1. 18, 19.

Christ is revealed to him, and how.

Act. 16. 30, 31.

2 Cor. 12. 9.

But I replyed, Lord, I am a great, a very great sinner; and he answered, *My grace is sufficient for thee.* Then I said But Lord, what is believing? And then I saw from that saying, [*He that cometh to me* John 6. 35. *shall never hunger, and he that believeth on me shall never thirst*] That believing and coming was all one, and that he that came, that is, run out in his heart and affections after Salvation by Christ, he indeed believed in Christ. Then the water stood in mine eyes, and I asked further, But Lord, may such a great sinner as I am, be indeed accepted of thee, and be saved by thee? And I heard him say, *And* John 6. 37. *him that cometh to me, I will in no wise cast out.* Then I said, But how, Lord, must I consider of thee in my coming to thee, that my Faith may be placed aright upon thee? Then he said, *Christ Jesus came into* 1 Tim. 1. *the World to save sinners. He is the end of the Law for righteousness to every* 15. *one that believes. He died for our sins, and rose again for our justification:* Rom. 10.4. *He loved us, and washed us from our sins in his own blood: He is Mediator* Heb. 7. 24, *between God and us. He ever liveth to make intercession for us.* From all 25. which I gathered, that I must look for righteousness in his person, and for satisfaction for my sins by his blood; that what he did in obedience to his Fathers Law, and in submitting to the penalty thereof, was not for himself, but for him that will accept it for his Salvation, and be thankful. And now was my heart full of joy, mine eyes full of tears, and mine affections running over with love, to the Name, People, and Ways of Jesus Christ.

Chr. *This was a Revelation of Christ to your soul indeed: But tell me particularly what effect this had upon your spirit?*

Hope. It made me see that all the World, notwithstanding all the righteousness thereof, is in a state of condemnation. It made me see that God the Father, though he be just, can justly justifie the coming sinner: It made me greatly ashamed of the vileness of my former life, and confounded me with the sence of mine own Ignorance; for there never came thought into mine heart before now, that shewed me so the beauty of Jesus Christ. It made me love a holy life, and long to do something for the Honour and Glory of the Name of the Lord Jesus. Yea I thought, that had I now a thousand gallons of blood in my body, I could spill it all for the sake of the Lord Jesus.

I then saw in my Dream, that *Hopeful* looked back and saw *Ignorance*, whom they had left behind, coming after. *Look*, said he, to *Christian*, how far yonder *Youngster loitereth behind.*

Chr. Ay, Ay, I see him; he careth not for our Company.

Hope. *But I tro, it would not have hurt him, had he kept pace with us hitherto.*

Chr. That's true, but I warrant you he thinketh otherwise.

Hope. *That I think he doth, but however let us tarry for him.* So they did.

Young Ignorance comes up again.

Then *Christian* said to him, *Come away man, why do you stay so behind?*

Their talk.

Ign. I take my pleasure in walking alone, even more a great deal then in Company, unless I like it the better.

Then said *Christian* to *Hopeful,* (but softly) *did I not tell you he cared not for our Company: But however, come up and let us talk away the time in this solitary place.* Then directing his Speech to *Ignorance,* he said, *Come, how do you? how stands it between God and your Soul now?*

Ignorance's hope, and the ground of it.

Ignor. *I hope well, for I am always full of good motions, that come into my mind to comfort me as I walk.

Chr. *What good motions? pray tell us.*

Ignor. Why, I think of God and Heaven.

Chr. *So do the Devils, and damned Souls.*

Ignor. But I think of them, and desire them.

Chr. *So do many that are never like to come there:* The Soul of the Sluggard†desires and hath nothing.

Ignor. But I think of them, and leave all for them.

Chr. *That I doubt, for leaving of all, is an hard matter, yea a harder matter then many are aware of. But why, or by what, art thou perswaded that thou hast left all for God and Heaven?*

Ignor. My heart tells me so.

Prov. 28. 26.

Chr. *The wise man sayes,* He that trusts his own heart is a fool.

Ignor. That is spoken of an evil heart, but mine is a good one.

Chr. *But how dost thou prove that?*

Ignor. It comforts me in the hopes of Heaven.

Chr. *That may be, through its deceitfulness, for a mans heart may minister comfort to him in the hopes of that thing, for which he yet has no ground to hope.*

Ignor. But my heart and life agree together, and therefore my hope is well grounded.

Chr. *Who told thee that thy heart and life agrees together?*

Ignor. My heart tells me so.

Chr. *Ask my fellow if I be a Thief: Thy heart tells thee so! Except the word of God beareth witness in this matter, other Testimony is of no value.*

Ignor. But is it not a good heart that has good thoughts? And is not that a good life, that is according to Gods Commandments?

Chr. *Yes, that is a good heart that hath good thoughts, and that is a good life that is according to Gods Commandments: But it is one thing indeed to have these, and another thing only to think so.*

Ignor. Pray, what count you good thoughts, and a life according to Gods Commandments?

Chr. *There are good thoughts of divers kinds, some respecting our selves, some God, some Christ, and some other things.*

Ignor. What be good thoughts respecting our selves?

Chr. *Such as agree with the Word of God.*

Ignor. When does our thoughts of our selves, agree with the Word of God?

What are good thoughts.

Chr. *When we pass the same Judgement upon our selves which the Word passes. To explain my self: The Word of God saith of persons in a natural condition, There is none Righteous, there is none that doth good. It saith also, That every imagination of the heart of man is only evil, and that continually. And again, The imagination of mans heart is evil from his Youth. Now then, when we think thus of our selves, having sense thereof, then are our thoughts good ones, because according to the Word of God.* Rom. 3. Gen. 6. 5.

Ignor. I will never believe that my heart is thus bad.

Chr. *Therefore thou never hadst one good thought concerning thy self in thy life. But let me go on: As the Word passeth a Judgement upon our HEART, so it passeth a Judgement upon our WAYS; and when our thoughts of our HEARTS and WAYS agree with the Judgement which the Word giveth of both, then are both good, because agreeing thereto.*

Ignor. Make out your meaning.

Chr. *Why, the Word of God saith, That mans ways are crooked ways, not good, but perverse: It saith, They are naturally out of the good way, that they have not known it. Now when a man thus thinketh of his ways, I say when he doth sensibly, and with heart-humiliation thus think, then hath he good thoughts of his own ways, because his thoughts now agree with the judgment of the Word of God.* Psal. 125. 5. Prov. 2. 15. Rom. 3.

Ignor. What are good thoughts concerning God?

Chr. Even (*as I have said concerning ourselves*) *when our thoughts of God do agree with what the Word saith of him. And that is, when we think of his Being and Attributes as the Word hath taught: Of which I cannot now discourse at large. But to speak of him with reference to us, Then we have right thoughts of God, when we think that he knows us better then we know our selves, and can see sin in us, when, and where we can see none in our selves; when we think he knows our in-most thoughts, and that our heart, with all its depths is alwayes open unto his eyes: Also when we think that all our Righteousness stinks in his Nostrils, and that therefore he cannot abide to see us stand before him in any confidence, even of all our best performances.*

Ignor. Do you think that I am such a fool, as to think God can see no further then I? or that I would come to God in the best of my performances?

Chr. Why, how dost thou think in this matter?

Ignor. Why, to be short, I think I must believe in Christ for Justification.

Chr. How! think thou must believe in Christ, when thou seest not thy need of him! Thou neither seest thy original, nor actual infirmities, but hast such an opinion of thy self, and of what thou doest, as plainly renders thee to be one that did never see a necessity of Christs personal righteousness to justifie thee before God: How then dost thou say, I believe in Christ?

Ignor. I believe well enough for all that.

Chr. How doest thou believe?

Ignor. I believe that Christ died for sinners, and that I shall be justified before God from the curse, through his gracious acceptance of my obedience to his Law: Or thus, Christ makes my Duties that are Religious, acceptable to his Father by vertue of his Merits; and so shall I be justified.

Chr. Let me give an answer to this confession of thy faith.

The Faith of Ignorance. 1. *Thou believest with a* Fantastical Faith, *for this faith is no where described in the Word.*

2. *Thou believest with a* False Faith, *because it taketh Justification from the personal righteousness of Christ, and applies it to thy own.*

3. *This faith maketh not Christ a Justifier of thy person, but of thy actions; and of thy person for thy actions sake, which is false.*

4. *Therefore this faith is deceitful, even such as will leave thee under*

wrath, in the day of God Almighty. For true Justifying Faith puts the soul (as sensible of its lost condition by the Law) upon flying for refuge unto Christs righteousness: (which righteousness of his, is, not an act of grace, by which he maketh for Justification thy obedience accepted with God, but his personal obedience to the Law in doing and suffering for us, what that required at our hands.) This righteousness, I say, true faith accepteth, under the skirt of which, the soul being shrouded, and by it presented as spotless before God, it is accepted, and acquit from condemnation.

Ignor. What! would you have us trust to what Christ in his own person has done without us? This conceit would loosen the reines of our lust, and tollerate us to live as we list: For what matter how we live if we may be Justified by Christs personal righteousness from all, when we believe it?†

Chr. Ignorance *is thy name, and as thy name is, so art thou; even this thy answer demonstrateth what I say.* Ignorant *thou art of what Justifying righteousness is, and, as* Ignorant *how to secure thy Soul through the faith of it from the heavy wrath of God. Yea, thou also art* Ignorant *of the true effects of saving faith in this righteousness of Christ, which is, to bow and win over the heart to God in Christ, to love his Name, his Word, Ways and People, and not as thou* ignorantly *imaginest.*

Hope. Ask him if ever he had Christ revealed to him from Heaven?

Ignor. What! you are a man for revelations! I believe that what both you, and all the rest of you say about that matter, is but the fruit of distracted braines.

Ignorance angles with them.

Hope. Why man! Christ is so hid in God from the natural apprehensions of all flesh, that he cannot by any man be savingly known, unless God the Father reveals him to them.

Ignor. That is your faith, but not mine; yet mine I doubt not, is as good as yours: though I have not in my head so many whimzies as you.

He speaks reproachfully of what he knows not. Mat. 11. 27. 1 Cor. 12. 3. Eph. 1. 18, 19.

Chr. Give me leave to put in a word: You ought not so slightly to speak of this matter: for this I will boldly affirm, (even as my good Companion hath done) that no man can know Jesus Christ but by the Revelation of the Father: yea, and faith too, by which the soul layeth hold upon Christ (if it be right) must be wrought by the exceeding greatness of his mighty power, the working of which faith, I perceive, poor *Ignorance*, thou art ignorant of. Be awakened then, see thine own wretchedness, and flie to the Lord Jesus; and by his righteousness, which is the righteousness of God, (for he himself is God) thou shalt be delivered from condemnation.

The talk broke up. *Ignor.* *You go so fast, I cannot keep pace with you; do you go on before, I must stay a while behind.*

Then they said,

> *Well* Ignorance, *wilt thou yet foolish be,*
> *To slight good Counsel, ten times given thee?*
> *And if thou yet refuse it, thou shalt know*
> *Ere long the evil of thy doing so:*
> *Remember man in time, stoop, do not fear,*
> *Good Counsel taken well, saves; therefore hear:*
> *But if thou yet shalt slight it, thou wilt be*
> *The loser* (Ignorance) *I'le warrant thee.*

Then *Christian* addressed thus himself to his fellow.

Chr. Well, come my good *Hopeful*, I perceive that thou and I must walk by our selves again.

So I saw in my Dream, that they went on a pace before, and *Ignorance* he came hobling after. Then said *Christian* to his Companion, *It pities me much for this poor man, it will certainly go ill with him at last.*

Hope. Alas, there are abundance in our Town in his condition; whole Families, yea, whole Streets, (and that of Pilgrims too) and if there be so many in our parts, how many, think you, must there be in the place where he was born?

Chr. *Indeed the Word saith,* He hath blinded their eyes, lest they should see, *&c. But now we are by our selves, what do you think of such men? Have they at no time, think you, convictions of sin, and so consequently fears that their state is dangerous?*

Hope. Nay, do you answer that question your self, for you are the elder man.

Chr. *Then, I say, sometimes (as I think) they may, but they being naturally ignorant, understand not that such convictions tend to their good; and therefore they do desperately seek to stifle them, and presumptuously continue to flatter themselves in the way of their own hearts.*

The good use of fear. *Hope.* I do believe as you say, that fear tends much to Mens good, and to make them right, at their beginning to go on Pilgrimage.

Chr. *Without all doubt it doth, if it be right: for so says the Word,* Job 28. 28.
The fear of the Lord is the beginning of Wisdom. Psal. 111.
10.

Hope. How will you describe right fear? Prov. 1. 7.
ch. 9. 10.

Chr. *True, or right fear, is discovered by three things.* *Right fear.*

1. By its rise. It is caused by saving convictions for sin.

2. It driveth the soul to lay fast hold of Christ for Salvation.

3. It begetteth and continueth in the soul a great reverence of
God, his word, and ways, keeping it tender, and making it afraid
to turn from them, to the right hand, or to the left, to any thing
that may dishonour God, break its peace, grieve the Spirit, or cause
the enemy to speak reproachfully.

Hope. Well said, I believe you have said the truth. Are we now
almost got past the Inchanted ground?

Chr. *Why, are you weary of this discourse?*

Hope. No verily, but that I would know where we are.

Chr. *We have not now above two Miles further to go thereon. But let us* *Why*
return to our matter. *Now the Ignorant know not that such convictions that* *ignorant
persons*
tend to put them in fear, are for their good, and therefore they seek to stifle *stifle con-*
them. *victions.*
* 1. In
general.*

Hope. How do they seek to stifle them?

Chr. *1. They think that those fears are wrought by the Devil * 2. In particular.
(though indeed they are wrought of God) and thinking so, they
resist them, as things that directly tend to their overthrow. 2. They
also think that these fears tend to the spoiling of their faith, (when
alas for them, poor men that they are! they have none at all) and
therefore they harden their hearts against them. 3. They presume
they ought not to fear, and therefore, in despite of them, wax presumptuously confident. 4. They see that these fears tend to take
away from them their pitiful old self-holiness, and therefore they
resist them with all their might.

Hope. I know something of this my self; for before I knew my self
it was so with me.

Chr. *Well, we will leave at this time our Neighbour* Ignorance *by himself, and fall upon another profitable question.*

Hope. With all my heart, but you shall still begin.

Chr. *Well then, Did you not know about ten years ago; one* Temporary *Talk about
one* Tem-
porary.
in your parts, who was a forward man in Religion then?

Hope. Know him! Yes, he dwelt in *Graceless*, a Town about two
Where he dwelt. miles off of *Honesty*, and he dwelt next door to one *Turn-back*.

Chr. Right, he dwelt under the same roof with him. Well, that man was
• He was much awakened once;* *I believe that then he had some sight of his sins, and*
towardly once. *of the wages that was due thereto.*

Hope. I am of your mind, for (my house not being above three
miles from him) he would oft times come to me, and that with many
tears. Truly I pitied the man, and was not altogether without hope
of him; but one may see, it is not every one that cries, *Lord, Lord.*†

*Chr. He told me once, That he was resolved to go on Pilgrimage, as we
do now; but all of a sudden he grew acquainted with one Save-self, and then
he became a stranger to me.*

Hope. Now since we are talking about him, let us a little enquire
into the reason of the suddain backsliding of him and such others.

Chr. It may be very profitable, but do you begin.

Hope. Well then, there are in my judgement four reasons for it.

Reason, why towardly ones go back. 1. Though the Consciences of such men are awakened, yet their
minds are not changed: therefore when the power of guilt weareth
away, that which provoked them to be Religious, ceaseth. Wherefore they naturally turn to their own course again: even as we see
the Dog that is sick of what he hath eaten, so long as his sickness
prevails, he vomits and casts up all; not that he doth this of a free
mind (if we may say a Dog has a mind) but because it troubleth his
Stomach; but now when his sickness is over, and so his Stomach
eased, his desires being not at all alienate from his vomit, he turns
2 Pet. 2. him about, and licks up all. And so it is true which is written, *The
22. Dog is turned to his own vomit again.* Thus, I say, being hot for Heaven,
by virtue only of the sense and fear of the torments of Hell, as their
sense of Hell, and the fears of damnation chills and cools, so their
desires for Heaven and Salvation cool also. So then it comes to pass,
that when their guilt and fear is gone, their desires for Heaven
and Happiness die; and they return to their course again.

2*ly.* Another reason is, They have slavish fears that do over-
Prov. 29. master them. I speak now of the fears that they have of men: *For
25. the fear of men bringeth a snare.* So then, though they seem to be hot
for Heaven, so long as the flames of Hell are about their ears, yet
when that terrour is a little over, they betake themselves to second

thoughts; namely, that 'tis good to be wise, and not to run (for they know not what) the hazard of loosing all; or at least, of bringing themselves into unavoidable and un-necessary troubles: and so they fall in with the world again.

3*ly*. The shame that attends Religion, lies also as a block in their way; they are proud and haughty, and Religion in their eye is low and contemptible: Therefore when they have lost their sense of Hell and wrath to come, they return again to their former course.

4*ly*, Guilt, and to meditate terrour, are grievous to them, they like not to see their misery before they come into it: Though perhaps the sight of it first, if they loved that sight, might make them flie whither the righteous flie and are safe; but because they do, as I hinted before, even shun the thoughts of guilt and terrour, therefore, when once they are rid of their awakenings about the terrors and wrath of God, they harden their hearts gladly, and chuse such ways as will harden them more and more.

Chr. You are pretty near the business, for the bottom of all is, for want of a change in their mind and will. And therefore they are but like the Fellon that standeth before the Judge: he quakes and trembles, and seems to repent most heartily; but the bottom of all is, the fear of the Halter, not of any detestation of the offence; as is evident, because, let but this man have his liberty, and he will be a Thief, and so a Rogue still; whereas, if his mind was changed, he would be otherwise.

Hope. Now I have shewed you the reasons of their going back, do you shew me the manner thereof.

Chr. So I will willingly.

1. They draw off their thoughts all that they may from the remembrance of God, Death, and Judgement to come. *How the Apostate goes back.*

2. Then they cast off by degrees private Duties, as Closet-Prayer, curbing their lusts, watching, sorrow for Sin, and the like.

3. Then they shun the company of lively and warm Christians.

4. After that, they grow cold to publick Duty, as Hearing, Reading, Godly Conference, and the like.

5. Then they begin to pick holes, as we say, in the Coats of some of the Godly, and that devilishly that they may have a seeming colour to throw Religion (for the sake of some infirmity they have spied in them) behind their backs.

6. Then they begin to adhere to, and associate themselves with carnal, loose, and wanton men.

7. Then they give way to carnal and wanton discourses in secret; and glad are they if they can see such things in any that are counted honest, that they may the more boldly do it through their example.

8. After this, they begin to play with little sins openly.

9. And then, being hardened, they shew themselves as they are. Thus being lanched again into the gulf of misery, unless a Miracle of Grace prevent it, they everlastingly perish in their own deceivings.

Now I saw in my Dream, that by this time the Pilgrims were got over the Inchanted Ground, and entering into the Country of *Beulah*, whose Air was very sweet and pleasant, the way lying directly through it, they solaced themselves there for a season. Yea, here they heard continually the singing of Birds, and saw every day the flowers appear in the earth: and heard the voice of the Turtle in the Land. In this Countrey the Sun shineth night and day; wherefore this was beyond the Valley of the *shadow of death*, and also out of the reach of Giant *Despair*; neither could they from this place so much as see *Doubting-Castle*. Here they were within sight of the City they were going to: also here met them some of the Inhabitants thereof. For in this Land the shining Ones commonly walked, because it was upon the Borders of Heaven. In this Land also the contract between the Bride and the Bridgroom was renewed: Yea here, *as the Bridegroom rejoyceth over the Bride, so did their God rejoyce over them.* Here they had no want of Corn and Wine; for in this place they met with abundance of what they had sought for in all their Pilgrimage. Here they heard voices from out of the City, loud voices, saying, *Say ye to the daughter of* Zion, *Behold thy Salvation cometh, behold his reward is with him.* Here all the Inhabitants of the Countrey called them, *The holy People, the redeemed of the Lord, Sought out,* &c.

Now as they walked in this Land they had more rejoycing then in parts more remote from the Kingdom, to which they were bound; and drawing near to the City, they had yet a more perfect view thereof. It was builded of Pearls and Precious Stones, also the Street thereof was paved with Gold, so that by reason of the natural glory

Marginal notes:

Isa. 62. 4.
Cant. 2. 10, 11, 12.

Angels.

Isa. 62. 5.
ver. 8.

ver. 11.
ver. 12.

of the City, and the reflection of the Sunbeams upon it, *Christian*, with desire fell sick, *Hopeful* also had a fit or two of the same Disease: Wherefore here they lay by it a while, crying out because of their pangs, *If you see my Beloved, tell him that I am sick of love.*†

But being a little strengthned, and better able to bear their sickness, they walked on their way, and came yet nearer and nearer, where were Orchards, Vineyards, and Gardens, and their Gates opened into the Highway. Now as they came up to these places, behold the Gardener stood in the way; to whom the Pilgrims said, Whose goodly Vineyards and Gardens are these? He answered, They are the Kings, and are planted here for his own delights, and also for the solace of Pilgrims, So the Gardiner had them into the Vineyards, and bid them refresh themselves with the Dainties; he also shewed them *there* the Kings Walks and the *Arbors* where he delighted to be: And here they tarried and slept. | Deut. 23. 24.

Now I beheld in my Dream, that they talked more in their sleep at this time, then ever they did in all their Journey; and being in a muse there-about, the Gardiner said even to me, Wherefore musest thou at the matter? It is the nature of the fruit of the Grapes of these Vineyards to go down so sweetly, as to cause the lips of them that are asleep to speak.

So I saw that when they awoke, they addressed themselves to go up to the City. But, as I said, the reflections of the Sun upon the City, (for the City was pure Gold) was so extreamly glorious, that they could not, as yet, with open face behold it, but through an *Instrument* made for that purpose. So I saw, that as they went on, there met them two men, in Raiment that shone like Gold, also their faces shone as the light. | Rev. 21. 18. 2 Cor. 3. 18.

These men asked the Pilgrims whence they came? and they told them; they also asked them, Where they had lodg'd, what difficulties, and dangers, what comforts and pleasures they had met in the way? and they told them. Then said the men that met them, You have but two difficulties more to meet with, and then you are in the City.

Christian then and his Companion asked the men to go along with them, so they told them they would; but, said they, you must obtain it by your own faith. So I saw in my Dream that they went on together till they came within sight of the Gate.

Death. Now I further saw, that betwixt them and the Gate was a River, but there was no Bridge to go over; the River was very deep; at the sight therefore of this River, the Pilgrims were much stounded, but the men that went with them, said, You must go through, or you cannot come at the Gate.

Death is not welcome to nature though by it we pass out of this World into glory.
1 Cor. 15. 51, 52.
The Pilgrims then began to enquire if there was no other way to the Gate; to which they answered, Yes; but there hath not any, save two, to wit, *Enoch* and *Elijah,* been permitted to tread that path, since the foundation of the World, nor shall, untill the last Trumpet shall sound. The Pilgrims then, especially *Christian,* began to dispond in his mind, and looked this way and that, but no way could be found by them, by which they might escape the River. Then they asked the men if the Waters were all of a depth. They said no; yet

Angels help us not comfortably through death.
they could not help them in that Case; for said they, *You shall find it deeper or shallower, as you believe in the King of the place.*

They then addressed themselves to the Water; and entring, *Christian* began to sink, and crying out to his good friend *Hopeful;* he said, I sink in deep Waters, the Billows go over my head, all his Waves go over me, *Selah.*†

Christians conflict at the hour of death.
Then said the other, Be of good chear, my Brother, I feel the bottom, and it is good. Then said *Christian,* Ah my friend, the sorrows of death have compassed me about, I shall not see the Land that flows with Milk and Honey.†And with that, a great darkness and horror fell upon *Christian,* so that he could not see before him; also here he in great measure lost his senses, so that he could neither remember nor orderly talk of any of those sweet refreshments that he had met with in the way of his Pilgrimage. But all the words that he spake, still tended to discover that he had horror of mind, and hearty fears that he should die in that River, and never obtain entrance in at the Gate: Here also, as they that stood by, perceived, he was much in the troublesome thoughts of the sins that he had committed, both since and before he began to be a Pilgrim. 'Twas also observed, that he was troubled with apparitions of Hob-goblins and Evil Spirits, For ever and anon he would intimate so much by words. *Hopeful* therefore here had much adoe to keep his Brothers

head above water, yea sometimes he would be quite gone down, and then ere a while he would rise up again half dead. *Hopeful* also would endeavour to comfort him, saying, Brother, I see the Gate, and men standing by it to receive us. But *Christian* would answer, 'Tis you, 'tis you they wait for, you have been *Hopeful* ever since I knew you: and so have you, said he to *Christian*. Ah Brother, said he, surely if I was right, he would now arise to help me; but for my sins he hath brought me into the snare, and hath left me. Then said *Hopeful*, My Brother, you have quite forgot the Text, where its said of the wicked, *There is no band in their death, but their strength is firm,* Psal. 73. *they are not troubled as other men, neither are they plagued like other men.* 4, 5. These troubles and distresses that you go through in these Waters, are no sign that God hath forsaken you, but are sent to try you, whether you will call to mind that which heretofore you have received of his goodness, and live upon him in your distresses.

Then I saw in my Dream that *Christian* was as in a muse a while; to whom also *Hopeful* added this word, Be of good cheer, *Jesus Christ* Christian *maketh thee whole:*†And with that, *Christian* brake out with a loud delivered *from his* voice, Oh I see him again! and he tells me, *When thou passest through* fears in *the waters, I will be with thee, and through the Rivers, they shall not over-* death. *flow thee.* Then they both took courage, and the enemy was after Isa. 43. 2. that as still as a stone,†until they were gone over. *Christian* therefore presently found ground to stand upon; and so it followed that the rest of the River was but shallow. Thus they got over. Now upon the bank of the River, on the other side, they saw the two shining men again, who there waited for them. Wherefore being come up out of the River, they saluted them, saying, *We are ministring Spirits,* The Angels *sent forth to minister for those that shall be Heirs of Salvation.* Thus they do wait for *them so soon* went along towards the Gate, now you must note that the City as they are *passed out* stood upon a mighty hill, but the Pilgrims went up that hill *with* of this *ease,* because they had these two men to lead them up by the Arms; world. also they had left their *Mortal* Garments behind them in the River: They have *put off* for though they went in with them, they came out without them. mortality. They therefore went up here with much agility and speed, though the foundation upon which the City was framed was higher then the Clouds. They therefore went up through the Regions of the Air, sweetly talking as they went, being comforted, because they safely got over the River, and had such glorious Companions to attend them.

The talk that they had with the shining Ones, was about the glory of the place, who told them, that the beauty, and glory of it was inexpressible. There, said they, is the Mount *Sion*, the heavenly *Jerusalem*, the inumerable company of Angels, and the Spirits of Just Men†made perfect: You are going now, said they, to the Paradice of God, wherein you shall see the Tree of Life, and eat of the never-fading fruits thereof: And when you come there, you shall have white Robes given you, and your walk and talk shall be every day with the King, even all the days of eternity. There you shall not see again, such things as you saw when you were in the lower Region upon the earth, to wit, sorrow, sickness, affliction, and death, *for the former things are passed away.* You are going now to *Abraham*, to *Isaac*, and *Jacob*, and to the Prophets; men that God hath taken away from the evil to come, and that are now resting upon their Beds, each one walking in his righteousness. The men then asked, What must we do in the holy place? To whom it was answered, You must there receive the comfort of all your toil, and have joy for all your sorrow; you must reap what you have sown, even the fruit of all your Prayers and Tears, and sufferings for the King by the way. In that place you must wear Crowns of Gold, and enjoy the perpetual sight and Visions of the *Holy One*, *for there you shall see him as he is.* There also you shall serve him continually with praise, with shouting and thanksgiving, whom you desired to serve in the World, though with much difficulty, because of the infirmity of your flesh. There your eyes shall be delighted with seeing, and your ears with hearing, the pleasant voice of the mighty One. There you shall enjoy your friends again, that are got thither before you; and there you shall with joy receive, even every one that follows into the Holy place after you. There also you shall be cloathed with Glory and Majesty, and put into an equipage fit to ride out with the King of Glory. When he shall come with sound of Trumpet in the Clouds, as upon the wings of the Wind, you shall come with him; and when he shall sit upon the Throne of Judgement, you shall sit by him; yea, and when he shall pass Sentence upon all the workers of Iniquity, let them be Angels or Men, you also shall have a voice

Marginal references:

Heb. 12. 22, 23, 24.
Rev. 2. 7.
Rev. 3. 4.

Rev. 21. 1.

Isa. 57. 1,
2.
Isa. 65. 14.

Gal. 6. 7.
1 John 3. 2.

1 Thes. 4. 13, 14, 15, 16.
Jude 14.
Da. 7. 9, 10.
1 Cor. 6. 2, 3.

in that Judgement, because they were his and your Enemies. Also when he shall again return to the City, you shall go too, with sound of Trumpet, and be ever with him.

Now while they were thus drawing towards the Gate, behold a company of the Heavenly Host came out to meet them: To whom it was said, by the other two shining Ones, These are the men that have loved our Lord, when they were in the World, and that have left all for his holy Name, and he hath sent us to fetch them, and we have brought them thus far on their desired Journey; that they may go in and look their Redeemer in the face with joy. Then the Heavenly Host gave a great shout, saying, *Blessed are they that are* Rev. 19. *called to the Marriage Supper of the Lamb.*

There came out also at this time to meet them, several of the Kings Trumpeters, cloathed in white and shining Rayment, who with melodious noises, and loud, made even the Heavens to eccho with their sound. These Trumpeters saluted *Christian* and his Fellow with ten thousand welcomes from the world: And this they did with shouting, and sound of Trumpet.

This done, they compassed them round on every side; some went before, some behind, and some on the right hand, some on the left (as 'twere to guard them through the upper Regions) continually sounding as they went, with melodious noise, in notes on high; so that the very sight was to them that could behold it, as if Heaven it self was come down to meet them. Thus therefore they walked on together, and as they walked, ever and anon, these Trumpeters, even, with joyful sound, would, by mixing their Musick, with looks and gestures, still signifie to *Christian* and his Brother, how welcome they were into their company, and with what gladness they came to meet them: And now were these two men, as 'twere, in Heaven, before they came at it; being swallowed up with the sight of Angels, and with hearing of their melodious notes. Here also they had the City it self in view, and they thought they heard all the Bells therein to ring, to welcome them thereto: but above all, the warm and joyful thoughts that they had about their own dwelling there, with such company, and that for ever and ever. Oh! by what tongue or pen can their glorious joy be expressed: and thus they came up to the Gate.

Now when they were come up to the Gate, there was written
Rev. 22. over it, in Letters of Gold, *Blessed are they that do his commandments,*
14. *that they may have right to the Tree of Life; and may enter in through the*
Gates into the City.

Then I saw in my Dream, that the shining men bid them call at
the Gate, the which when they did, some from above looked over
the Gate; to wit, *Enoch, Moses,* and *Elijah, &c.* to whom it was said,
These Pilgrims are come from the City of *Destruction,* for the love
that they bear to the King of this place: and then the Pilgrims gave
in unto them each man his Certificate, which they had received in
the beginning; those therefore were carried into the King, who
when he had read them, said, Where are the men? to whom it was
answered, They are standing without the Gate. The King then
Isa. 26. 2. commanded to open the Gate; *That the righteous Nation,* said he, *that*
keepeth Truth may enter in.

Now I saw in my Dream, that these two men went in at the Gate;
and loe, as they entered, they were transfigured, and they had
Raiment put on that shone like Gold. There was also that met them
with Harps and Crowns, and gave them to them; The Harp to
praise withal, and the Crowns in token of honor: Then I heard in
my Dream, that all the Bells in the City Rang again for joy; and
that it was said unto them, *Enter ye into the joy of your Lord.* I also
heard the men themselves, that they sang with a loud voice, saying,
Rev. 5. 13, *Blessing, Honour, Glory, and Power, be to him that sitteth upon the Throne,*
14. *and to the Lamb for ever and ever.*

Now just as the Gates were opened to let in the men, I looked
after them; and behold, the City shone like the Sun, the Streets also
were paved with Gold, and in them walked many men, with Crowns
on their heads, Palms in their hands, and golden Harps to sing
praises withall.

There were also of them that had wings, and they answered one
another without intermission, saying, *Holy, Holy, Holy, is the Lord.*
And after that, they shut up the Gates: which when I had seen,
I wished my self among them.

Now while I was gazing upon all these things, I turned my head
to look back, and saw *Ignorance* come up to the River side: but he
soon got over, and that without half that difficulty which the other

two men met with. For it happened, that there was then in that place one *Vain-hope* a Ferry-man, that with his Boat helped him over: so he, as the other I saw, did ascend the Hill to come up to the Gate, only he came alone; neither did any man meet him with the least incouragement. When he was come up to the Gate, he looked up to the writing that was above; and then began to knock, supposing that entrance should have been quickly administred to him: But he was asked by the men that lookt over the top of the Gate, Whence came you? and what would you have? He answered, I have eat and drank in the presence of the King, and he has taught in our Streets. Then they asked him for his Certificate, that they might go in and shew it to the King. So he fumbled in his bosom for one, and found none. Then said they, Have you none? But the man answered never a word. So they told the King but he would not come down to see him; but commanded the two shining Ones that conducted *Christian* and *Hopeful* to the City to go out and take *Ignorance* and bind him hand and foot, and have him away. Then they took him up, and carried him through the air to the door that I saw in the side of the Hill, and put him in there. Then I saw that there was a way to Hell, even from the Gates of Heaven, as well as from the City of *Destruction*. So I awoke, and behold it was a Dream.

FINIS

THE CONCLUSION

NOW Reader, I have told my Dream to thee;
See if thou canst Interpret it to me;
Or to thy self, or Neighbour: but take heed
Of mis-interpreting: for that, instead
Of doing good, will but thy self abuse:
By mis-interpreting evil issues.

Take heed also, that thou be not extream,
In playing with the out-side of my Dream:
Nor let my figure, or similitude,
Put thee into a laughter or a feud;
Leave this for Boys and Fools; but as for thee,
Do thou the substance of my matter see.

Put by the Curtains, look within my Vail;
Turn up my Metaphors and do not fail:
There, if thou seekest them, such things to find,
As will be helpful to an honest mind.

What of my dross *thou findest there, be bold*
To throw away, but yet preserve the Gold.
What if my Gold be wrapped up in Ore?
None throws away the Apple for the Core:
But if thou shalt cast all away as vain,
I know not but 'twill make me Dream again.

THE END

THE AUTHOR'S WAY
OF SENDING FORTH HIS
SECOND PART OF
THE PILGRIM

GO, now my little Book, to every place,
Where my first Pilgrim has but shewn his Face,
Call at their door: If any say, who's there?
Then answer thou, Christiana is here.
If they bid thee come in, then enter thou
With all thy boys. And then, as thou know'st how,
Tell who they are, also from whence they came,
Perhaps they'l know them, by their looks, or name:
But if they should not, ask them yet again
If formerly they did not Entertain
One Christian a Pilgrim; If they say
They did: And was delighted in his way:
Then let them know that those related were
Unto him: Yea, his Wife and Children are.

Tell them that they have left their House and Home,
Are turned Pilgrims, seek a World to come:
That they have met with hardships in the way,
That they do meet with troubles night and Day;
That they have trod on Serpents, fought with Devils,
Have also overcome a many evils.
Yea tell them also of the rest, who have
Of love to Pilgrimage been stout and brave
Defenders of that way, and how they still
Refuse this World, to do their Fathers will.

Go, tell them also of those dainty things,
That Pilgrimage unto the Pilgrim brings,
Let them acquainted be, too, how they are
Beloved of their King, under his care;

What goodly Mansions *for them he Provides.*
Tho they meet with rough Winds, and swelling Tides,
How brave a calm they will enjoy at last,
Who to their Lord, and by his ways hold fast.

   Perhaps with heart and hand they will imbrace
Thee, as they did my firstling,†and will Grace
Thee, and thy fellows with such chear and fair,
As shew will, they of Pilgrims *lovers are.*

1 *Object*

But how if they will not believe of me
That I am truly thine, 'cause some there be
That Counterfeit the Pilgrim, and his name,
Seek by disguise to seem the very same.
And by that means have wrought themselves into
The Hands and Houses of I know not who.

Answer

'Tis true, some have of late, to Counterfeit
My Pilgrim,†*to their own, my Title set;*
Yea others, half my Name and Title too;
Have stitched to their Book, to make them do;
But yet they by their Features *do declare*
Themselves not mine to be, whose ere they are.

   If such thou meetst with, then thine only way
Before them all, is, to say out thy say,
In thine own native Language, which no man
Now useth, nor with ease dissemble can.

   If after all, they still of you shall doubt,
Thinking that you like Gipsies *go about,*
In naughty-wise the Countrey to defile,
Or that you seek good People to beguile
With things unwarrantable: Send for me
And I will Testifie, you Pilgrims *be;*
Yea, I will Testifie that only you
My Pilgrims *are; And that alone will do.*

2 Object

But yet, perhaps, I may enquire for him,
Of those that wish him Damned life and limb,
What shall I do, when I at such a door,
For *Pilgrims* ask, and they shall rage the more?

Answer

Fright not thy self my Book, for such Bugbears
Are nothing else but ground for groundless fears,
My Pilgrims Book has travel'd Sea and Land,
Yet could I never come to understand,
That it was slighted, or turn'd out of Door
By any Kingdom, were they Rich or Poor.

In France *and Flanders where men kill each other*
My Pilgrim is esteem'd a Friend, a Brother.

In Holland *too, 'tis said, as I am told,*
My Pilgrim is with some, worth more than Gold.

Highlanders, and Wild-Irish can agree,
My Pilgrim should familiar with them be.

'Tis in New-England *under such advance,*
Receives there so much loving Countenance,
As to be Trim'd, new Cloth'd & Deckt with Gems,†
That it might shew its Features, and its Limbs,
Yet more; so comely doth my Pilgrim walk,
That of him thousands daily Sing and talk.

If you draw nearer home, it will appear
My Pilgrim knows no ground of shame, or fear;
City, and Countrey will him Entertain,
With welcome Pilgrim. Yea, they can't refrain
From smiling, if my Pilgrim be but by,
Or shews his head in any Company.

Brave Galants do my Pilgrim hug and love,
Esteem it much, yea value it above
Things of a greater bulk, yea, with delight,
Say my Larks leg is beter than a Kite.

Young Ladys, and young Gentle-women too,
Do no small kindness to my Pilgrim shew;

Their Cabinets, their Bosoms, and their Hearts
My Pilgrim *has, 'cause he to them imparts*
His pretty riddles in such wholsome straines
As yields them profit double to their paines
Of reading. Yea, I think I may be bold
To say some prize him far above their Gold.

The very Children that do walk the street,
If they do but my holy Pilgrim meet,
Salute him will, will wish him well and say,
He is the only Stripling *of the Day.*

They that have never seen him, yet admire
What they have heard of him, and much desire
To have his Company, and hear him tell
Those Pilgrim *storyes which he knows so well.*

Yea, some who did not love him at the first,
But cal'd him Fool, *and* Noddy, *say they must*
Now they have seen & heard him, him commend,
And to those whom they love, they do him send.

Wherefore my Second Part, *thou needst not be*
Afraid to shew thy Head: None can hurt thee,
That wish but well to him, that went before,
'Cause thou com'st after with a Second store,
Of things as good, as rich, as profitable,
For Young, for Old, for Stag'ring and for stable.

3 Object

But some there be that say he laughs too loud;
And some do say his Head is in a Cloud.
Some say, his Words and Storys are so dark,
They know not how, by them, to find his mark.

Answer

One may (I think) *say both his laughs & cryes,*
May well be guest at by his watry Eyes.
Some things are of that Nature as to make
Ones fancie Checkle while his Heart doth ake,

When Jacob *saw his* Rachel *with the Sheep,*†
He did at the same time both kiss and weep.

  Whereas some say a Cloud is in his Head,
That doth but shew how Wisdom's covered
With its own mantles: And to stir the mind
To a search after what it fain would find,
Things that seem to be hid in words obscure,
Do but the Godly mind the more alure;
To study what those Sayings should contain,
That speak to us in such a Cloudy strain.

  I also know, a dark Similitude
Will on the Fancie more it self intrude,
And will stick faster in the Heart and Head,
Then things from Similies not borrowed.

  Wherefore, my Book, let no discouragement
Hinder thy travels. Behold, thou art sent
To Friends, not foes: to Friends that will give place
To thee, thy Pilgrims, *and thy words imbrace.*

  Besides, what my first Pilgrim *left conceal'd,*
Thou my brave Second Pilgrim *hast reveal'd;*
What Christian *left lock't up and went his way,*
Sweet Christiana *opens with her Key.*†

4 *Object*

But some love not the method of your first,
Romance they count it, throw't away as dust,
If I should meet with such, what should I say?
Must I slight them as they slight me, or nay?

Answer

My Christiana, *if with such thou meet,*
By all means in all Loving-wise, them greet;
Render them not reviling for revile:
But if they frown, I prethee on them smile.
Perhaps 'tis Nature, or some ill report
Has made them thus *dispise, or* thus *retort.*

  Some love no Cheese, some love no Fish, &c *some*

Love not their Friends, nor their own House or home;
Some start at Pigg, slight Chicken, love not Fowl,
More then they love a Cuckoo or an Owl.
Leave such, my Christiana, *to their choice,*
And seek those, who to find thee will rejoyce;
By no means strive, but in all humble wise,
Present thee to them in thy Pilgrims *guise.*

   Go then, my little Book and shew to all
That entertain, and bid thee welcome shall,
What thou shalt keep close, shut up from the rest,
And wish what thou shalt shew them may be blest
To them for good, may make them chuse to be
Pilgrims, better by far, then thee or me.

   Go then, I say, tell all men who thou art,
Say, I am Christiana, *and my part*
Is now with my four Sons, to tell you what
It is for men to take a Pilgrims *lot.*

   Go also tell them who, and what they be,
That now do go on Pilgrimage with thee;
Say, here's my neighbour Mercy, *she is one,*
That has long-time with me a Pilgrim gone:
Come see her in her Virgin *Face, and learn*
Twixt Idle ones, and Pilgrims *to discern.*
Yea, let young Damsels learn of her to prize
The World which is to come, in any wise;
When little Tripping *Maidens follow God,*
And leave old doting Sinners to his Rod;
'Tis like those Days wherein the young ones cri'd
Hosanah *to whom old ones did deride.*

   Next tell them of old Honest, *who you found*
With his white hairs treading the Pilgrims ground,
Yea, tell them how plain hearted this *man was,*
How after his good Lord he bare his Cross:
Perhaps with some gray Head this may prevail,
With Christ to fall in Love, and Sin bewail.

   Tell them also how Master Fearing *went*
On Pilgrimage, and how the time he spent

In Solitariness, with Fears and Cries,
And how at last, he won the Joyful Prize.
He was a good man, though much down in Spirit,
He is a good Man, and doth Life inherit.

Tell them of Master Feeblemind also,
Who, not before, but still behind would go;
Show them also how he had like been slain,
And how one Great-Heart did his life regain:
This man was true of Heart, tho weak in grace,
One might true Godliness read in his Face.

Then tell them of Master Ready-to-halt,
A Man with Crutches, but much without fault:
Tell them how Master Feeblemind, and he
Did love, and in Opinions much agree.
And let all know, tho weakness was their chance,
Yet sometimes one could Sing, the other Dance.

Forget not Master Valiant-for-the-Truth,
That Man of courage, tho a very Youth.
Tell every one his Spirit was so stout,
No Man could ever make him face about,
And how Great-Heart, & he could not forbear
But put down Doubting Castle, slay Despair.

Overlook not Master Despondancie,
Nor Much-a-fraid, his Daughter, tho they ly
Under such Mantles as may make them look
(With some) as if their God had them forsook.
They softly went, but sure, and at the end,
Found that the Lord of Pilgrims was their Friend.
When thou hast told the World of all these things,
Then turn about, my book, and touch these strings,
Which, if but touched will such Musick make,
They'l make a Cripple dance, a Gyant quake.
Those Riddles that lie couch't within thy breast,
Freely propound, expound: and for the rest
Of thy misterious lines, let them remain,
For those whose nimble Fancies shall them gain.

Now may this little Book a blessing be,

To those that love this little Book and me,
And may its buyer have no cause to say,
His Money is but lost or thrown away.
Yea, may this Second Pilgrim yield that Fruit,
As may with each good Pilgrims fancie sute,
And may it perswade some that go astray,
To turn their Foot and Heart to the right way.

Is the Hearty Prayer
of the Author

JOHN BUNYAN

THE PILGRIM'S PROGRESS
IN THE SIMILITUDE
OF A DREAM

THE SECOND PART

COURTEOUS Companions, sometime since, to tell you my Dream that I had of *Christian* the Pilgrim, and of his dangerous Journey toward the Celestial Countrey; was pleasant to me, and profitable to you. I told you then also what I saw concerning his *Wife* and *Children*, and how unwilling they were to go with him on Pilgrimage: Insomuch that he was forced to go on his Progress without them, for he durst not run the danger of that destruction which he feared would come by staying with them, in the City of Destruction: Wherefore, as I then shewed you, he left them and departed.

Now it hath so happened,†thorough the Multiplicity of Business, that I have been much hindred, and kept back from my wonted Travels into those Parts whence he went, and so could not till now obtain an opportunity to make further enquiry after whom he left behind, that I might give you an account of them. But having had some concerns that way of late, I went down again thitherward. Now, having taken up my Lodgings in a Wood about a mile off the Place, as I slept I dreamed again.

And as I was in my Dream, behold, an aged Gentleman came by where I lay; and because he was to go some part of the way that I was travelling, me thought I got up and went with him. So as we walked, and as Travellers usually do, it was as if we fell into discourse, and our talk happened to be about *Christian* and his Travels: For thus I began with the Old-man.

Sir, said I, *what Town is that there below, that lieth on the left hand of our way?*

Then said Mr. *Sagasity*, for that was his name, it is the City of *Destruction*, a populous place, but possessed with a very ill conditioned, and idle sort of People.

I thought that was that City, quoth I, *I went once my self through that Town, and therefore know that this report you give of it is true.*

Sag. Too true, I wish I could speak truth in speaking better of them that dwell therein.

Well, Sir, quoth I, *Then I perceive you to be a well meaning man: and so one that takes pleasure to hear and tell of that which is good; pray did you never hear what happened to a man sometime ago in this Town (whose name was* Christian*) that went on Pilgrimage up towards the higher Regions?*

Sag. Hear of him! Aye, and I also heard of the Molestations, Troubles, Wars, Captivities, Cries, Groans, Frights and Fears that he met with, and had in his Journey. Besides, I must tell you, all our Countrey rings of him, there are but few Houses that have heard of him and his doings, but have sought after and got the *Records* of his Pilgrimage; yea, I think I may say, that that his hazzardous *Journey* has got a many wel-wishers to his ways: For though when he was here, he was *Fool* in every mans mouth, yet now he is gone, he is highly commended of all. For 'tis said he lives bravely where he is: Yea, many of them that are resolved never to run his hazzards, yet have their mouths water at his gains.

Christians are well spoken of when gone, tho' called Fools while they are here.

They may, quoth I, *well think, if they think any thing that is true, that he liveth well where he is, for he now lives at and in the Fountain of Life, and has what he has without Labour and Sorrow, for there is no grief mixed therewith.*

Sag. Talk! The People talk strangely about him. Some say that he *now walks in White*, that he has a Chain of Gold about his Neck, that he has a Crown of Gold, beset with Pearls upon his Head: Others say, that the shining ones that sometimes shewed themselves to him in his Journey, are become his Companions, and that he is as familiar with them in the place where he is, as here one Neighbour is with another. Besides, 'tis confidently affirmed concerning him, that the King of the place where he is, has bestowed upon him already, a very rich and pleasant Dwelling at Court, and that he every day eateth and drinketh, and walketh, and talketh with him, and receiveth of the smiles and favours of him that is Judg of all there. Moreover, it is expected of some that his Prince, the Lord of that Countrey, will shortly come into *these* parts, and will know the reason, if they can give any, why his Neighbours set so

Revel. 3. 4.
chap. 6. 11.

Zech. 3. 7.

Luke 14. 15.

Jude 14. 15.

little by him, and had him so much in derision when they perceived that he would be a Pilgrim. *For they say, that now he is so in the Affections of his Prince, and that his *Soveraign* is so much concerned with the *Indignities* that was cast upon *Christian* when he became a Pilgrim, that he will look upon all as if done unto himself; and no marvel, for 'twas for the love that he had to his Prince, that he ventured as he did.

I dare say, quoth I, *I am glad on't, I am glad for the poor mans sake, for that now he has rest from his labour, and for that he now reapeth the benefit of his Tears with Joy; and for that he is got beyond the Gun-shot of his Enemies, and is out of the reach of them that hate him. I also am glad for that a Rumour of these things is noised abroad in this Countrey; Who can tell but that it may work some good effect on some that are left behind? But, pray Sir, while it is fresh in my mind, do you hear any thing of his Wife and Children? poor hearts, I wonder in my mind what they do.*

Sag. Who! *Christiana,*†and her Sons! *They are like to do as well as did *Christian* himself, for though they all plaid the Fool at the first, and would by no means be perswaded by either the tears or intreaties of *Christian*, yet second thoughts have wrought wonderfully with them, so they have packt up and are also gone after him.

Better, and better, quoth I, *But what! Wife and Children and all?*

Sag. 'Tis true, I can give you an account of the matter, for I was upon the spot at the instant, and was throughly acquainted with the whole affair.

Then, said I, *a man it seems may report it for a truth?*

Sag. You need not fear to affirm it, I mean that they are all gon on Pilgrimage, both the good Woman and her four Boys. And being we are, as I perceive, going some considerable way together, I will give you an account of the whole of the matter.

This *Christiana* (for that was her name from the day that she with her Children betook themselves to a *Pilgrims* Life,) after her Husband was gone *over the River*, and she could hear of him no more, her thoughts began to work in her mind; First, for that she had lost her Husband, and for that the loving bond of that Relation was utterly broken betwixt them. For you know, said he to me, nature can do no less but entertain the living with many a heavy Cogitation in the remembrance of the loss of loving Relations. This

*Christians
King will
take Chris-
tians part.
Luke 10.
16.

*Revel. 14.
13.
Psal. 126.
5, 6.

*Good
tidings of
Christians
Wife and
Children.

1Part, pag.
129.

Mark this, you that are Churles to your godly Relations. therefore of her Husband did cost her many a Tear. But this was not all, for *Christiana* did also begin to consider with her self, whether her unbecoming behaviour towards her Husband, was not one cause that she saw him no more, and that in such sort he was taken away from her. And upon this, came into her mind by *swarms*, all her unkind, unnatural, and ungodly Carriages to her dear Friend: Which also clogged her Conscience, and did load her with guilt. She was moreover much broken with recalling to remembrance the restless Groans, brinish Tears and self-bemoanings of her Husband, and how she did harden her heart against all his entreaties, and loving perswasions (of her and her Sons) to go with him, yea, there was not any thing that *Christian* either said to her, or did before her, all the while that his burden did hang on his back, but it returned upon her like a flash of lightning, and rent the Caul of her Heart in sunder. Specially that bitter out-cry of his, *What shall I do to be saved,* did ring in her ears most dolefully.

1 Part, pag. 9.

Then said she to her Children, Sons, we are all undone. I have sinned away your Father, and he is gone; he would have had us with him; but I would not go my self; I also have hindred you of Life. With that the Boys fell all into Tears, and cryed out to go after their Father. Oh! Said *Christiana*, that it had been but our lot to go with him, then had it fared well with us beyond what 'tis like to do now. For tho' I formerly foolishly imagin'd concerning the Troubles of your Father, that they proceeded of a foolish fancy that he had, or for that he was over run with Melancholy Humours; yet now 'twill not out of my mind, but that they sprang from another cause, to wit, for that the Light of Light was given him, by the help of which, as I perceive, he has escaped the Snares of Death. Then they all wept again, and cryed out: Oh, Wo, worth the day.†

James 1. 23, 24, 25.

Christiana's Dream. The next night *Christiana* had a Dream, and behold she saw as if a broad Parchment was opened before her, in which were recorded the sum of her ways, and the times, as she thought, look'd *very black upon her.* Then she cryed out aloud in her sleep, Lord have mercy upon me a Sinner, and the little Children heard her.

Luke 18. 13.

* *Mark this, this is the quintescence of Hell.* After this she thought she saw two very ill favoured ones standing by her Bed-side, and saying, **What shall we do with this Woman? For she cryes out for Mercy waking and sleeping: If she be suffered to go on*

as she begins, we shall lose her as we have lost her Husband. Wherefore we must by one way or other, seek to take her off from the thoughts of what shall be hereafter: else all the World cannot help it, but she will become a Pilgrim.

Now she awoke in a great Sweat, also a trembling was upon her, but after a while she fell to sleeping again. *And then she thought she saw *Christian* her Husband in a place of Bliss among many *Immortals*, with an *Harp* in his Hand, Standing and playing upon it before one that sate on a Throne with a Rainbow about his Head. She saw also as if he bowed his Head with his Face to the Pav'd-work that was under the Princes Feet, saying, *I heartily thank my Lord and King for bringing of me into this Place.* Then shouted a Company of them that stood round about, and harped with their Harps: but no man living could tell what they said, but *Christian* and his Companions.

• *Help against Discourage-ment.*

Revel. 14. 2, 3.

Next Morning when she was up, had prayed to God, and talked with her Children a while, one knocked hard at the door; to whom she spake out saying, *If thou comest in Gods Name, come in.* So he said *Amen*, and opened the Door, and saluted her with *Peace be to this House.* *The which when he had done, he said, *Christiana*, knowest thou wherefore I am come? Then she blush'd and trembled, also her Heart began to wax warm with desires to know whence he came, and what was his Errand to her. So he said unto her; my name is *Secret*, I dwell with those that are high. It is talked of where I dwell, as if thou had'st a desire to go thither; also there is a report that thou art aware of the evil thou hast formerly done to thy Husband in hardening of thy Heart against his way, and in keeping of these thy Babes in their Ignorance. *Christiana*, the merciful one has sent me to tell thee that he is a God ready to forgive, and that he taketh delight to multiply pardon to offences. He also would have thee know that he inviteth thee to come into his presence, to his Table, and that he will feed thee with the Fat of his House, and with the Heritage of *Jacob* thy Father.

• *Convictions seconded with fresh Tidings of Gods readiness to Pardon.*

There is *Christian* thy Husband, *that was*, with Legions more his Companions, ever beholding that face that doth minister Life to beholders: and they will all be glad when they shall hear the sound of thy feet step over thy Fathers Threshold.

Christiana at this was greatly abashed in her self, and bowing her head to the ground, this *Visitor* proceeded and said, *Christiana!* Here is also a Letter for thee which I have brought from thy Husbands King. So she took it and opened it, but it smelt after the manner of the best Perfume, also it was Written in Letters of Gold. The Contents of the Letter was, *That the King would have her do as did Christian her Husband; For that was the way to come to his City, and to dwell in his Presence with Joy, forever.* At this the good Woman was quite overcome: So she cried out to her *Visitor, Sir, will you carry me and my children with you, that we also may go and Worship this King?*

Then said the Visitor, *Christiana! The bitter is before the sweet:* Thou must through Troubles, as did he that went before thee, enter this Celestial City. Wherefore I advise thee, to do as did *Christian* thy Husband: go to the *Wicket Gate* yonder, over the Plain, for that stands in the head of the way up which thou must go, and I wish thee all good speed. Also I advise that thou put this Letter in thy Bosome, that thou read therein to thy self and to thy Children, until you have got it by root-of-Heart.†For it is one of the Songs that thou must Sing while thou art in this House of thy Pilgrimage. Also this thou must deliver in at the *further* Gate.

Now I saw in my Dream that this Old Gentleman, as he told me this Story, did himself seem to be greatly affected therewith. He moreover proceeded and said, So *Christiana* called her Sons together, and began thus to Address her self unto them. *My Sons, I have, as you may perceive, been of late under much exercise in my Soul about the Death of your Father; not for that I doubt at all of his Happiness: For I am satisfied now that he is well. I have also been much affected with the thoughts of mine own State and yours, which I verily believe is by nature miserable: My Carriages also to your Father in his distress, is a great load to my Conscience. For I hardened both mine own heart and yours against him, and refused to go with him on Pilgrimage.

The thoughts of these things would now kill me outright; but that for a Dream which I had last night, and but that for the incouragement that this Stranger has given me this Morning. Come, my Children, let us pack up, and be gon to the Gate that leads to the Celestial Countrey, that we may see your Father, and be with

Marginal notes:

Song. 1. 3.

Christiana quite overcome.

Further Instruction to Christiana.

Psal. 119. 54.

•*Christiana prays well for her Journey.*

him and his Companions in Peace according to the Laws of that Land.

Then did her Children burst out into Tears for Joy that the Heart of their Mother was so inclined: So their *Visitor* bid them farewel: and they began to prepare to set out for their Journey.

But while they were thus about to be gon, two of the Women that were *Christiana's* Neighbours, came up to her House and knocked at her Dore. To whom she said as before. *If you come in Gods Name, come in.* *At this the Women were stun'd, for this kind of Language they used not to hear, or to perceive to drop from the Lips of *Christiana.* Yet they came in; but behold they found the good Woman a preparing to be gon from her House.

So they began and said, *Neighbour, pray what is your meaning by this?*

Christiana answered and said to the eldest of them, whose name was Mrs. *Timorous,* I am preparing for a Journey (This *Timorous* was Daughter to him that met *Christian* upon the Hill *Difficulty;* and would a had him gone back for fear of the Lyons.)

Timorous. For what Journey I pray you?

Chris. Even to go after my good Husband, and with that she fell aweeping.

Timo. I hope not so, good Neighbour, pray for your poor Childrens sake, do not so unwomanly cast away your self.

Chris. Nay, my Children shall go with me; not one of them is willing to stay behind.

Timo. I wonder in my very Heart, what, or who, has brought you into this mind.

Chris. Oh, Neighbour, knew you but as much as I do, I doubt not but that you would go with me.

Timo. Prithee what new knowledg hast thou got that so worketh off thy mind from thy Friends, and that tempteth thee to go no body knows where?

Chris. Then *Christiana* reply'd, I have been sorely afflicted since my Husbands departure from me; but specially since he went *over the River.* But that which troubleth me most, is, my churlish Carriages to him when he was under his distress. Besides, I am *now,* as he was *then;* nothing will serve me but going on Pilgrimage. I was a dreamed last night, that I saw him. O that my Soul was with him. He dwelleth in the presence of the King of the Country, he sits and eats with

[margin notes:]
* Christiana's *new* language *stunds* her old Neighbours.

1 *Part, pag.* 35.

Timorous *comes to* visit Christiana, *with* Mercie *one of her* Neighbours.

over the Death.

2 Cor. 5. 1, 2, 3, 4. him at his Table, he is become a Companion of *Immortals*, and has a House now given him to dwell in, to which, the best Palaces on Earth, if compared, seems to me to be but as a Dunghil. The Prince of the Place has also sent for me, with promise of entertainment if I shall come to him; his messenger was here even now, and has brought me a Letter, which Invites me to come. And with that she pluck'd out her Letter, and read it, and said to them, what now will you say to this?

Timo. Oh the madness that has possessed thee and thy Husband, to run your selves upon such difficulties! You have heard, I am sure, what your Husband did meet with, even in a manner at the first step, that he took on his *1 Part, pag. 11-13.* *way, as our Neighbour* Obstinate *yet can testifie; for he went along with him, yea and* Plyable *too, until they, like wise men, were afraid to go any further. We also heard over and above, how he met with the Lyons,* Apollyon, *the Shadow of Death, and many other things: Nor is the danger he met with* *The reasonings of the flesh.* *at* Vanity fair *to be forgotten by thee. For if he, tho' a man, was so hard put to it, what canst thou being but a poor Woman do? Consider also that these four sweet Babes are thy Children, thy Flesh and thy Bones. Wherefore, though thou shouldest be so rash as to cast away thy self, yet for the sake of the Fruit of thy Body, keep thou at home.*

But *Christiana* said unto her, tempt me not, my Neighbour: I have now a price put into mine hand to get gain, and I should be a Fool of the greatest size, if I should have no heart to strike in with the opportunity. And for that you tell me of all these Troubles that *• A pertinent reply to fleshly reasonings.* I am like to meet with in the way, *they are so far off from being to me a discouragement, that they shew I am in the right. *The bitter must come before the sweet*, and that also will make the sweet the sweeter. Wherefore, since you came not to my House, *in Gods name*, as I said, I pray you to be gon, and not to disquiet me further.

Then *Timorous* all to revil'd her, and said to her Fellow, come Neighbour *Mercie*, lets leave her in her own hands, since she scorns our Counsel and Company. But *Mercie* was at a stand, and could not so readily comply with her Neighbour: and that for a twofold *Mercies Bowels yearn over Christiana.* reason. First, her Bowels yearned†over *Christiana*: so she said with in her self, If my Neighbour will needs be gon, I will go a little way with her, and help her. Secondly, her Bowels yearned over her own Soul, (for what *Christiana* had said, had taken some hold upon her mind.) Wherefore she said within her self again, I will yet have more talk with this *Christiana*, and if I find Truth and Life in what she shall say, my self with my Heart shall also go with her. Wherefore *Mercie* began thus to reply to her Neighbour *Timorous*.

Mercie. Neighbour, *I did indeed come with you, to see* Christiana *this Morning, and since she is, as you see, a taking of her last farewel of her Country, I think to walk this Sun-shine Morning, a little way with her to help her on the way.* But she told her not of her second Reason, but kept that to her self. Timorous *forsakes her; but* Mercie *cleaves to her.*

Timo. Well, I see you have a mind to go a fooling too; but take heed in time, and be wise: while we are out of danger we are out; but when we are in, we are in. So Mrs. *Timorous* returned to her House, and *Christiana* betook herself to her Journey. But when *Timorous* was got home to her House, she sends for some of her Neighbours, to wit, Mrs. *Bats-eyes,*†Mrs. *Inconsiderate*, Mrs. *Light-mind*, and Mrs. *Know-nothing*. So when they were come to her House, she falls to telling of the story of *Christiana*, and of her intended Journey. And thus she began her Tale. Timorous *acquaints her Friends what the good* Christiana *intends to do.*

Timo. Neighbours, having had little to do this Morning, I went to give *Christiana* a Visit, and when I came at the Door, I knocked, as you know 'tis our Custom: And she answered, *If you come in God's Name, come in.* So in I went, thinking all was well: But when I came in, I found her preparing her self to depart the Town, she and also her Children. So I asked her what was her meaning by that? and she told me in short, That she was now of a mind to go on Pilgrimage, as did her Husband. She told me also of a Dream that she had, and how the King of the Country where her Husband was, had sent her an inviting Letter to come thither.

Then said Mrs. Know-nothing. *And what! do you think she will go?* Mrs. Know-nothing.

Timo. Ayé, go she will, what ever come on't; and methinks I know it by this; for that which was my great Argument to perswade her to stay at home, (to wit, the Troubles she was like to meet with in the way) is one great Argument with her to put her forward on her Journey. For she told me in so many words, *The bitter goes before the sweet.* Yea, and for as much as it so doth, it makes the sweet the sweeter.

Mrs.
Bats-eyes.
Mrs. *Bats-eyes.* Oh this blind and foolish Woman, said she, Will she not take warning by her Husband's Afflictions? For my part, I say if he was here again he would rest him content in a whole Skin, and never run so many hazards for nothing.

Mrs. In-
considerate.
Mrs. *Inconsiderate* also replied, saying, away with such Fantastical Fools from the Town; a good riddance, for my part I say, of her. Should she stay where she dwels, and retain this her mind, who could live quietly by her? for she will either be dumpish or unneighbourly, or talk of such matters as no wise body can abide: Wherefore, for my part, I shall never be sorry for her departure; let her go, and let better come in her room: 'twas never a good World since these whimsical Fools dwelt in it.

Mrs.
Light-
mind.
Madam
Wanton,
she that had
like to a bin
too hard for
Faithful in
time past.
Then Mrs. *Light-mind* added as followeth: Come put this kind of Talk away. I was Yesterday at Madam *Wantons*, where we were as merry as the Maids. For who do you think should be there, but I, and Mrs. *Love-the-flesh*, and three or four more, with Mr. *Lechery*, Mrs. *Filth*, and some others. So there we had Musick and dancing, and what else was meet to fill up the pleasure. And I dare say my Lady her self is an admirably well-bred Gentlewoman, and Mr. *Lechery* is as pretty a Fellow.

1 Part, pag.
56.

Discourse
betwixt
Mercie
and good
Christiana.
By this time *Christiana* was got on her way, and *Mercie* went along with her. So as they went, her Children being there also, *Christiana* began to discourse. And, *Mercie*, said *Christiana*, I take this as an unexpected favour, that thou shouldest set foot out of Doors with me to accompany me a little in my way.

Mercie
inclines
to go.
Mercie. Then said young *Mercie* (*for she was but young,*) If I thought it would be to purpose to go with you, I would never go near the Town any more.

Christiana
would have
her Neigh-
bour with
her.
Chris. Well *Mercie*, said *Christiana*, cast in thy Lot with me. I well know what will be the end of our Pilgrimage, my Husband is where he would not but be, for all the Gold in the *Spanish* Mines. Nor shalt thou be rejected, tho thou goest but upon *my Invitation*. The King, who hath sent for me and my Children, is one that delighteth in *Mercie*. Besides, if thou wilt, I will hire thee, and thou shalt go along with me as my servant. Yet we will have all things in common betwixt thee and me, only go along with me.

Mercie
doubts of
acceptance.
Mercie. But how shall I be ascertained that I also shall be entertained?

Had I but this hope from one that can tell, I would make no stick at all, but would go, being helped by him that can help, tho' the way was never so tedious.

Christiana. Well, loving *Mercie*, I will tell thee what thou shalt do; go with me to the *Wicket Gate*, and there I will further enquire for thee, and if there thou shalt not meet with incouragement, I will be content that thou shalt return to thy place. I also will pay thee for thy Kindness which thou shewest to me and my Children, in thy accompanying of us in our way as thou doest. *Christiana alures her to the Gate which is Christ, and promiseth there to enquire for her.*

Mercie. *Then will I go thither, and will take what shall follow, and the Lord grant that my Lot may there fall even as the King of Heaven shall have his heart upon me.* *Mercie prays.*

Christiana then was glad at her heart, not only that she had a Companion, but also for that she had prevailed with this poor Maid to fall in love with her own Salvation. So they went on together, and *Mercie* began to weep. Then said *Christiana*, wherefore weepeth my Sister so? *Christiana glad of Mercie's company.*

Mer. *Alas! said she, who can but lament that shall but rightly consider what a State and Condition my poor Relations are in, that yet remain in our sinful Town: and that which makes my Grief the more heavy, is, because they have no Instructor, nor any to tell them what is to come.* *Mercie grieves for her carnal Relations.*

Chris. Bowels becometh Pilgrims. And thou dost for thy Friends, as my good *Christian* did for me when he left me; he mourned for that I would not heed nor regard him, but his Lord and ours did gather up his Tears and put them into his Bottle, and now both I, and thou, and these my sweet Babes, are reaping the Fruit and Benefit of them. I hope, *Mercie*, these Tears of thine will not be lost, for the truth hath said, *That they that sow in Tears shall reap in Joy, in singing. And he that goeth forth and weepeth, bearing precious seed, shall doubtless come again with rejoicing, bringing his Sheaves with him.* Then said *Mercie*, *Christian's Prayers were answered for his Relations after he was dead.* *Psal. 126. 5, 6.*

> *Let the most blessed be my guide,*
> *If't be his blessed Will,*
> *Unto his Gate, into his Fould,*
> *Up to his Holy Hill.*

> And let him never suffer me
> To swarve, or turn aside
> From his free grace, and Holy ways,
> What ere shall me betide.

> And let him gather them of mine,
> That I have left behind.
> Lord make them pray they may be thine,
> With all their heart and mind.

1 Part, pag.   Now my old Friend proceeded, and said, But when *Christiana*
12-13. came up to the Slow of *Despond*, she began to be at a stand; for, said
she, This is the place in which my dear Husband had like to a been
smuthered with Mud. She perceived also, that notwithstanding the
Command of the King to make this place for Pilgrims good; yet it
was rather worse than formerly. So I asked if that was true? Yes,
said the Old Gentleman, too true. For that many there be that
pretend to be the Kings Labourers;†and that say they are for mend-
Their own ing the Kings High-way, that bring *Dirt* and *Dung* instead of Stones,
Carnal
Conclusions, and so marr, instead of mending. Here *Christiana* therefore, with her
instead of Boys, did make a stand: but said *Mercie*, *come let us venture, only
the word
of life. let us be wary. Then they looked well to the *Steps*, and made a shift
* *Mercie* to get staggeringly over. Yet *Christiana* had like to a been in, and
the boldest
at the that not once nor twice. Now they had no sooner got over, but they
Slow of
Despond. thought they heard words that said unto them, *Blessed is she that*
believeth, for there shall be a performance of the things that have been told
Luke 1. 45. *her from the Lord.*

Then they went on again; and said *Mercie* to *Christiana*, Had I as
good ground to hope for a loving Reception at the *Wicket-Gate*, as
you, I think no Slow of *Despond* would discourage me.

Well, said the other, you know *your sore*, and I know *mine*; and,
good friend, we shall all have enough evil before we come at our
Journeys end.

For can it be imagined, that the people that design to attain such
excellent Glories *as we do*, and that are so envied that Happiness *as
we are*; but that we shall meet with what Fears and Scares, with
what Troubles and Afflictions they can possibly assault us with,
that hate us?

And now Mr. *Sagacity* left me to Dream out my Dream by my self. Wherefore me-thought I saw *Christiana*, and *Mercie* and the Boys go all of them up to the Gate. To which when they were come, they betook themselves to a short debate about *how* they must manage their calling at the Gate, and what should be said to him that did open to them. So it was concluded, since *Christiana* was the eldest, that she should knock for entrance, and that she should speak to him that did open, for the rest. So *Christiana* began to knock, and as her poor Husband did, she *knocked* and *knocked* again. But instead of any that answered, they all thought that they heard, as if a Dog†came barking upon them. A Dog, and a great one too, and this made the Women and Children afraid. Nor durst they for a while dare to knock any more, for fear the *Mastiff* should fly upon them. *Now therefore they were greatly tumbled up and down in their minds, and knew not what to do. Knock they durst not, for fear of the Dog: go back they durst not, for fear that the Keeper of that Gate should espy them, as they so went, and should be offended with them. At last they thought of knocking again, and knocked more vehemently then they did at the first. Then said the Keeper of the Gate, who is there? So the *Dog* left off to bark, and he opened unto them.

Then *Christiana* made low obeysance, and said, Let not our Lord be offended with his Handmaidens, for that we have knocked at his Princely Gate. Then said the Keeper, Whence come ye, and what is that you would have?

Christiana answered, We are come from whence *Christian* did come, and upon the same *Errand* as he; to wit, to be, if it shall please you, graciously admitted by this Gate, into the way that leads to the Celestial City. And I answer, my Lord, in the next place, that I am *Christiana*, once the Wife of *Christian*, that now is gotten above.

With that the Keeper of the Gate did marvel, saying, *What! is she become now a Pilgrim, that but awhile ago abhorred that Life?* Then she bowed her Head, and said, yes; and so are these my sweet Babes also.

Then he took her by the hand, and led her in, and said also, *Suffer the little Children to come unto me,*†and with that he shut up the Gate.

Prayer should be made with Consideration, and Fear: As well as in Faith and Hope.

1 Part, pag. 21.

The Dog, the Devil, an Enemy to Prayer.

• *Christiana and her companions perplexed about Prayer.†*

*How
Christiana
is enter-
tained at
the Gate.
Luke 15. 7.* This don, he called to a Trumpeter that was above over the Gate, to entertain *Christiana* with shouting and sound of Trumpet for joy. So he obeyed and sounded, and filled the Air with his melodious Notes.

Now all this while, poor *Mercie* did stand without, trembling and crying for fear that she was rejected. But when *Christiana* had gotten admittance for her self and her Boys: then she began to make Intercession for *Mercy*.

Chris. *And she said, my Lord, I have a Companion of mine that stands* ***Chris-*** *yet without, that is come hither upon the same account as my self.* ***One that*** *tiana's* *is much dejected in her mind, for that she comes, as she thinks, without sending* *Prayer for* *for,* *whereas I was sent to, by my Husbands King, to come.* *her friend*
Mercie.
The delays Now *Mercie* began to be very impatient, for each *minute* was as *make the* long to her as an Hour, wherefore she prevented *Christiana* from a *hungring* fuller interceding for her, by knocking at the Gate her self. And she *Soul the* knocked then so loud, that she made *Christiana* to start. Then said *ferventer.* the Keeper of the Gate, Who is there? And said *Christiana*, It is my Friend.

Mercie So he opened the Gate, and looked out; *but *Mercie* was fallen *faints.* down without in a Swoon, for she fainted, and was afraid that no Gate should be opened to her.

Then he took her by the hand, and said, *Damsel*, I bid thee arise. O Sir, said she, I am faint, there is scarce Life left in me. But he *Jonah 2. 7.* answered, That *Jonah* once said, *When my Soul fainted within me, I remembred the Lord, and my prayer came in unto thee, into thy Holy Temple.* Fear not, but stand upon thy Feet, and tell me wherefore thou art come.

Mer. I am come, for *that*, unto which I was invited, as my ***The cause*** Friend *Christiana* was. **Hers* was from the King, and *mine* was but *of her* from *her*: Wherefore I fear I presume. *fainting.*
Did she desire thee to come with her to this Place?

Mer. Yes. And as my Lord sees, I am come. And if there is any Grace and forgiveness of Sins to spare, I beseech that I thy poor Handmaid may be partaker thereof.

Then he took her again by the Hand, and led her gently in, and ***mark this.*** said: *I pray for all them that believe on me, by what means soever they come unto me. Then said he to those that stood by: Fetch something, and give it *Mercie* to smell on, thereby to stay her fainting. So they fetcht her a *Bundle* of *Myrrh*, and a while after she was revived.

And now was *Christiana*, and her Boys, and *Mercie*, *received* of the Lord at the head of the way, and spoke kindly unto by him.

Then said they yet further unto him, We are sorry for our Sins, and beg of our Lord his Pardon, and further information what we must do.

I grant Pardon, said he, by word, and deed; by word in the promise of forgiveness: by deed in the way I obtained it. Take the first from my Lips with a kiss, and the other, as it shall be revealed. *Song. 1. 2. John 20. 20.*

Now I saw in my Dream that he spake many good words unto them, whereby they were greatly gladed. He also had them up to the top of the Gate and shewed them by what *deed* they were saved, and told them withall that that sight they would have again as they went along in the way, to their comfort. *Christ Crucified seen afar off.*

So he left them a while in a Summer-Parler below, where they entred into talk by themselves. And thus *Christiana* began, *O Lord! How glad am I, that we are got in hither!* *• Talk between the Christians.*

Mer. *So you well may; but I, of all have cause to leap for joy.*

Chris. *I thought, one time, as I stood at the Gate (because I had knocked and none did answer) that all our Labour had been lost: Specially when that ugly Curr made such a heavy barking against us.*

Mer. *But my worst Fears was after I saw that you was taken into his favour, and that I was left behind: Now thought I, 'tis fulfiled which is Written, Two Women shall be Grinding together; the one shall be taken, and the other left. I had much ado to forbear crying out, Undone, undone.* *Mat. 24. 41.*

And afraid I was to knock any more; but when I looked up, to what was Written over the Gate, I took Courage. I also thought that I must either knock again or dye. So I knocked; but I cannot tell how, for my spirit now struggled betwixt life and death. *1 Part, pag. 13.*

Chris. *Can you not tell how you knocked? I am sure your knocks were so earnest, that the very sound of them made me start. I thought I never heard such knocking in all my Life. I thought you would a come in by violent hand, or a took the Kingdom by storm.* *Christiana thinks her Companion prays better then she. Mat. 11. 12.*

Mer. *Alas, to be in my Case, who that so was, could but a done* so? You saw that the Door was shut upon me, and that there was a most cruel *Dog* thereabout. Who, I say, that was so faint hearted as I, that would not a knocked with all their might? But pray what said my Lord to my rudeness, was he not angry with me?

• Christ pleased with loud and restless praises.

Chris. *When he heard your lumbring noise, he gave a wonderful Innocent smile. I believe what you did pleas'd him well enough, for he shewed no sign to the contrary. But I marvel in my heart why he keeps such a Dog; had I known that afore, I fear I should not have had heart enough to a ventured my self in this manner. But now we are in, we are in, and I am glad with all my heart.*

Mer. I will ask if you please next time he comes down, why he keeps such a filthy Cur in his Yard. I hope he will not take it amiss.

• The Children are afraid of the dog.

Ay do, said the Children, and perswade him to hang him, for we are afraid that he will bite us when we go hence.

So at last he came down to them again, and *Mercie* fell to the Ground on her Face before him and worshipped, and said, Let my Lord accept of the Sacrifice of praise which I now offer unto him, with the calves of my Lips.†

So he said to her, peace be to thee, stand up.

Jer. 12. 1, 2.

But she continued upon her Face and said, *Righteous art thou O Lord when I plead with thee, yet let me talk with thee of thy Judgments:*

• Mercie expostulates about the dog.
• Devill.

Wherefore dost thou keep so cruel a Dog in thy Yard, at the sight of which, such Women and Children as we, are ready to fly from thy Gate for fear?

He answered, and said: *That Dog* has another *Owner, he also is kept close in an other man's ground; only my Pilgrims hear his

1 Part, pag. 21.

barking. He belongs to the Castle which you see there at a distance: but can come up to the Walls of this Place. He has frighted many an honest Pilgrim from worse to better, by the great voice of his roaring. Indeed he that oweth him, doth not keep him of any good will to me or mine; but with intent to keep the Pilgrims from coming to me, and that they may be afraid to knock at this Gate for entrance. Sometimes also he has broken out and has *worried* some that I love; but I take all at present patiently: I also give my Pilgrims timely help, so they are not delivered up to his power to do to them

• A Check to the carnal fear of the Pilgrims.

what his Dogish nature would prompt him to. *But what! My purchased one,†I tro, hadst thou known never so much before hand, thou wouldst not a bin afraid of a *Dog*.

The Beggers that go from Door to Door, will, rather then they will lose a supposed Alms, run the hazzard of the bauling, barking, and biting too of a Dog: and shall a Dog, a Dog in an other Mans Yard: a Dog, whose barking I turn to the Profit of Pilgrims, keep any from coming to me? I deliver them from the *Lions,* their Darling from the power of the Dog.

Mer. Then said *Mercie,* *I confess my Ignorance: I spake what I under- stood not: I acknowledg that thou doest all things well.

• Christians
when wise
enough
acquiesce in
the wisdom
of their
Lord.
1 Part, pag.
23.

Chris. Then *Christiana* began to talk of their Journey, and to en-quire after the way. So he fed them, and washed their feet, and set them in the way of his Steps, according as he had dealt with her Husband before.

So I saw in my Dream, that they walkt on in their way, and had the weather very comfortable to them.

Then *Christiana* began to sing, saying,

> *Bless't be the Day that I began*
> *A Pilgrim for to be;*
> *And blessed also be that man*
> *That thereto moved me.*

> *'Tis true, 'twas long ere I began*
> *To seek to live for ever:*
> *But now I run fast as I can,*
> *'Tis better late then never.*

Mat. 20. 6.

> *Our Tears to joy, our fears to Faith*
> *Are turned, as we see:*
> *Thus our beginning, (as one saith,)*
> *Shews what our end will be.*

Now there was, on the other side of the Wall that fenced in the way up which *Christiana* and her Companions was to go, a *Garden; and that Garden belonged to him whose was that *Barking Dog,* of whom mention was made before. And some of the Fruit-Trees that grew in that Garden shot their Branches over the Wall, and being mellow, they that found them did gather them up and oft eat of them to their hurt. So *Christiana's* Boys, as Boys are apt to do, being pleas'd with the Trees, and with the Fruit that did hang thereon,

• The devils
garden.

The Chil- did *Plash* them, and began to eat. Their mother did also chide them
dren eat of for so doing; but still the Boys went on.
the Enemies
Fruit.   Well, said she, my Sons, you Transgress, for that Fruit is none
of ours: but she did not know that they did belong to the Enemy;
Ile warrant you if she had, she would a been ready to die for fear.
But that passed, and they went on their way. Now by that they
Two ill were gon about two Bows-shot from the place that let them into the
favoured way: they espyed two very *ill-favoured ones* coming down a pace to
ones. meet them. With that *Christiana*, and *Mercie* her Friend, covered
themselves with their Vails, and so kept on their Journey: The
Children also went on before, so at last they met together. Then
they that came down to meet them, came just up to the Women,
They as if they would imbrace them; but *Christiana* said, Stand back, or
assault go peaceably by as you should. Yet these two, as men that are deaf,
Christiana. regarded not *Christiana's* words; but began to lay hands upon them;
The at that *Christiana* waxing very wroth, spurned at them with her feet.
pilgrims *Mercie* also, as well as she could, did what she could to shift them.
struggle *Christiana* again, said to them, Stand back and be gon, for we have
with them. no Money to loose being Pilgrims as ye see, and such too as live
upon the Charity of our Friends.

Ill-fa. Then said one of the two of the Men, We make no assault
upon you for Money; but are come out to tell you, that if you will
but grant one small request which we shall ask, we will make
Women of you for ever.

Christ. Now *Christiana*, imagining what they should mean, made
answer again, *We will neither hear nor regard, nor yield to what you shall
ask. We are in haste, cannot stay, our Business is a Business of Life and
Death.* So again she and her Companions made a fresh assay to go
past them. But they letted them in their way.

Ill-fa. And they said, we intend no hurt to your lives, 'tis an other
thing we would have.

Christ. Ay, quoth *Christiana*, you would have us Body and Soul,
for I know 'tis for that you are come; but we will die rather upon
the spot, then suffer our selves to be brought into such Snares as
shall hazzard our well being hereafter. And with that they both
She cryes Shrieked out, and cryed Murder, Murder: and so put themselves
out.

under those Laws that are provided for the Protection of Women. *Deut. 22.*
But the men still made their approach upon them, with design to *25, 26, 27.*
prevail against them: They therefore cryed out again.

*Now they being, as I said, not far from the Gate in at which they *• 'Tis good*
came, their voice was heard from where they was, thither: Where- *to cry out*
when
fore some of the House came out, and knowing that it was *Chris-* *we are*
tiana's Tongue: they made haste to her relief. But by that they was *assaulted.*
got within sight of them, the Women was in a very great scuffle,
the Children also stood crying by. Then did he that came in for *The*
their relief, call out to the Ruffins saying, What is that thing that *Reliever*
you do? Would you make my Lords People to transgress? He also *comes.*
attempted to take them; but they did make their escape over the *The Ill-*
ones fly to
Wall into the Garden of the Man, to whom the great Dog belonged, *the devill*
so the Dog became their Protector. This *Reliever* then came up to *for releif.*
the Women, and asked them how they did. So they answered, we
thank thy Prince, pretty well, only we have been somewhat
affrighted; we thank thee also for that thou camest in to our help,
for otherwise we had been overcome.

Reliever. So after a few more words, this *Reliever* said as followeth:
I marvelled much when you was entertained at the Gate above, being ye knew *The Re-*
that ye were but weak Women, that you petitioned not the Lord there for a *liever talks*
to the
Conductor: Then might you have avoided these Troubles, and Dangers: For *Women.*
he would have granted you one.

Christ. *Alas said *Christiana*, we were so taken with our present *• mark this.*
blessing, that Dangers to come were forgotten by us; besides, who
could have thought that so near the Kings Palace there should have
lurked such naughty ones? indeed it had been well for us had we
asked our Lord for one; but since our Lord knew 'twould be for our
profit, I wonder he sent not one along with us.

Relie. *It is not always necessary to grant things not asked for, lest by so* *We lose for*
doing they become of little esteem; but when the want of a thing is felt, it *want of*
asking for.
then comes under, in the Eyes of him that feels it, that estimate, that properly
is its due, and so consequently will be thereafter used. Had my Lord granted
you a Conductor, you would not neither so have bewailed that oversight of
yours in not asking for one, as now you have occasion to do. So all things work
for good, and tend to make you more wary.

Christ. Shall we go back again to my Lord, and confess our folly and ask one?

Relie. Your Confession of your folly, I will present him with: To go back again, you need not. For in all places where you shall come, you will find no want at all, for in every of my Lord's Lodgings, which he has prepared for the reception of his Pilgrims, there is sufficient to furnish them against all attempts whatsoever. But, as I said, he will be enquired of by them to do it for them: *and 'tis a poor thing that is not worth asking for.* When he had thus said, he went back to his place, and the Pilgrims went on their way.

Ezek. 36. 37.

The mistake of Mercie.

Mer. Then said *Mercie,* what a sudden blank is here? I made account we had now been past all danger, and that we should never see sorrow more.

Christiana's Guilt.

Christ. Thy *Innocency,* my Sister, said *Christiana* to *Mercie,* may excuse thee much; but as for me, my fault is so much the greater, for that I saw this danger before I came out of the Doors, and yet did not provide for it where provision might a been had. I am therefore much to be blamed.

Mer. Then said Mercie, *how knew you this before you came from home? pray open to me this Riddle.*

Christ. Why, I will tell you. Before I set Foot out of Doors, one Night, as I lay in my Bed, I had a Dream about this. For methought I saw two men, as like these as ever the World they could look, stand at my *Beds-feet,* plotting how they might prevent my Salvation. I will tell you their very words. They said, ('twas when I was in my Troubles,) *What shall we do with this Woman? for she cries out* Christiana's Dream repeated. *waking and sleeping for forgiveness. If she be suffered to go on as she begins, we shall lose her as we have lost her Husband.* This you know might a made me take heed, and have provided when Provision might a been had.

Mercie makes good use of their neglect of duty.

Mer. Well, said *Mercie, as by this neglect, we have an occasion ministred unto us to behold our own imperfections: So our Lord has taken occasion thereby, to make manifest the Riches of his Grace. For he, as we see, has followed us with un-asked kindness, and has delivered us from their hands that were stronger then we, of his meer good pleasure.*

Thus now when they had talked away a little more time, they drew nigh to an House which stood in the way, which House was built for the relief of Pilgrims as you will find more fully related in the first part of these Records of the *Pilgrims Progress*. So they drew on towards the House (the House of the Interpreter) and when they came to the Door, they heard a great talk in the House; they then gave ear, and heard, as they thought, *Christiana* mentioned by name. For you must know that there went along, even before her, a talk of her and her Childrens going on Pilgrimage. And this thing was the more pleasing to them, because they had heard that she was *Christian's* Wife; that Woman who was sometime ago so unwilling to hear of going on Pilgrimage. Thus therefore they stood still and heard the good people within commending her, who they little thought stood at the Door. *At last *Christiana* knocked as she had done at the Gate before. Now when she had knocked, there came to the Door a young Damsel and opened the Door and looked, and behold two Women was there.

Damsel. Then said the Damsel to them, *With whom would you speak in this Place?*

Christ. *Christiana* answered, we understand that this is a priviledged place for those that are become Pilgrims, and we now at this Door are such: Wherefore we pray that we may be partakers of that for which we at this time are come; for the day, as thou seest, is very far spent, and we are loth to night to go any further.

Damsel. Pray what may I call your name, that I may tell it to my Lord within?

Christ. My name is *Christiana*, I was the Wife of that Pilgrim that some years ago did Travel this way, and these be his four Children. This Maiden also is my Companion, and is going on Pilgrimage too.

Innocent. Then ran *Innocent* in (for that was her name) and said to those within, Can you think who is at the Door? There is *Christiana* and her Children, and her Companion, all waiting for entertainment here. *Then they leaped for Joy, and went and told their Master. So he came to the Door, and looking upon her, he said, *Art thou that Christiana, whom Christian, the Good-man, left behind him, when he betook himself to a Pilgrims Life?*

Christ. I am that Woman that was so hard-hearted as to slight my Husbands Troubles, and that left him to go on in his Journey alone, and these are his four Children; but now I also am come, for I am convinced that no way is right but this.

1 *Part, pag.* 23-31.

Talk in the Interpreter's house about Christiana's going on pilgrimage.

* *She knocks at the Door.*

The door is opened to them by Innocent.

* *Joy in the house of the Interpreter that Christiana is turned Pilgrim.*

Mat. 21. Inter. *Then is fulfilled that which also is written of the Man that said*
29. *to his Son, go work to day in my Vineyard, and he said to his Father, I will*
not; but afterwards repented and went.

Christ. Then said *Christiana*, So be it, *Amen*, God make it a true
saying upon me, and grant that I may be found at the last, of him
in peace without spot and blameless.

Inter. *But why standest thou thus at the Door, come in thou Daughter*
of *Abraham,*†*we was talking of thee but now: For tidings have come to us*
before, how thou art become a Pilgrim. Come Children, come in; come
Maiden, come in; so he had them all into the House.

So when they were within, they were bidden sit down and rest
them, the which when they had done, those that attended upon the
Pilgrims in the House, came into the Room to see them. And one
Old Saints smiled, and another smiled, and they all smiled for Joy that *Chris-*
glad to see *tiana* was become a Pilgrim. They also looked upon the Boys, they
the young
ones walk in stroaked them over the Faces with the Hand, in token of their kind
Gods ways. reception of them: they also carried it lovingly to *Mercie*, and bid
them all welcome into their Masters House.

• The After a while, because Supper was not ready, *the *Interpreter* took
Significant
Rooms. them into his *Significant* Rooms, and shewed them what *Christian,*
Christiana's Husband had seen sometime before. Here therefore they
saw the *Man* in the *Cage*, the man and his Dream, the man that cut
his way thorough his Enemies, and the Picture of the biggest of
them all: together with the rest of those things that were then so
profitable to *Christian.*

This done, and after these things had been somewhat digested by
Christiana, and her Company: the *Interpreter* takes them apart again:
and has them first into a Room, *where was a man that could look no way*
The man *but downwards, with a Muckrake in his hand. There stood also one over his*
with the
Muck-rake *head with a Celestial Crown in his Hand, and proffered to give him that*
expounded. Crown for his Muck-rake; but the man did neither look up, nor regard; but
raked to himself the Straws, the small Sticks, and Dust of the Floar.

Then said *Christiana*, I perswade my self that I know somewhat the
meaning of this: For this is a Figure of a man of this World: Is it not, good
Sir ?

Inter. Thou hast said the right, said he, and his *Muck-rake* doth
show his Carnal mind. And whereas thou seest him rather give heed

to rake up Straws and Sticks, and the Dust of the Floar, then to what he says that calls to him from above with the Celestial Crown in his Hand; it is to show, that Heaven is but as a Fable to some, and that things here are counted the only things substantial. Now whereas it was also shewed thee, that the man could look no way but downwards: It is to let thee know that earthly things when they are with Power upon Mens minds, quite carry their hearts away from God.

Chris. Then said Christiana, O! *deliver me from this Muck-rake.*   * Christiana's prayer against the Muck-rake. Pro. 30. 8.

Inter. That Prayer said the *Interpreter,* has lain by till 'tis almost rusty: *Give me not Riches,* is scarce the Prayer of one of ten thousand. Straws, and Sticks, and Dust, with most, are the great things now looked after.

With that *Mercie,* and *Christiana* wept, and said, It is alas! too true.

When the *Interpreter* had shewed them this, he has them into the very best Room in the house, (a very brave Room it was) so he bid them look round about, and see if they could find any thing profitable there. Then they looked round and round: For there was nothing there to be seen but a very great *Spider* on the Wall: and that they overlook't.

Mer. Then said Mercie, *Sir, I see nothing; but* Christiana *held her peace.*

Inter. But said the *Interpreter,* look again: she therefore lookt again and said, Here is not any thing, but an *ugly Spider,*†who hangs by her Hands upon the Wall. Then said he, Is there but one *Spider* in all this spacious Room? Then the water stood in *Christiana's* Eyes, for she was a Woman quick of apprehension: and she said, Yes Lord, there is more here then one. Yea, and *Spiders* whose Venom is far more destructive then that which is in her. The *Interpreter* then looked pleasantly upon her, and said, Thou hast said the Truth. This made *Mercie* blush, and the Boys to cover their Faces. For they all began now to understand the Riddle.   *Of the Spider.*   *Talk about the Spider.*

Then said the *Interpreter* again, *The Spider taketh hold with her hands, as you see, and is in Kings Pallaces.* And wherefore is this recorded; but to show you, that how full of the Venome of Sin soever you be, yet you may by the hand of Faith lay hold of, and dwell in the best Room that belongs to the Kings House above?   Pro. 30. 28.   *The Interpretation.*

Chris. I thought, said *Christiana*, of something of this; but I could not imagin it all. I thought that we were like *Spiders*, and that we looked like ugly Creatures, in what fine Room soever we were: But that by this *Spider*, this venomous and ill favoured Creature, we were to learn *how to act Faith*, that came not into my mind. And yet she has taken hold with her hands as I see, and dwells in the best Room in the House. God has made nothing in vain.

Then they seemed all to be glad; but the water stood in their Eyes: Yet they looked one upon another, and also bowed before the *Interpreter*.

He had them then into another Room where was a Hen and *Of the* Chickens, and bid them observe a while. So one of the Chickens *Hen and* went to the Trough to drink, and every time she drank she lift up *Chickens.* her head and her eyes towards Heaven. See, said he, what this little Chick doth, and learn of her to acknowledge whence your Mercies come, by receiving them with looking up. Yet again, said he, observe and look: So they gave heed, and perceived that the Hen did walk in a fourfold Method towards her Chickens. 1. She had a *common call*, and that she hath all day long. 2. She had a *special call*, and that she had but sometimes. 3. She had a *brooding note*. And 4. she had an *out-cry*.

Now, said he, compare this *Hen* to your King, and these Chickens to his Obedient ones. For answerable to her, himself has his Methods, which he walketh in towards his People. By his common call, *he gives nothing*, by his special call, he always *has something to give*, Mat. 23. he has also a brooding voice, *for them that are under his Wing*. And 37. he has an out-cry, to give *the Alarm when he seeth the Enemy come*. I chose, my Darlings, to lead you into the Room where such things are, because you are Women, and they are easie for you.

Chris. And Sir, said *Christiana*, pray let us see some more: So he *Of the* had them into the Slaughter-house, where was a *Butcher* a killing of *Butcher* a Sheep: And behold the Sheep was quiet, and took her Death *and the* *Sheep.* patiently. Then said the *Interpreter*: You must learn of this Sheep, to suffer: And to put up wrongs without murmurings and complaints. Behold how quietly she takes her Death, and without objecting she suffereth her Skin to be pulled over her Ears. Your King doth call you his Sheep.

After this, he led them into his Garden, where was great variety *Of the* of Flowers: and he said, do you see all these? So *Christiana* said, yes. *Garden.* Then said he again, Behold the Flowers are divers in *Stature*, in *Quality*, and *Colour*, and *Smell*, and *Virtue*, and some are better then some: Also where the Gardiner has set them, there they stand, and quarrel not one with another.

Again he had them into his Field, which he had sowed with *Of the* Wheat and Corn: but when they beheld, the tops of all was cut off, *Field.* only the Straw remained. He said again, this Ground was Dunged, and Plowed, and Sowed; but what shall we do with the Crop? Then said *Christiana*, burn some and make muck of the rest. Then said the *Interpreter* again, Fruit you see is that thing you look for, and for want of that you condemn it to the Fire, and to be trodden under foot of men: Beware that in this you condemn not your selves.

Then, as they were coming in from abroad, they espied a little *Of the* *Robbin* with a great *Spider* in his mouth. So the *Interpreter* said, look *Robbin and* here. So they looked, and *Mercie* wondred; but *Christiana* said, what *the* Spider. a disparagement is it to such a little pretty Bird as the *Robbin-red-breast* is, he being also a Bird above many, that loveth to maintain a kind of Sociableness with man? I had thought they had lived upon crums of Bread, or upon other such harmless matter. I like him worse then I did.

The *Interpreter* then replied, This *Robbin* is an Emblem very apt to set forth some Professors by; for to sight they are as this *Robbin*, pretty of Note, Colour and Carriages, they seem also to have a very great Love for Professors that are sincere; and above all other to desire to sociate with, and to be in their Company, as if they could live upon the good Mans Crums. They pretend also that therefore it is, that they frequent the House of the Godly, and the appointments of the Lord: but when they are by themselves, *as the Robbin*, they can catch and gobble up *Spiders*, they can change their Diet, drink *Iniquity*, and swallow down *Sin* like Water.

So when they were come again into the House, because Supper *Pray, and* as yet was not ready, *Christiana* again desired that the *Interpreter* *you will get* would either *show* or *tell* of some other things that are Profitable. *at that* *which yet* Then the *Interpreter* began and said, *The fatter the Sow is, the more* *lies un-* *she desires the Mire; the fatter the Ox is, the more gamesomly he goes to the* *revealed.*

Slaughter; and the more healthy the lusty man is, the more prone he is unto Evil.

There is a desire in Women, to go neat and fine, and it is a comely thing to be adorned with that, that in Gods sight is of great price.

'Tis easier watching a night or two, then to sit up a whole year together: So 'tis easier for one to begin to profess well, then to hold out as he should to the end.

Every Ship-Master, when in a Storm, will willingly cast that over Board that is of the smallest value in the Vessel; but who will throw the best out first? none but he that feareth not God.

One leak will sink a Ship, and one Sin will destroy a Sinner.

He that forgets his Friend, is ungrateful unto him; but he that forgets his Saviour is unmerciful to himself.

He that lives in Sin, and looks for Happiness hereafter, is like him that soweth Cockle, and thinks to fill his Barn with Wheat, or Barley.

If a man would live well, let him fetch his last day to him, and make it always his company-Keeper.

Whispering and change of thoughts, proves that Sin is in the World.

If the world which God sets light by, is counted a thing of that worth with men: what is Heaven which God commendeth?

If the Life that is attended with so many troubles, is so loth to be let go by us, What is the Life above?

Every Body will cry up the Goodness of Men; but who is there that is, as he should, affected with the Goodness of God?

We seldom sit down to Meat; but we eat, and leave. So there is in Jesus Christ more Merit and Righteousness then the whole World has need of.

Of the Tree that is rotten at heart. When the *Interpreter* had done, he takes them out into his Garden again, and had them to a Tree whose *inside* was all rotten, and gone, and yet it grew and had Leaves. Then said *Mercie*, what means this? This Tree, said he, whose *out-side* is fair, and whose *inside* is rotten; is it to which many may be compared that are in the Garden of God: Who with their mouths speak high in behalf of God, but indeed will do nothing for him: Whose Leaves are fair; but their heart Good for nothing, but to be *Tinder* for the Devils *Tinder-box*.

They are at Supper. Now Supper was ready, the Table spread, and all things set on the Board; so they sate down and did eat when one had given thanks. And the *Interpreter* did usually entertain those that lodged with him with Musick at Meals, so the Minstrels played. There was also one that did Sing. And a very fine voice he had.

His Song was this.

> *The Lord is only my support,*
> *And he that doth me feed:*
> *How can I then want any thing*
> *Whereof I stand in need?*

When the Song and Musick was ended, the *Interpreter* asked *Christiana*, what it was that at first did move her to betake her self to a Pilgrims Life?

Christiana answered: *First*, the loss of my Husband came into my mind, at which I was heartily grieved: but all that was but natural Affection. Then after that, came the Troubles, and Pilgrimage of my Husband into my mind, and also how like a Churl I had carried it to him as to that. So guilt took hold of my mind, and would have drawn me into the *Pond*; but that opportunely I had a Dream of the well-being of my Husband, and a Letter sent me by the King of that Country where my Husband dwells, to come to him. The Dream and the Letter together so wrought upon my mind, that they forced me to this way. *Talk at Supper.*

A Repetition of Christiana's Experience.

Inter. *But met you with no opposition afore you set out of Doors?*

Chris. Yes, a Neighbour of mine, one Mrs. *Timerous.* (She was a kin to him that would have perswaded my Husband to go back for fear of the Lions.) She all-to-be-fooled me for, as she called it, my intended desperate adventure; she also urged what she could, to dishearten me to it, the hardships and Troubles that my Husband met with in the way; but all this I got over pretty well. But a Dream that I had, of two ill-lookt ones, that I thought did Plot how to make me miscarry in my Journey, that hath troubled me much: Yea, it still runs in my mind, and makes me afraid of every one that I meet, lest they should meet me to do me a mischief, and to turn me out of the way. Yea, I may tell my Lord, tho' I would not have every body know it, that between this and the Gate by which we got into the way, we were both so sorely assaulted, that we were made to cry out Murder, and the two that made this assault upon us, were like the two that I saw in my Dream.

A question   Then said the *Interpreter*, Thy beginning is good, thy latter end
put to   shall greatly increase. So he addressed himself to *Mercie*: and said
Mercie. unto her, *And what moved thee to come hither, sweet-heart?*

Mercie. Then *Mercie* blushed and trembled, and for a while continued silent.

Interpreter. *Then said he, be not afraid, only believe, and speak thy
mind.*

Mercys   *Mer.* So she began and said, Truly Sir, my want of Experience, is
answer. that that makes me covet to be in silence, and that also that fills me
with fears of coming short at last. I cannot tell of Visions, and
Dreams as my friend *Christiana* can; nor know I what it is to mourn
for my refusing of the Counsel of those that were good Relations.

Interpreter. *What was it then, dear-heart, that hath prevailed with thee
to do as thou hast done?*

Mer. Why, when our friend here, was packing up to be gone from
our Town, I and another went accidentally to see her. So we
knocked at the Door and went in. When we were within, and seeing
what she was doing, we asked what was her meaning. She said, she
was sent for to go to her Husband, and then she up and told us, how
she had seen him in a Dream, dwelling in a curious place among
Immortals wearing a Crown, playing upon a Harp, eating and drinking at his Princes Table, and singing Praises to him for bringing him
thither, *&c.* Now methought, while she was telling these things
unto us, my heart burned within me. And I said in my Heart, if this
be true, I will leave my Father and my Mother, and the Land of my
Nativity, and will, if I may, go along with *Christiana*.

So I asked her further of the truth of these things, and if she
would let me go with her: For I saw now that there was no dwelling,
but with the danger of ruin, any longer in our Town. But yet I came
away with a heavy heart, not for that I was unwilling to come away;
but for that so many of my Relations were left behind. And I am
come with all the desire of my heart, and will go if I may with
Christiana unto her Husband and his King.

Inter. Thy setting out is good, for thou hast given credit to the
Ruth 2. 11, truth, Thou art a *Ruth*, who did for the love that she bore to *Naomi*,
12. and to the Lord her God, leave Father and Mother, and the land of
her Nativity to come out, and go with a People that she knew not

heretofore. *The Lord recompence thy work, and a full reward be given thee of the Lord God of Israel, under whose Wings thou art come to trust.*

Now Supper was ended, and Preparations was made for Bed, the Women were laid singly alone, and the Boys by themselves. Now when *Mercie* was in Bed, she could not sleep for joy, for that now her doubts of missing at last, were removed further from her than ever they were before. So she lay blessing and Praising God who had had such favour for her.

In the Morning they arose with the *Sun*, and prepared themselves for their departure: But the *Interpreter* would have them tarry a while, for, said he, you must orderly go from hence. Then said he to the Damsel that at first opened unto them, Take them and have them into the Garden, to the *Bath*,† and there wash them, and make them clean from the soil which they have gathered by travelling. Then *Innocent* the Damsel took them and had them into the Garden, and brought them to the *Bath*, so she told them that there they must wash and be clean, for so her Master would have the Women to do that called at his House as they were going on *Pilgrimage*. They then went in and washed, yea they and the Boys and all, and they came out of that *Bath* not only sweet, and clean; but also much enlivened and strengthened in their Joynts: So when they came in, they looked fairer a deal, then when they went out to the washing.

When they were returned out of the Garden from the *Bath*, the *Interpreter* took them and looked upon them and said unto them, fair as the *Moon*.† Then he called for the *Seal*† wherewith they used to be *Sealed* that were washed in his *Bath*. So the *Seal* was brought, and he set his Mark upon them, that they might be known in the Places whither they were yet to go: Now the seal was the contents and sum of the Passover which the Children of *Israel* did eat when they came out from the Land of *Egypt*: and the mark was set between their Eyes. This seal greatly added to their Beauty, for it was an Ornament to their Faces. It also added to their gravity, and made their Countenances more like them of Angels.

Then said the *Interpreter* again to the Damsel that waited upon these Women, Go into the Vestry and fetch out Garments for these People: So she went and fetched out white Rayment, and laid it down before him; so he commanded them to put it on. *It was fine*

They address themselves for bed.

Mercy's good nights rest.

The Bath Sanctification.

They wash in it.

They are sealed.

Exo. 13. 8, 9, 10.

They are **Linnen, white and clean.** When the Women were thus adorned they
clothed. seemed to be a Terror one to the other, for that they could not see
that glory each one on her self, which they could see in each other.
True Now therefore they began to esteem each other better then them-
humility. selves. For, You are fairer then I am, said one, and, You are more
comely then I am, said another. The Children also stood amazed to
see into what fashion they were brought.

The *Interpreter* then called for a *Man-servant* of his, one *Great-heart,*†
and bid him take *Sword,* and *Helmet* and *Shield,* and take these my
Daughters, said he, and conduct them to the House called *Beautiful,*
at which place they will rest next. So he took his Weapons, and
went before them, and the *Interpreter* said, God speed. Those also
that belonged to the Family sent them away with many a good
wish. So they went on their way, and Sung.

> *This place has been our second Stage,*
> *Here we have heard and seen*
> *Those good things that from Age to Age,*
> *To others hid have been.*
>   *The Dunghil-raker, Spider, Hen,*
> *The Chicken too to me*
> *Hath taught a Lesson, let me then*
> *Conformed to it be.*
>   *The Butcher, Garden and the Field,*
> *The Robbin and his bait,*
> *Also the Rotten-tree doth yield*
> *Me Argument of weight*
>   *To move me for to watch and pray,*
> *To strive to be sincere,*
> *To take my Cross up day by day,*
> *And serve the Lord with fear.*

1 Part, pag. Now I saw in my Dream that they went on, and *Great-heart* went
31. before them, so they went and came to the place where *Christians*
Burthen fell off his Back, and tumbled into a Sepulchre. Here then
they made a pause, and here also they blessed God. Now said
Christiana, it comes to my mind what was said to us at the Gate, to
wit, that we should have Pardon, by *Word* and *Deed;* by word, that

is, by the promise; by *Deed*, to wit, in the way it was obtained. What the promise is, of that I know something: But what is it to have Pardon by deed, or in the way that it was obtained, Mr. *Great-heart*, I suppose you know; wherefore if you please let us hear you discourse thereof.

Great-heart. Pardon by the deed done, is Pardon obtained by some one, for another that hath need thereof: Not by the Person pardoned, but in the way, *saith another*, in which I have obtained it. So then to speak to the question more large, The pardon that you and *Mercie* and these Boys have *attained*, was *obtained* by another, to wit, by him that let you in at the Gate: And he hath obtain'd it in this double way. He has performed Righteousness to cover you, and spilt blood to wash you in. *A comment upon what was said at the Gate, or a discourse of our being justified by Christ.†*

Chris. But if he parts with his Righteousness to us: What will he have for himself?

Great-heart. He has more Righteousness then you have need of, or then he needeth himself.

Chris. Pray make that appear.

Great-heart. With all my heart, but first I must premise that he of whom we are now about to speak, is one that has not his Fellow. He has two Natures in one Person, plain to be *distinguished*, impossible to be *divided*. Unto each of these Natures a Righteousness belongeth, and each Righteousness is essential to that Nature. So that one may as easily cause the Nature to be extinct, as to separate its Justice or Righteousness from it. Of *these* Righteousnesses therefore, we are not made partakers so, as that they, or any of them, should be put upon us that we might be made just, and live thereby. Besides these there is a Righteousness which this Person has, as these two Natures are joyned in one. And this is not the Righteousness of the *Godhead*, as distinguished from the *Manhood*; nor the Righteousness of the *Manhood*, as distinguished from the *Godhead*; but a Righteousness which standeth in the Union of both Natures: and may properly be called, the Righteousness that is essential to his being prepared of God to the capacity of the Mediatory Office which he was to be intrusted with. If he parts with his first Righteousness, he parts with his *Godhead*; if he parts with his second Righteousness, he parts with the purity of his *Manhood*; if he parts with this third, he parts with that perfection that capacitates him to the Office of Mediation. He has therefore another Righteousness which standeth in *performance*, or obedience to a revealed Will: And that is it that he puts upon Sinners, and that by which their Sins are covered. Wherefore he saith, *as by one mans disobedience many were made Sinners: So by the obedience of one shall many be made Righteous.*

Rom. 5. 19.

Chris. But are the other Righteousnesses of no use to us?

Great heart. Yes, for though they are essential to his Natures and Office, and so cannot be communicated unto another, yet it is by Virtue of them that the Righteousness that justifies, is for that purpose efficacious. The *Righteousness* of his *God-head* gives *Virtue* to his Obedience; the *Righteousness* of his *Manhood* giveth capability to his obedience to justifie, and the Righteousness that standeth in the Union of these two Natures to his Office, giveth Authority to that Righteousness to do the work for which it is ordained.

So then, here is a Righteousness that Christ, as God, has no need of, for he is God without it: here is a Righteousness that Christ, as Man, has no need of to make him so, for he is perfect Man without it. Again, here is a Righteousness that Christ as God-man has no need of, for he is perfectly so without it. Here then is a Righteousness that Christ, as God, as Man, as God-man has no need of, with Reference to himself, and therefore he can spare it, a justifying Righteousness, that he for himself wanteth not, and therefore he giveth it away. Hence 'tis called the *gift of Righteousness*. This Righteousness, since Christ Jesus the Lord, has made himself under the Law, *must* be given away: For the Law doth not only bind him that is under it, *to do justly*; but to use Charity: Wherefore he *must*, he *ought* by the Law, if he hath two Coats,† to give one to him that has none. Now our Lord hath indeed *two Coats*, one for himself, and one to spare: Wherefore he freely bestows one upon those that have none. And thus *Christiana*, and *Mercie*, and the rest of you that are here, doth your Pardon come by *deed*, or by the work of another man. Your Lord Christ is he that has worked, and given away what he wrought for to the next poor Beggar he meets.

Rom. 5. 17.

But again, in order to Pardon by *deed*, there must something be paid to God as a price, as well as something prepared to cover us withal. Sin has delivered us up to the just Curse of a Righteous

Law: Now from this Curse we must be justified by way of Redemption, a price being paid for the harms we have done, and this is by the Blood of your Lord, who came and stood in your place, and stead, and died your Death for your Transgressions. Thus has he ransomed you from your Transgressions by Blood, and covered your poluted and deformed Souls with Righteousness: For the sake of which, God passeth by you, and will not hurt you, when he comes to Judge the World. Rom. 4. 24.

Gala. 3. 13.

Chris. This is brave. Now I see that there was something to be learnt by our being pardoned by word and deed. Good Mercie, let us labour to keep this in mind, and my Children do you remember it also. But, Sir, was not this it that made my good Christians Burden fall from off his Shoulder, and that made him give three leaps for Joy?

Christiana affected with this way of Redemption.

Great-heart. *Yes, 'twas the belief of this, that cut those Strings that could not be cut by other means, and 'twas to give him a proof of the Virtue of this, that he was suffered to carry his Burden to the Cross.

• How the Strings that bound Christians burden to him were cut.

Chris. I thought so, for tho' my heart was lightful and joyous before, yet it is ten times more lightsome and joyous now. And I am perswaded by what I have felt, tho' I have felt but little as yet, that if the most burdened Man in the Worla was here, and did see and believe, as I now do, 'twould make his heart the more merry and blithe.

Great-heart. There is not only comfort, and the ease of a Burden brought to us, by the sight and Consideration of these; but an indeared Affection begot in us by it: For who can, if he doth but once think that Pardon comes, not only by promise, but thus; but be affected with the way and means of his Redemption, and so with the man that hath wrought it for him?

How affection to Christ is begot in the Soul.

Chris. True, methinks it makes my Heart bleed to think that he should bleed for me. Oh! thou loving one, Oh! thou Blessed one. Thou deservest to have me, thou hast bought me: Thou deservest to have me all, thou hast paid for me ten thousand times more than I am worth. No marvel that this made the Water stand in my Husbands Eyes, and that it made him trudg so nimbly on: I am perswaded he wished me with him; but vile wretch, that I was, I let him come all alone. O Mercie, that thy Father and Mother were here, yea, and Mrs. Timorous also. Nay I wish now with all my Heart, that here was Madam Wanton too. Surely, surely, their Hearts would be affected, nor

1 Part, pag. 31. Cause of admiration.

could the fear of the one, nor the powerful Lusts of the other, prevail with them to go home again, and to refuse to become good Pilgrims.

Great-heart. You speak now in the warmth of your Affections, will it, think you, be always thus with you? Besides, this is not communicated to every one, nor to every one that did see your Jesus bleed. There was that stood by, and that saw the Blood run from his Heart to the Ground, and yet was so far off this, that instead of lamenting, they laughed at him, and instead of becoming his Disciples, did harden their Hearts against him. So that all that you have, my Daughters, you have by a peculiar impression made by a Divine contemplating upon what I have spoken to you. Remember that 'twas told you, that the *Hen* by her common call, gives no meat to her *Chickens*. This you have therefore by a special Grace.

To be affected with Christ and with what he has don is a thing special.

Now I saw still in my Dream, that they went on until they were come to the place that *Simple*, and *Sloth* and *Presumption* lay and slept in, when *Christian* went by on Pilgrimage. And behold they were hanged up in Irons a little way off on the other-side.

Note: marginal note reads: Simple and Sloth and Presumption hanged, and why.

Mercie. Then said *Mercie* to him that was their Guide, and Conductor, *What are those three men? and for what are they hanged here?*

Great-heart. These three men, were Men of very bad Qualities, they had no mind to be Pilgrims themselves, and whosoever they could they hindred; they were for Sloth and Folly themselves, and whoever they could perswade with, they made so too, and withal taught them to presume that they should do well at last. They were asleep when *Christian* went by, and now you go by they are hanged.

Mercie. But could they perswade any to be of their Opinion?

Great-heart. Yes, they turned several out of the way. There was *Slow-pace* that they perswaded to do as they. They also prevailed with one *Short-wind*, with one *No-heart*, with one *Linger-after-lust*, and with one *Sleepy-head*, and with a young Woman her name was *Dull*, to turn out of the way and become as they. Besides, they brought up an ill report of your Lord, perswading others that he was a task-Master. They also brought up an evil report of the good Land, saying, 'twas not half so good as some pretend it was: They also began to villifie his Servants, and to count the very best of them meddlesome, troublesome busie-Bodies: Further, they would call

Note: marginal note reads: Their Crimes. Who they prevailed upon to turn out of the way.

the Bread of God, *Husks*; the *Comforts* of his Children, *Fancies*; the Travel and Labour of Pilgrims, things to no purpose.

Chris. Nay, said Christiana, *if they were such, they shall never be bewailed by me, they have but what they deserve, and I think it is well that they hang so near the High-way that others may see and take warning. But had it not been well if their Crimes had been ingraven in some Plate of Iron or Brass, and left here, even where they did their Mischiefs, for a caution to other bad Men?*

Great-heart. So it is, as you well may perceive if you will go a little to the Wall.

Mercie. No no, let them hang and their Names Rot, and their Crimes live for ever against them; I think it a high favour that they were hanged afore we came hither, who knows else what they might a done to such poor Women as we are? Then she turned it into a Song, saying,

> *Now then, you three, hang there and be a Sign*
> *To all that shall against the Truth combine:*
> *And let him that comes after, fear this end,*
> *If unto Pilgrims he is not a Friend.*
> *And thou my Soul of all such men beware,*
> *That unto Holiness Opposers are.*

Thus they went on till they came at the foot of the Hill *Difficulty*, where again their good Friend, Mr. *Great-heart*, took an occasion to tell them of what happened there when *Christian* himself went by. So he had them first to the Spring. *Lo*, saith he, *This is the Spring that* *Christian* *drank of*, before he went up this Hill, and then 'twas clear and good; but now 'tis Dirty with the feet of some that are not desirous that Pilgrims here should quench their Thirst: Thereat *Mercie* said, *And why so envious tro?* But said their Guide, It will do, if taken up, and put into a Vessel that is sweet and good; for then the Dirt will sink to the bottom, and the Water come out by it self more clear. Thus therefore *Christiana* and her Companions were compelled to do. They took it up, and put it into an Earthen-pot and so let it stand till the Dirt was gone to the bottom, and then they drank thereof.

Next he shewed them the two *by-ways* that were at the foot of the

1 Part, pag. 34. Ezek. 34. 18.

'Tis difficult getting of good Doctrine in erroneous Times.

Hill, where *Formality* and *Hypocrisie*, lost themselves. And, said he, these are dangerous Paths: Two were here cast away when *Christian*

• By paths tho barred up will not keep all from going in them.

came by. *And although, as you see, these ways are since stopt up with *Chains*, *Posts* and a *Ditch*: Yet there are that will chuse to adventure here, rather then take the pains to go up this Hill.

1 Part, pag. 33-4. Pro. 13. 15.

Christiana. *The way of Transgressors is hard.* 'Tis a wonder that they can get into those ways, without danger of breaking their Necks.

Great-heart. They will venture, yea, if at any time any of the Kings Servants doth happen to see them, and doth call unto them, and tell them that *they* are in the wrong ways, and do bid them beware the danger. Then they will railingly return them answer and say,

Jer. 44. 16, 17.

As for the Word that thou hast spoken unto us in the name of the King, we will not hearken unto thee; but we will certainly do whatsoever thing goeth out of our own Mouths, &c. Nay if you look a little farther, you shall see that these ways, are made cautionary enough, not only by these *Posts* and *Ditch* and *Chain*; but also by being hedged up. Yet they will chuse to go there.

• The reason why some do chuse to go in by-waies. Pro. 15. 19.

Christiana. *They are Idle, they love not to take Pains, up-hill-way is unpleasant to them. So it is fulfilled unto them as it is Written, The way of the slothful man is a Hedg of Thorns.* Yea, they will rather chuse to walk upon a Snare, then to go up this Hill, and the rest of this way to the City.

The Hill puts the Pilgrims to it.

Then they set forward and began to go up the Hill, and up the Hill they went; but before they got to the top, *Christiana* began to Pant, and said, I dare say this is a breathing Hill, no marvel if they that love their ease more than their Souls, chuse to themselves a smoother way. Then said *Mercie*, I must sit down, also the least of

They sit in the Arbour.

the Children began to cry. Come, come, said *Great-heart*, sit not down here, for a little above is the Princes *Arbour*. Then took he the little Boy by the Hand, and led him up thereto.

1 Part, pag. 35-7. Mat. 11. 28.

When they were come to the *Arbour* they were very willing to sit down, for they were all in a pelting heat. Then said *Mercie*, *How sweet is rest to them that Labour!* And how good is the Prince of Pilgrims, to provide such resting places for them! Of *this Arbour* I have heard much; but I never saw it before. But here let us beware of sleeping: For as I have heard, for that it cost poor *Christian* dear.

Then said Mr. *Great-heart* to the little ones, Come my pretty The little
Boys, how do you do? what think you now of going on Pilgrimage? Boys answer to
Sir, said the least, I was almost beat out of heart; but I thank you the guide,
for lending me a hand at my need. And I remember now what my and also to Mercie.
Mother has told me, namely, That the way to Heaven is as up a
Ladder, and the way to Hell is as down a Hill. But I had rather go
up the Ladder to Life, then down the Hill to Death.

Then said *Mercie*, But the Proverb is, *To go down the Hill is easie*: Which is hardest up
But *James* said (for that was his Name) The day is coming when in Hill or
my Opinion, *going down Hill will be the hardest of all*. 'Tis a good Boy, down Hill?
said his Master, thou hast given her a right answer. Then *Mercie*
smiled, but the little Boy did blush.

Chris. Come, said *Christiana*, will you eat a bit, a little to sweeten They refresh
your Mouths, while you sit here to rest your Legs? For I have here themselves.
a piece of Pomgranate which Mr. *Interpreter* put in my Hand, just
when I came out of his Doors; he gave me also a piece of an Honey-
comb, and a little Bottle of Spirits. I thought he gave you something, said *Mercie*, because he called you a to-side. Yes, so he did,
said the other, But *Mercie*, It shall still be as I said it should, when
at first we came from home: Thou shalt be a sharer in all the good
that I have, because thou so willingly didst become my Companion.
Then she gave to them, and they did eat, both *Mercie*, and the Boys.
And said *Christiana* to Mr. *Great-heart*, Sir will you do as we? But he
answered, You are going on Pilgrimage, and presently I shall return;
much good may what you have, do to you. At home I eat the same
every day. Now when they had eaten and drank, and had chatted
a little longer, their guide said to them, The day wears away, if you
think good, let us prepare to be going. So they got up to go, and the
little Boys went before; but *Christiana* forgat to take her Bottle of
Spirits with her, so she sent her little Boy back to fetch it. Then
said *Mercie*, I think this is a *losing* Place. Here *Christian* lost his *Role*, Christiana
and here *Christiana* left her Bottle behind her: Sir, what is the cause forgets her
of this? so their guide made answer and said, The cause is *sleep*, or Bottle of
forgetfulness; some *sleep*, when they should keep *awake*; and some Spirits.
forget, when they should *remember*; and this is the very cause, why
often at the resting places, some Pilgrims in some things come off
losers. Pilgrims should watch and remember what they have already Mark this.

received under their greatest enjoyments: But for want of doing so, oft times their rejoicing ends in Tears, and their Sun-shine in a Cloud: Witness the story of *Christian* at this place.

1 Part, pag.
35–6.

When they were come to the place where *Mistrust* and *Timorous* met *Christian* to perswade him to go back for fear of the Lions, they perceived as it were a Stage, and before it towards the Road, a broad plate with a Copy of Verses Written thereon, and underneath, the reason of the raising up of that Stage in that place, rendered. The Verses were these.

> *Let him that sees this Stage take heed,*
> *Unto his Heart and Tongue:*
> *Lest if he do not, here he speed*
> *As some have long agone.*

The words underneath the Verses were. *This Stage was built to punish such upon, who through* Timorousness, *or* Mistrust, *shall be afraid to go further on Pilgrimage. Also on this Stage both* Mistrust, *and* Timorous *were burned thorough the Tongue with an hot Iron, for endeavouring to hinder* Christian *in his Journey.*

Then said *Mercie*. This is much like to the saying of the beloved,

Psal. 120.
3, 4.

What shall be given unto thee? or what shall be done unto thee thou false Tongue? sharp Arrows of the mighty, with Coals of Juniper.

1 Part, pag.
37.
An Emblem of those that go on bravely, when there is no danger; but shrink when troubles come.

So they went on till they came within sight of the Lions. Now Mr. *Great-heart* was a strong man, so he was not afraid of a Lion. But yet when they were come up to the place where the Lions were, the Boys that went before, were now glad to cringe behind, for they were afraid of the Lions, so they stept back and went behind. At this their guide smiled, and said, How now my Boys, do you love to go before when no danger doth approach, and love to come behind so soon as the Lions appear?

Of Grim *the Giant, and of his backing the Lions.*

Now as they went up, Mr. *Great-heart* drew his Sword with intent to make a way for the Pilgrims in spite of the Lions. Then there appeared one, that it seems, had taken upon him to back the Lions. And he said to the Pilgrims guide, What is the cause of your coming hither? Now the name of that man was *Grim* or *Bloody man,* because of his slaying of Pilgrims, and he was of the race of the *Gyants.*

Great-heart. Then said the *Pilgrims* guide, these Women and

Children, are going on Pilgrimage, and this is the way they must go, and go it they shall in spite of thee and the Lions.

Grim. This is not their way, neither shall they go therein. I am come forth to withstand them, and to that end will back the Lions.

Now to say truth, by reason of the fierceness of the Lions, and of the *Grim*-Carriage of him that did back them, this way had of late lain much un-occupied, and was almost all grown over with Grass.

Christiana. Then said *Christiana*, Tho' the Highways have been unoccupied heretofore, and tho' the Travellers have been made in time past, to walk thorough by-Paths, it must not be so now I am risen, *Now I am Risen a Mother in* Israel.   Judg. 5. 6,

Grim. Then he swore *by the Lions*, but it should; and therefore bid 7. them turn aside, for they should not have passage there.

Great-heart. But their guide made first his Approach unto *Grim*, and laid so heavily at him with his Sword, that he forced him to a retreat.

Grim. Then said he (that attempted to back the Lions) will you slay me upon mine own Ground?

Great-heart. 'Tis the Kings High-way that we are in, and in this *A fight* way it is that thou hast placed thy Lions; but these Women and *betwixt* these Children, tho' weak, shall hold on their way in spite of thy *Great-* Lions. And with that he gave him again a down-right blow, and *heart.* brought him upon his Knees. With this blow he also broke his Helmet, and with the next he cut off an Arm. Then did the *Giant Roar* so hideously, that his Voice frighted the Women, and yet they *The* were glad to see him lie sprawling upon the Ground. Now the Lions *Victory.* were chained, and so of themselves could do nothing. Wherefore when old *Grim* that intended to back them was dead, Mr. *Great-heart* said to the Pilgrims, Come now and follow me, and no hurt shall happen to you from the Lions. They therefore went on; but *They pass* the Women trembled as they passed by them, the Boys also look't *by the* as if they would die; but they all got by without further hurt. *Lyons.*

Now then they were within sight of the *Porters* Lodg, and they *They come* soon came up unto it; but they made the more haste after this to *to the Por-* go thither, because 'tis dangerous travelling there in the Night. So *ters Lodge.* when they were come to the Gate, the guide knocked, and the

Porter cried, *who is there*; but as soon as the Guide had said *it is I*, he knew his Voice and came down (For the Guide had oft before that, came thither as a Conductor of Pilgrims). When he was came down, he opened the Gate, and seeing the Guide standing just before it (for he saw not the Women, for they were behind) he said unto him, How now Mr. *Great-heart*, what is your business here so late to Night? I have brought, said he, some Pilgrims hither, where by my Lords Commandment they must Lodg. I had been here some time ago, had I not been opposed by the Giant that did use to back the Lyons. But I after a long and tedious combate with him, have cut him off, and have brought the Pilgrims hither in safety.

Great-
heart
attempts to
go back.
The
Pilgrims
implore his
company
still.
Porter. *Will you not go in, and stay till Morning?*

Great-heart. No, I will return to my Lord to night.

Christiana. Oh Sir, I know not how to be willing you should leave us in our Pilgrimage, you have been so faithful, and so loving to us, you have fought so stoutly for us, you have been so hearty in counselling of us, that I shall never forget your favour towards us.

Mercie. Then said *Mercie*, O that we might have thy Company to our Journeys end! How can such poor Women as we, hold out in a way so full of Troubles as this way is, without a Friend, and Defender?

James. Then said *James*, the youngest of the Boys, Pray Sir be perswaded to go with us and help us, because we are so weak, and the way so dangerous as it is.

Help lost
for want of
asking for.
Great-heart. I am at my Lords Commandment. If he shall allot me to be your Guide quite thorough, I will willingly wait upon you; but here you failed at first; for when he bid me come thus far with you, then you should have begged me of him to have gon quite thorough with you, and he would have granted your request. However, at present I must withdraw, and so good *Christiana*, *Mercie*, and my brave Children, Adieu.

1 Part, pag.
38.
Then the Porter, Mr. *Watchfull*, asked *Christiana* of her Country, Christiana
makes her
self known
to the Por-
ter, he tells
it to a dam-
sel. and of her Kindred, and she said, *I came from the City of* Destruction, *I am a Widdow Woman, and my Husband is dead, his name was* Christian *the Pilgrim.* How, said the Porter, was he your Husband? Yes, said she, and these are his Children: and this, pointing to *Mercie*, is one of my Towns-Women. Then the Porter rang his Bell, as at such

times he is wont, and there came to the Door one of the Damsels, whose Name was *Humble-mind*. And to her the Porter said, Go tell it within that *Christiana* the Wife of *Christian* and her Children are come hither on Pilgrimage. She went in therefore and told it. But Oh what a Noise for gladness was there within, when the Damsel did but drop that word out of her Mouth! *Joy at the noise of the Pilgrims coming.*

So they came with haste to the Porter, for *Christiana* stood still at the Door; then some of the most grave, said unto her, *Come in Christiana, come in thou Wife of that Good Man, come in thou Blessed Woman, come in with all that are with thee.* So she went in, and they followed her that were her Children, and her Companions. Now when they were gone in, they were had into a very large Room, where they were bidden to sit down: So they sat down, and the chief of the House was called to see, and welcome the Guests. Then they came in, and understanding who they were, did Salute each one with a kiss, and said, Welcome ye Vessels of the Grace of God, welcome to us your Friends. *Christians love is kindled at the sight of one another.*

Now because it was somewhat late, and because the Pilgrims were weary with their Journey, and also made faint with the sight of the Fight, and of the terrible Lyons, therefore they desired as soon as might be, to prepare to go to Rest. Nay, said those of the Family, refresh your selves first with a Morsel of Meat. For they had prepared for them a Lamb, with the accustomed Sauce belonging thereto. For the Porter had heard before of their coming, and had told it to them within. So when they had Supped, and ended their Prayer with a Psalm, they desired they might go to rest. But let us, said *Christiana*, if we may be so bold as to chuse, be in that Chamber that was my Husbands, when he was here. So they had them up thither, and they lay all in a Room. When they were at Rest, *Christiana* and *Mercie* entred into discourse about things that were convenient. *Exo. 12. 3, 8.* *Joh. 1. 29.* *1 Part, pag. 44.*

Chris. Little did I think once, that when my Husband went on Pilgrimage I should ever a followed.

Mercie. And you as little thought of lying in his Bed, and in his Chamber to Rest, as you do now.

Chris. And much less did I ever think of seeing his Face with Comfort, and of Worshipping the Lord the King with him, and yet now I believe I shall. *Christs Bosome is for all Pilgrims.*

Mercie. Hark, don't you hear a Noise?

Musick. *Christiana.* Yes, 'tis as I believe a Noise of Musick, for Joy that we are here.

Mer. Wonderful! Musick in the House, Musick in the Heart, and Musick also in Heaven, for joy that we are here.

Thus they talked a while, and then betook themselves to sleep; so in the morning, when they were awake, *Christiana* said to *Mercy.*

Mercy *did* *laugh in* *her sleep.* Chris. *What was the matter that you did laugh in your sleep to Night? I suppose you was in a Dream?*

Mercy. So I was, and a sweet Dream it was; but are you sure I laughed?

Christiana. Yes, you laughed heartily; But prethee Mercy *tell me thy Dream?*

Mercy's *Dream.* *Mercy.* I was a Dreamed that I sat all alone in a Solitary place, and was bemoaning of the hardness of my Heart. Now I had not sat there long, but methought many were gathered about me to see me, and to hear what it was that I said. So they harkened, and I went on bemoaning the hardness of my Heart. At this, some of them laughed at me, some called me Fool, and some began to thrust me *What her* *dream was.* about. With that, methought I looked up, and saw one coming with Wings towards me. So he came directly to me, and said *Mercy*, what aileth thee? Now when he had heard me make my complaint; he said, *Peace be to thee:* he also wiped mine Eyes with his Hankerchief, Ezek. 16. 8, 9, 10, 11. and *clad* me in *Silver* and *Gold*; he put a Chain about my Neck, and Ear-rings in mine Ears, and a beautiful Crown upon my Head. Then he took me by my Hand, and said, *Mercy*, come after me. So he went up, and I followed, till we came at a Golden Gate. Then he knocked, and when they within had opened, the man went in and I followed him up to a Throne, upon which one sat, and he said to me, *welcome Daughter.* The place looked bright, and twinkling like the Stars, or rather like the *Sun*, and I thought that I saw your Husband there, so I awoke from my Dream. But did I laugh?

Christiana. Laugh! Ay, and well you might to see your self so well. For you must give me leave to tell you, that I believe it was a good Dream, and that as you have begun to find the first part true, so you shall find the second Job 33. 14, 15. *at last.* God speaks once, yea twice, yet man perceiveth it not. In a Dream, in a Vision of the Night, when deep sleep falleth upon men, in slumbering upon the Bed. *We need not, when a-Bed, lie awake to talk with God; he can visit us while we sleep, and cause us then to hear his Voice. Our Heart oft times wakes when we sleep, and God can speak to that, either by Words, by Proverbs, by Signs, and Similitudes, as well as if one was awake.*

Mercy. Well, I am glad of my Dream, for I hope ere long to see it fulfilled, to the making of me laugh again. Mercy glad of her dream.

Christiana. *I think it is now time to rise, and to know what we must do.*

Mercy. Pray, if they invite us to stay a while, let us willingly accept of the proffer. I am the willinger to stay awhile here, to grow better acquainted with these Maids; methinks *Prudence, Piety* and *Charity*, have very comly and sober Countenances.

Chris. *We shall see what they will do.* So when they were up and ready, they came down. And they asked one another of their rest, and if it was Comfortable, or not?

Mer. *Very good, said* Mercy. *It was one of the best Nights Lodging that ever I had in my Life.*

Then said *Prudence*, and *Piety*, If you will be perswaded to stay here a while, you shall have what the House will afford. They stay here some time.

Charity. *Ay, and that with a very good will, said* Charity. So they consented, and stayed there about a Month or above: and became very Profitable one to another. And because *Prudence* would see how *Christiana* had brought up her Children, she asked leave of her to Catechise them: So she gave her free consent. Then she began at the youngest whose Name was *James*. Prudence desires to catechise Christianas Children.

Pru. *And she said, Come* James, *canst thou tell who made thee?* James Catechised.

Jam. God the Father, God the Son, and God the Holy-Ghost.

Pru. *Good Boy. And canst thou tell who saves thee?*

Jam. God the Father, God the Son, and God the Holy Ghost.

Pru. *Good Boy still. But how doth God the Father save thee?*

Jam. By his Grace.

Pru. *How doth God the Son save thee?*

Jam. By his Righteousness, Death, and Blood, and Life.

Pru. *And how doth God the Holy Ghost save thee?*

Jam. By his *Illumination*, by his *Renovation*, and by his *Preservation*.

Then said *Prudence* to *Christiana*, You are to be commended for thus bringing up your Children. I suppose I need not ask the rest these Questions, since the youngest of them can answer them so well. I will therefore now apply my self to the Youngest next.

Prudence. Then she said, Come *Joseph,* (for his Name was *Joseph*) will you let me Catechise you?

Joseph. With all my Heart.

Pru. *What is Man?*

Joseph. A Reasonable Creature, so made by God, as my Brother said.

Pru. *What is supposed by this Word, saved?*

Joseph. That man by Sin has brought himself into a State of Captivity and Misery.

Pru. *What is supposed by his being saved by the Trinity?*

Joseph. That Sin is so great and mighty a Tyrant, that none can pull us out of its clutches but God, and that God is so good and loving to man, as to pull him indeed out of this Miserable State.

Pru. *What is Gods design in saving of poor Men?*

Joseph. The glorifying of his Name, of his Grace, and Justice, *&c.* And the everlasting Happiness of his Creature.

Pru. *Who are they that must be saved?*

Joseph. Those that accept of his Salvation.

Good Boy *Joseph,* thy Mother has taught thee well, and thou hast harkened to what she has said unto thee.

Then said *Prudence* to *Samuel,* who was the eldest but one.

Prudence. Come *Samuel,* are you willing that I should Catechise you also?

Sam. Yes, forsooth, if you please.

Pru. *What is Heaven?*

Sam. A place, and State most blessed, because God dwelleth there.

Pru. *What is Hell?*

Sam. A Place and State most woful, because it is the dwelling place of Sin, the Devil, and Death.

Prudence. *Why wouldest thou go to Heaven?*

Sam. That I may see God, and serve him without weariness; that I may see Christ, and love him everlastingly; that I may have that fulness of the Holy Spirit in me, that I can by no means here enjoy.

Pru. *A very good Boy also, and one that has learned well.*

Then she addressed her self to the eldest, whose Name was Mathew, and she said to him, Come *Mathew*, shall I also Catechise you? Mathew *Catechised.*

Mat. *With a very good will.*

Pru. *I ask then, if there was ever any thing that had a being, Antecedent to, or before God?*

Mat. No, for God is Eternal, nor is there any thing excepting himself, that had a being until the beginning of the first day. *For in six days the Lord made Heaven and Earth, the Sea and all that in them is.*

Pru. *What do you think of the Bible?*

Mat. It is the Holy Word of God.

Pru. *Is there nothing Written therein, but what you understand?*

Mat. Yes, a great deal.

Pru. *What do you do when you meet with such places therein, that you do not understand?*

Mat. I think God is wiser then I. I pray also that he will please to let me know all therein that he knows will be for my good.

Pru. *How believe you as touching the Resurrection of the Dead?*

Mat. I believe they shall rise, the same that was buried: The same in *Nature*, tho' not in Corruption. And I believe this upon a double account. First, because God has promised it. Secondly, because he is able to perform it.

Then said *Prudence* to the Boys, You must still harken to your Mother, for she can learn you more. You must also diligently give ear to what good talk you shall hear from others, for for your sakes do they speak good things. Observe also and that with carefulness, what the Heavens and the Earth do teach you; but especially be much in the Meditation of that Book that was the cause of your Fathers becoming a Pilgrim. I for my part, my Children, will teach you what I can while you are here, and shall be glad if you will ask me Questions that tend to Godly edifying. Prudences *conclusion upon the* Catechising *of the Boys.*

Now by that these Pilgrims had been at this place a week, *Mercie* had a Visitor that pretended some good Will unto her, and his name was Mr. *Brisk*.† A man of some breeding, and that pretended to Religion; but a man that stuck very close to the World. So he came once or twice, or more to *Mercie*, and offered love unto her. Now *Mercie* was of a fair Countenance, and therefore the more alluring. Mercie *has a sweet heart.*

Mercies temper. Her mind also was, to be always busying of her self in doing, for when she had nothing to do for her self, she would be making of Hose and Garments for others, and would bestow them upon them that had need. And Mr. *Brisk* not knowing where or how she disposed of what she made, seemed to be greatly taken, for that he found her never Idle. I will warrant her a good Huswife, quoth he to himself.

* Mercie enquires of the Maids concerning Mr. Brisk. **Mercie* then revealed the business to the Maidens that were of the House, and enquired of them concerning him: for they did know him better then she. So they told her that he was a very busie Young-Man, and one that pretended to Religion; but was as they feared, a stranger to the Power of that which was good.

Nay then, said Mercie, *I will look no more on him, for I purpose never to have a clog to my Soul.*

Prudence then replied, That there needed no great matter of discouragement to be given to him, her continuing so as she had began to do for the Poor, would quickly cool his Courage.

Talk betwixt Mercie and Mr. Brisk. *1 Tim. 6. 17, 18, 19.* *He forsakes her, and why.* So the next time he comes, he finds her at her old work, a making of things for the Poor. Then said he, What always at it? Yes, said she, either for my self, or for others. And what canst thee *earn* a day, quoth he? I do these things, said she, *That I may be Rich in good Works, laying up in store a good Foundation against the time to come, that I may lay hold on Eternal Life:* Why prethee what dost thou with them? said he; Cloath the naked, said she. With that his Countenance fell. So he forbore to come at her again. And when he was asked the reason why, he said, *That* Mercie *was a pretty lass; but troubled with ill Conditions.*

Mercie in the practice of Mercie rejected; While Mercie in the Name of Mercie is liked. When he had left her, *Prudence* said, Did I not tell thee that Mr. *Brisk* would soon forsake thee? yea, he will raise up an ill report of thee: For notwithstanding his pretence to Religion, and his seeming love to *Mercie,* yet *Mercie* and he are of tempers so different, that I believe they will never come together.

Mercie. I might a had Husbands afore now, tho' I spake not of it to any; but they were such as did not like my Conditions, though never did any of them find fault with my Person:†So they and I could not agree.

Prudence. Mercie in our days is little set by, any further then as to its Name: the Practice, which is set forth by thy Conditions, there are but few that can abide.

Mercie. *Well, said Mercie, if no body will have me, I will dye a Maid, or my Conditions shall be to me as a Husband. For I cannot change my Nature, and to have one that lies cross to me in this, that I purpose never to admit of, as long as I live. I had a Sister named* Bountiful *that was married to one of these Churles; but he and she could never agree; but because my Sister was resolved to do as she had began, that is, to show Kindness to the Poor, therefore her Husband first cried her down at the Cross,†and then turned her out of his Doors.*

Mercie's resolution.

How Mercie's Sister was served by her Husband.

Pru. And yet he was a Professor, I warrant you?

Mer. *Yes, such a one as he was, and of such as he, the World is now full; but I am for none of them all.*

Now Mathew *the eldest Son of* Christiana *fell Sick, and his Sickness was sore upon him, for he was much pained in his Bowels, so that he was with it, at times, pulled as 'twere both ends together. There dwelt also not far from thence, one Mr.* Skill, *an Antient, and well approved Physician. So* Christiana *desired it, and they sent for him, and he came. When he was entred the Room, and had a little observed the Boy, he concluded that he was sick of the Gripes. Then he said to his Mother, What Diet has* Mathew *of late fed upon? Diet, said* Christiana, *nothing but that which is wholsome.* *The Physician answered, This Boy has been tampering with something which lies in his Maw undigested, and that will not away without means. And I tell you he must be purged or else he will dye.*

• *Mathew falls sick.*

Gripes of Conscience.

• *The Physicians Judgment.*

Samuel. *Then said* Samuel, *Mother, Mother, what was that which my Brother did gather up and eat, so soon as we were come from the Gate, that is at the head of this way? You know that there was an Orchard on the left hand, on the otherside of the Wall, and some of the Trees hung over the Wall, and my Brother did plash and did eat.*

• *Samuel puts his Mother in mind of the fruit his Brother did eat.*

Christiana. True my Child, said Christiana, he did take thereof and did eat; naughty Boy as he was, I did chide him, and yet he would eat thereof.

Skill. *I knew he had eaten something that was not wholsome Food. And that Food, to wit, that Fruit is even the most hurtful of all. It is the Fruit of* Belzebubs *Orchard. I do marvel that none did warn you of it; many have died thereof.*

Christiana. Then Christiana began to cry, and she said, O naughty Boy, and O careless Mother, what shall I do for my Son?

Skill. Come, do not be too much Dejected; the Boy may do well again; but he must purge and Vomit.

Christiana. Pray Sir try the utmost of your Skill with him whatever it costs.

Heb. 10. 1,
2, 3, 4.

• *Potion prepared.*

John 6. 54,
55, 56, 57.
Mark 9. 49.
The Lattine
I borrow.
Heb. 9. 14.

• *The boy
loth to take
the Physick.*
Zech. 12.
10.

The
Mother
tasts it, and
perswades
him.

A word of
God in the
hand of his
Faith.

Heb. 13.
11, 12, 13,
14, 15.

This Pill an
Universal
Remedy.

Skill. Nay, I hope I shall be reasonable: So he made him a Purge; but it was too weak. 'Twas said, it was made of the Blood of a Goat, the Ashes of an Heifer, and with some of the Juice of Hyssop, &c. *When Mr. *Skill* had seen that that Purge was too weak, he made him one to the purpose. 'Twas made *ex Carne & Sanguine Christi.* (You know Physicians give strange Medicines to their Patients) and it was made up into Pills with a Promise or two, and a proportionable quantity of Salt. Now he was to take them three at a time fasting in half a quarter of a Pint of the Tears of Repentance. When this potion was prepared, and brought to the Boy; *he was loth to take it, tho' torn with the Gripes, as if he should be pulled in pieces. *Come, come, said the Physician, you must take it.* It goes against my Stomach, said the Boy. *I must have you take it, said his Mother.* I shall Vomit it up again, said the Boy. Pray Sir, said *Christiana* to Mr. *Skill,* how does it taste? It has no ill taste, said the Doctor, and with that she touched one of the pills with the tip of her Tongue. Oh *Mathew,* said she, this potion is sweeter then Hony. If thou lovest thy Mother, if thou lovest thy Brothers, if thou lovest *Mercie,* if thou lovest thy Life, take it. So with much ado, after a short Prayer for the blessing of God upon it, he took it; and it wrought kindly with him. It caused him to Purge, it caused him to sleep, and rest quietly, it put him into a fine heat and breathing sweat, and did quite rid him of his Gripes.

So in little time he got up, and walked about with a Staff, and would go from Room to Room, and talk with *Prudence, Piety,* and *Charity* of his Distemper, and how he was healed.

So when the Boy was healed, *Christiana* asked Mr. *Skill,* saying, Sir, what will content you for your pains and care to and of my Child? And he said, you must pay the *Master of the Colledge* of Physicians, according to rules made, in that case, and provided.

Chris. But Sir, said she, what is this Pill good for else?

Skill. It is an universal Pill, 'tis good against all the Diseases that Pilgrims are incident to, and when it is well prepared it will keep good, *time* out of *mind.*

Christiana. Pray Sir, make me up twelve Boxes of them: For if I can get these, I will never take other Physick.

Skill. These *Pills* are good to prevent Diseases, as well as to *cure* when one is Sick. Yea, I dare say it, and stand to it, that if a man will but use this Physick as he should, *it will make him live for ever.* But, good *Christiana*, thou must give these Pills, *no other way;* *but as I have prescribed: For if you do, they will do no good. So he gave unto *Christiana* Physick for her self, and her Boys, and for *Mercie:* and bid *Mathew* take heed how he eat any more *Green Plums*, and kist them and went his way.

* *In a Glass of the Tears of Repentance.*

It was told you before, That *Prudence* bid the Boys, that if at any time they would, they should ask her some Questions, that might be profitable, and she would say something to them.

Mat. Then *Mathew* who had been sick, asked her, *Why for the most part Physick should be bitter to our Palats?*

Of Physick.

Pru. To shew how unwelcome the word of God and the Effects thereof are to a Carnal Heart.

Mathew. *Why does Physick, if it does good, Purge, and cause that we Vomit?*

Of the Effects of Physick.

Prudence. To shew that the Word when it works effectually, cleanseth the Heart and Mind. For look what the one doth to the Body, the other doth to the Soul.

Mathew. *What should we learn by seeing the Flame of our Fire go upwards? and by seeing the Beams, and sweet Influences of the Sun strike downwards?*

Of Fire and of the Sun.

Prudence. By the going up of the Fire, we are taught to ascend to Heaven, by fervent and hot desires. And by the Sun his sending his Heat, Beams, and sweet Influences downwards; we are taught, that the Saviour of the World, tho' high, reaches down with his Grace and Love to us below.

Mathew. *Where have the Clouds their Water?*

Of the Clouds.

Pru. Out of the Sea.

Mathew. *What may we learn from that?*

Pru. That Ministers should fetch their Doctrine from God.

Mat. *Why do they empty themselves upon the Earth?*

Pru. To shew that Ministers should give out what they know of God to the World.

Of the Mat. *Why is the Rainbow caused by the Sun?*
Rainbow. *Prudence.* To shew that the Covenant of Gods Grace is confirmed
to us in Christ.

Mat. *Why do the Springs come from the Sea to us, thorough the Earth?*

Prudence. To shew that the Grace of God comes to us thorough
the Body of Christ.

Of the Mat. *Why do some of the Springs rise out of the tops of high Hills?*
Springs. *Prudence.* To shew that the Spirit of Grace shall spring up in *some*
that are Great and Mighty, as well as in *many* that are Poor and low.

Of the Mat. *Why doth the Fire fasten upon the Candle-wick?*
Candle. *Pru.* To shew that unless Grace doth kindle upon the Heart, there
will be no true Light of Life in us.

Mathew. *Why is the Wick and Tallow and all, spent to maintain the
light of the Candle?*

Prudence. To shew that Body and Soul and all, should be at the
Service of, and spend themselves to maintain in good Condition
that Grace of God that is in us.

Of the Mat. *Why doth the Pelican pierce her own Brest with her Bill?*†
Pelican. *Pru.* To nourish her Young ones with her Blood, and thereby to
shew that Christ the blessed, so loveth his Young, his People, as to
save them from Death by his Blood.

Of the Cock. Mat. *What may one learn by hearing the Cock to Crow.*

Prudence. Learn to remember *Peter*'s Sin, and *Peter*'s Repentance.
The Cocks crowing, shews also that day is coming on, let then the
crowing of the Cock put thee in mind of that last and terrible Day
of Judgment.

The weak Now about this time their month was out, wherefore they signi-
may some- fied to those of the House that 'twas convenient for them to up and
times call be going. Then said *Joseph* to his Mother, It is convenient that you
the strong to forget not to send to the House of Mr. *Interpreter*, to pray him to
Prayers. grant that Mr. *Great-heart* should be sent unto us, that he may be
our Conductor the rest of our way. Good *Boy*, said she, I had almost
forgot. So she drew up a Petition, and prayed Mr. *Watchful* the
Porter to send it by some fit man to her good Friend Mr. *Interpreter*;
who when it was come, and he had seen the contents of the Petitions,
They pro- said to the Messenger, Go tell them that I will send him.
vide to be When the Family where *Christiana* was, saw that they had a pur-
gone on
their way.

pose to go forward, they called the whole House together to give thanks to their King, for sending of them such profitable Guests as these. Which done, they said to *Christiana*, And shall we not shew thee something, according as our Custom is to do to Pilgrims, on which thou mayest meditate when thou art upon the way? So they took *Christiana*, her Children and *Mercy* into the Closet, and shewed them one of the *Apples* that *Eve* did eat of, and that she also did give to her Husband, and that for the eating of which they both were turned out of Paradice, and asked her what she thought that was? Then *Christiana* said, 'Tis Food, or Poyson, I know not which; so they opened the matter to her, and she held up her hands and wondered. *Eves Apple.*
A sight of Sin is amazing. Gen. 3. 6. Rom. 7. 24.

Then they had her to a place, and shewed her *Jacob's Ladder*. Now at that time there were some Angels ascending upon it. So *Christiana* looked and looked, to see the Angels go up, and so did the rest of the Company. Then they were going into another place to shew them something else: But *James* said to his Mother, pray bid them stay here a little longer, for this is a curious sight. So they turned again, and stood feeding their Eyes with this *so pleasant a Prospect*. After this they had them into a place where did hang up a *Golden Anchor*, so they bid *Christiana* take it down; for, said they, you shall have it with you, for 'tis of absolute necessity that you should, that you may lay hold of that within the vail, and stand stedfast, in case you should meet with turbulent weather: So they were glad thereof. Then they took them, and had them to the mount upon which *Abraham* our Father, had offered up *Isaac* his Son, and shewed them the *Altar*, the *Wood*, the *Fire*, and the *Knife*, for they remain to be seen to this very Day. When they had seen it, they held up their hands and blest themselves, and said, Oh! What a man, for love to his Master and for denial to himself, was *Abraham*: After they had shewed them all these things, *Prudence* took them into the Dining-Room, where stood a pair of Excellent Virginals, so she played upon them, and turned what she had shewed them into this excellent Song, saying:
Jacob's Ladder. Gen. 28. 12.
A sight of Christ is taking.
Golden Anchor. Joh. 1. 51. Heb. 6. 12, 19.
Gen. 22.
Of Abraham offering up Isaac.
Prudences Virginals.

> Eve's *Apple* we have shewed you,
> *Of that be you aware:*
> You have seen Jacobs *Ladder too,*
> *Upon which Angels are.*

An Anchor you received have;
But let not these suffice,
Until with Abra'm *you have gave,*
Your best, a Sacrifice.

Mr. Great-
heart *come*
again.

Now about this time one knocked at the Door, So the Porter opened, and behold Mr. *Great-heart* was there; but when he was come in, what Joy was there? For it came now fresh again into their minds, how but a while ago he had slain old *Grim Bloody-man*, the Giant, and had delivered them from the Lions.

He brings
a token from
his Lord
with him.

Then said Mr. *Great-heart* to *Christiana*, and to *Mercie*, My Lord has sent each of you a Bottle of Wine, and also some parched Corn, together with a couple of Pomgranates. He has also sent the Boys some Figs, and Raisins to refresh you in your way.

Then they addressed themselves to their Journey, and *Prudence*, and *Piety* went along with them. When they came at the Gate, *Christiana* asked the Porter, if any of late went by. He said, No, only one some time since: who also told me that of late there had been

Robbery. a great Robbery committed on the Kings High-way, as you go: But he saith, the Thieves are taken, and will shortly be Tryed for their Lives. Then *Christiana*, and *Mercie*, was afraid; but *Mathew* said, Mother fear nothing, as long as Mr. *Great-heart* is to go with us, and to be our Conductor.

Christiana
takes her
leave of the
Porter.

Then said *Christiana* to the Porter, Sir, I am much obliged to you for all the Kindnesses that you have shewed me since I came hither, and also for that you have been so loving and kind to my Children. I know not how to gratifie your Kindness: Wherefore pray as a token of my respects to you, accept of this small mite: So she put a Gold Angel in his Hand, and he made her a low obeisance, and

The
Porters
blessing.

said, Let thy Garments be always White, and let thy Head want no Ointment. Let *Mercie* live and not die, and let not her Works be few.†And to the Boys he said, Do you fly Youthful lusts, and follow after Godliness with them that are Grave, and Wise, so shall you put Gladness into your Mothers Heart, and obtain Praise of all that are sober minded. So they thanked the Porter and departed.

Now I saw in my Dream, that they went forward until they were come to the Brow of the Hill, where *Piety* bethinking her self cryed out, *Alas!* I have forgot what I intended to bestow upon *Christiana*, and her Companions. I will go back and fetch it. So she ran, and fetched it. While she was gone, *Christiana* thought she heard in a Grove a little way off, on the Right-hand, a most curious melodious Note, with Words much like these,

> *Through all my Life thy favour is* †
> *So frankly shew'd to me,*
> *That in thy House for evermore*
> *My dwelling place shall be.*

And listning still she thought she heard another answer it, saying,

> *For why, the Lord our God is good,*
> *His Mercy is forever sure:*
> *His truth at all times firmly stood:*
> *And shall from Age to Age endure.*

So *Christiana* asked *Prudence*, what 'twas that made those curious Notes? They are, said she, our Countrey Birds: They sing these Notes but seldom, except it be at the Spring, when the Flowers appear, and the Sun shines warm, and then you may hear them all day long. I often, said she, go out to hear them, we also oft times keep them tame in our House. They are very fine Company for us when we are *Melancholy*, also they make the Woods and Groves, and Solitary places, places desirous to be in. *Song 2.* *11, 12.*

By this time *Piety* was come again, So she said to *Christiana*, Look here, I have brought thee a *Scheme* of all those things that thou hast seen at our House: Upon which thou mayest look when thou findest thy self forgetful, and call those things again to remembrance for thy Edification, and comfort. *Piety bestoweth something on them at parting.*

Now they began to go down the Hill into the Valley of *Humiliation*. It was a steep Hill, & the way was slippery; but they were very careful, so they got down pretty well. When they were down in the Valley, *Piety* said to *Christiana*, This is the place where *Christian* your Husband met with the foul Fiend *Apollyon*, and where they had that dreadful fight that they had. I know you cannot but have heard *1 Part, pag. 46.*

thereof. But be of good Courage, as long as you have here Mr. *Great-heart* to be your Guide and Conductor, we hope you will fare the better. So when these two had committed the Pilgrims unto the Conduct of their Guide, he went forward, and they went after.

Mr. Great-heart at the Valley of Humiliation.

Great-heart. Then said Mr. *Great-heart*, We need not be so afraid of this Valley: For here is nothing to hurt us, unless we procure it to our selves. 'Tis true, *Christian* did here meet with *Apollion*, with whom he also had a sore Combate; but that *frey*, was the fruit of those slips that he got in his going down the Hill. For they that get slips there, must look for *Combats* here. And hence it is that this Valley has got so hard a name. For the common people when they hear that some frightful thing has befallen such an one in such a place, are of an Opinion that that place is haunted with some foul Fiend, or evil Spirit; when alas it is for the fruit of their doing, that such things do befal them there.

1 Part, pag. 46-7.

The reason why Christian was so beset here.

This Valley of *Humiliation* is of it self as fruitful a place, as any the Crow flies over; and I am perswaded if we could hit upon it, we might find somewhere here abouts something that might give us an Account why *Christian* was so hardly beset in this place.

A Pillar with an Inscription on it.

Then *James* said to his Mother, Lo, yonder stands a Pillar, and it looks as if something was Written thereon: let us go and see what it is. So they went, and found there Written, *Let* Christian's *slips before he came hither, and the Battels that he met with in this place, be a warning to those that come after.* Lo, said their Guide, did not I tell you, that there was something here abouts that would give Intimation of the reason why *Christian* was so hard beset in this place? Then turning himself to *Christiana*, he said: No disparagement to *Christian* more than to many others whose Hap and Lot his was. For 'tis easier going *up*, then *down this* Hill; and that can be said but of few Hills in all these parts of the World. But we will leave the good Man, he is at rest, he also had a brave Victory over his Enemy; let him grant that dwelleth above, that we fare no worse when we come to be tryed then he.

This Valley a brave place.

But we will come again to this Valley of *Humiliation*. It is the best, and most fruitful piece of Ground in all those parts. It is fat Ground, and as you see, consisteth much in Meddows: and if a man was to come here in the Summer-time, as we do now, if he knew not any thing before thereof, and if he also delighted himself in the sight of his Eyes, he might see that that would be delightful to him. Behold, how green this Valley is, also how beautified *with Lillies.* I have also known many labouring Men that have got good Estates in this Valley of *Humiliation.* (For God resisteth the Proud; but gives *more, more* Grace to the Humble;) for indeed it is a very fruitful Soil, and doth bring forth by handfuls. Some also have wished that the next way to their Fathers House were here, that they might be troubled no more with either Hills or Mountains to go over; but the way is the way, and there's an end.

Song 2. 1.
Jam. 4. 6.
1 Pet. 5. 5.
Men thrive in the Valley of Humiliation.

Now as they were going along and talking, they espied a Boy feeding his Fathers Sheep. The Boy was in very mean Cloaths, but of a very fresh and well-favoured Countenance, and as he sate by himself he Sung. Hark, said Mr. *Great-heart,* to what the Shepherds Boy saith. So they hearkned, and he said,

> He that is down, needs fear no fall,
> He that is low, no Pride:
> He that is humble, ever shall
> Have God to be his Guide.

Philip. 4. 12, 13.

> I am content with what I have,
> Little be it, or much:
> And, Lord, Contentment still I crave,
> Because thou savest such.

> Fulness to such a burden is
> That go on Pilgrimage:
> Here little, and hereafter Bliss,
> Is best from Age to Age.

Heb. 13. 5.

Then said their *Guide,* Do you hear him? I will dare to say, that this Boy lives a merrier Life, and wears more of that Herb called *Hearts-ease* in his Bosom, then he that is clad in Silk and Velvet; but we will proceed in our Discourse.

In this Valley our Lord formerly had his *Countrey-House,* he loved much to be here; He loved also to walk these Medows, for he found the Air was pleasant: Besides here a man shall be free from the Noise, and from the hurryings of this Life; all States are full of Noise

Christ, when in the Flesh, had his Countrey-House in the Valley of Humiliation.

and Confusion, only the Valley of *Humiliation* is that empty and Solitary Place. Here a man shall not be so let and hindred in his Contemplation, as in other places he is apt to be. This is a Valley that no body walks in, but those that love a Pilgrims Life. And though *Christian* had the hard hap to meet here with *Apollyon*, and to enter with him a brisk encounter: Yet I must tell you, that in former times men have met with Angels here, have found Pearls here, and have in this place found the words of Life.

Hos. 12. 4, 5.

Did I say, our Lord had here in former Days his Countrey-house, and that he loved here to walk? I will add, in this Place, and to the People that live and trace these Grounds, he has left a yearly revenue to be faithfully payed them at certain Seasons, for their maintenance by the way, and for their further incouragement to go on in their Pilgrimage.

Mat. 11. 29.

Samuel. Now as they went on, *Samuel* said to Mr. *Great-heart: Sir, I perceive that in this Valley, my Father and* Apollyon *had their Battel; but whereabout was the Fight, for I perceive this Valley is large?*

Great-heart. Your Father had that Battel with *Apollyon* at a place yonder, before us, in a narrow Passage just beyond *Forgetful-Green.* And indeed that place is the most dangerous place in all these Parts. For if at any time the Pilgrims meet with any brunt, it is when they forget what Favours they have received, and how unworthy they are of them. This was the Place also where others have been hard put to it. But more of the place when we are come to it; for I perswade my self, that to this day there remains either some sign of the Battel, or some Monument to testifie that such a Battle there was fought.

Forgetful-Green.

Mercie. Then said *Mercie*, I think I am as well in this Valley, as I have been any where else in all our Journey: The place methinks suits with my Spirit. I love to be in such places where there is no ratling with Coaches, nor rumbling with Wheels: Methinks here one may without much molestation be thinking what he is, whence he came, what he has done, and to what the King has called him: Here one may think, and break at Heart, and melt in ones Spirit, until ones Eyes become like the *Fish Pools of Heshbon.* They that go rightly thorow this Valley of Baca make it a Well, the Rain that God sends down from Heaven upon them that are here also *filleth*

Humility a sweet Grace.

Song 7. 4.
Psal. 84. 5, 6, 7.

the Pools. This Valley is that from whence also the King will give to *Hos. 2. 15.* his their Vineyards, and they that go through it, shall sing, (as *Christian* did, for all he met with *Apollyon*.)

Great-heart. 'Tis true, said their Guide, I have gone thorough this *An Experiment of it.* Valley many a time, and never was better then when here.

I have also been a Conduct to several Pilgrims, and they have confessed the same; *To this man will I look, saith the King, even to him that is Poor, and of a contrite Spirit, and that trembles at my Word.*[†]

Now they were come to the place where the afore mentioned Battel was fought. Then said the Guide to *Christiana*, her Children, *The place* and *Mercie*: This is the place, on this Ground *Christian* stood, and *where* up there came *Apollyon* against him. And look, did not I tell you, *Christian* here is some of your Husbands Blood upon these Stones to this day: *Fiend did* Behold also how here and there are yet to be seen upon the place, *signs of the* some of the Shivers of *Apollyon*'s Broken *Darts*: See also how they *remains.* did beat the Ground with their Feet as they fought, to make good their Places against each other, how also with their by-blows they did split the very stones in pieces. Verily *Christian* did here play the Man, and shewed himself as stout, as could, had he been here, even *Hercules*[†] himself. When *Apollyon* was beat, he made his retreat to the next Valley, that is called The Valley of the Shadow of Death, unto which we shall come anon.

Lo yonder also stands a Monument, on which is Engraven this *A Monument of the* Battle, and *Christian*'s Victory to his Fame throughout all Ages: So *Battel.* because it stood just on the way-side before them, they stept to it and read the Writing, which word for word was this:

> *Hard by, here was a Battle fought,*
> *Most strange, and yet most true.*
> Christian *and* Apollyon *sought*
> *Each other to subdue.*
>
> *The Man so bravely play'd the Man,*
> *He made the* Fiend *to fly:*
> *Of which a Monument I stand,*
> *The same to testifie.*

A Monument of Christians Victory.

When they had passed by this place, they came upon the Borders *1 Part, pag.* of the Shadow of Death, and this Valley was longer then the other, *51.*

a place also most strangely haunted with evil things, as many are able to testifie: But these Women and Children went the better thorough it, because they had day-light, and because Mr. *Great-heart* was their Conductor.

Groanings heard. When they were entred upon this Valley, they thought that they heard a groaning as of dead men: a very great groaning. They thought also they did hear Words of Lamentation spoken, as of some in extream Torment. These things made the Boys to quake, the Women also looked pale and wan; but their Guide bid them be of Good Comfort.

The Ground shakes. So they went on a little further, and they thought that they felt the Ground begin to shake under them, as if some hollow place was there; they heard also a kind of a hissing as of Serpents; but nothing as yet appeared. Then said the Boys, Are we not yet at the end of this doleful place? But the Guide also bid them be of good Courage, and look well to their Feet, lest haply, said he, you be taken in some Snare.

James sick with fear. Now *James* began to be Sick; but I think the cause thereof was Fear, so his Mother gave him some of that Glass of Spirits that she had given her at the *Interpreters* House, and three of the Pills that Mr. *Skill* had prepared, and the Boy began to revive. Thus they went on till they came to about the middle of the Valley, and then *The Fiend appears.* *Christiana* said, Methinks I see something yonder upon the Road before us, a thing of a shape such as I have not seen. Then said *The Pilgrims are afraid.* *Joseph*, Mother, what is it? An ugly thing, Child; an ugly thing, said she. But Mother, what is it like, said he? 'Tis like I cannot tell what, said she. And now it was but a little way off. Then said she, it is nigh.

Great-heart incourages them. Well, well, said Mr. *Great-heart*, let them that are most afraid keep close to me: So the *Fiend* came on, and the Conductor met it; but when it was just come to him, it vanished to all their sights. Then remembred they what had been said sometime agoe. *Resist the Devil, and he will fly from you.*[†]

They went therefore on, as being a little refreshed; but they had not gone far, before *Mercie* looking behind her, saw as she thought, *A Lion.* something most like a Lion, and it came a great padding pace after; and it had a hollow Voice of Roaring, and at every Roar that it gave, it made all the Valley Eccho, and their Hearts to ake, save the Heart of him that was their Guide. So it came up, and Mr. *Great-heart* went behind, and put the Pilgrims all before him. The Lion also came on apace, and Mr. *Great-heart* addressed himself to give him Battel. But when he saw that it was determined that resistance should be made, he also drew back and came no further.

1 Pet. 5. 8, 9.

Then they went on again, and their Conductor did go before them, till they came at a place where was cast up a pit, the whole breadth of the way, and before they could be prepared to go over that, a great mist and a darkness fell upon them, so that they could not see: Then said the Pilgrims, Alas! now what shall we do? But their Guide made answer; Fear not, stand still and see what an end will he put to this also; so they stayed there because their Path was marr'd. They then also thought that they did hear more apparently the noise and rushing of the Enemies, the fire also and the smoke of the Pit was much easier to be discerned. Then said *Christiana* to *Mercie*, Now I see what my poor Husband went through. I have heard much of this place, but I never was here afore now; poor man, he went here all alone in the night; he had night almost quite through the way, also these Fiends were busie about him, as if they would have torn him in pieces. Many have spoke of it, but none can tell what the Valley of the Shadow of Death should mean, until they come in it themselves. *The heart knows its own bitterness, and a stranger intermeddleth not with its Joy.*[†] To be here is a fearful thing.

A pit and darkness.

Christiana now knows what her Husband felt.

Greath. This is like doing business in great Waters, or like going down into the deep; this is like being in the heart of the Sea, and like going down to the Bottoms of the Mountains: Now it seems as if the Earth with its bars were about us for ever. *But let them that walk in darkness and have no light, trust in the name of the Lord, and stay upon their God.*[†] For my part, as I have told you already, I have gone often through this Valley, and have been much harder put to it than now I am, and yet you see I am alive. I would not boast, for that I am not mine own Saviour. But I trust we shall have a good deliverance. Come let us pray for light to him that can lighten our darkness, and that can rebuke, not only these, but all the Satans in Hell.

Great-heart's Reply.

They pray. So they cryed and prayed, and God sent light and deliverance, for there was now no lett in their way, no not there, where but now they were stopt with a pit.

Yet they were not got through the Valley; so they went on still, and behold great stinks and loathsome smells, to the great *Mercie to* annoyance of them. Then said *Mercie* to *Christiana*, there is not such *Christiana.* pleasant being here as at the *Gate*, or at the Interpreters, or at the House where we lay last.

One of the O but, said one of the Boys, *it is not so bad to go through here, as it is Boys Reply.* *to abide here always, and for ought I know, one reason why we must go this way to the House prepared for us, is, that our home might be made the sweeter to us.*

Well said, *Samuel*, quoth the *Guide*, thou hast now spoke like a man. Why, if ever I get out here again, said the *Boy*, I think I shall prize light and good way better than ever I did in all my life. Then said the *Guide*, we shall be out by and by.

So on they went, and *Joseph* said, *Cannot we see to the end of this Valley as yet?* Then said the *Guide*, Look to your feet, for you shall presently be among the Snares.† So they looked to their feet and went on; but they were troubled much with the Snares. Now when *Heedless is* they were come among the Snares, they espyed a Man cast into the *slain, and* Ditch on the left hand, with his flesh all rent and torn. Then said *Takeheed* *preserved.* the *Guide*, that is one *Heedless*, that was agoing this way; he has lain there a great while. There was one *Takeheed* with him, when he was taken and slain, but *he* escaped their hands. You cannot imagine how many are killed here about, and yet men are so foolishly venturous, as to set out lightly on Pilgrimage, and to come without a *Guide*. Poor *Christian*, it was a wonder that he here escaped, but he was beloved of his God, also he had a good heart of his own, or else *1 Part, pag.* he could never a-done it. Now they drew towards the end of the *54.* way, and just there where *Christian* had seen the Cave when he went *Maull* by, out thence came forth *Maull* a Gyant.† This *Maull* did use to *a Gyant.* spoyl young Pilgrims with Sophistry, and he called *Great-heart* by his name, and said unto him, how many times have you been for-*He quarrels* bidden to do these things? Then said Mr. *Great-heart*, what things? *with* What things, quoth the Gyant, you know what things; but I will *Great-* put an end to your trade. But pray, said Mr. *Great-heart*, before we *heart.*

fall to it, let us understand wherefore we must fight; (now the Women and Children stood trembling, and knew not what to do); quoth the Gyant, You rob the Countrey, and rob it with the worst of Thefts. These are but Generals, said Mr. *Great-heart*, come to particulars, man.

Then said the *Gyant*, thou practises the craft of a *Kidnapper*, thou gatherest up Women and Children, and carriest them into a strange Countrey, to the weaking of my Masters Kingdom. But now *Great-heart* replied, I am a Servant of the God of Heaven, my business is to perswade sinners to Repentance, I am commanded to do my endeavour to turn Men, Women and Children, from darkness to light, and from the power of Satan to God, and if this be indeed the ground of thy quarrel, let us fall to it as soon as thou wilt. *God's Ministers counted as Kidnappers.*

Then the *Giant* came up, and Mr. *Great-heart* went to meet him, and as he went, he drew his *Sword*, but the *Giant* had a *Club*: So without more ado they fell to it, and at the first blow the *Giant* stroke Mr. *Great-heart* down upon one of his knees; with that the Women, and Children cried out. So Mr. *Great-heart* recovering himself, laid about him in full lusty manner, and gave the *Giant* a wound in his arm; thus he fought for the space of an hour to that height of heat, that the breath came out of the *Giants* nostrils, as the heat doth out of a boiling Caldron. *The Gyant and Mr. Great-heart must fight. Weak folks prayers do sometimes help strong folks cries.*

Then they sat down to rest them, but Mr. *Great-heart* betook him to prayer; also the Women and Children did nothing but sigh and cry all the time that the Battle did last.

When they had rested them, and taken breath, they both fell to it again, and Mr. *Great-heart* with a full blow fetch't the *Giant* down to the ground. Nay hold, and let me recover, quoth he. So Mr. *Great-heart* fairly let him get up; So to it they went again; and the *Giant* mist but little of all-to-breaking Mr. *Great-heart*'s Scull with his Club. *The Gyant struck down.*

Mr. *Great-heart* seeing that, runs to him in the full heat of his Spirit, and pierceth him under the fifth rib; with that the *Giant* began to faint, and could hold up his Club no longer. Then Mr. *Great-heart* seconded his blow, and smit the head of the *Giant* from his shoulders. Then the Women and Children rejoyced, and Mr. *Great-heart* also praised God, for the deliverance he had wrought. *He is slain, and his head disposed of.*

When this was done, they amongst them erected a Pillar, and fastned the *Gyant's* head thereon, and wrote underneath in letters that Passengers might read,

> *He that did wear this head, was one*
> *That Pilgrims did misuse;*
> *He stopt their way, he spared none,*
> *But did them all abuse;*
> *Until that I,* Great-heart, *arose,*
> *The Pilgrims* Guide *to be;*
> *Until that I did him oppose,*
> *That was their Enemy.*

1 Part, pag. 55.

Now I saw, that they went to the Ascent that was a little way off cast up to be a Prospect for Pilgrims. (That was the place from whence *Christian* had the first sight of *Faithful* his Brother.) Wherefore here they sat down, and rested, they also here did eat and drink, and make merry; for that they had gotten deliverance from this so dangerous an Enemy. As they sat thus and did eat, *Christiana* asked the Guide, *if he had caught no hurt in the battle.* Then said Mr. *Great-heart, No,* save a little on my flesh; yet that also shall be so far from being to my determent, that it is at present a proof of my love to my Master and you, and shall be a means by Grace to encrease my reward at last.

2 Cor. 4.

Discourse of the fight.

But was you not afraid, good Sir, when you see him come with his Club?

It is my duty, said he, to distrust mine own ability, that I may have reliance on him that is stronger then all. *But what did you think when he fetched you down to the ground at the first blow?* Why I thought, quoth he, that so my master himself was served, and yet he it was that conquered at the last.

Mat. here admires Goodness.

Mat. *When you all have thought what you please, I think God has been wonderful good unto us, both in bringing us out of this Valley, and in delivering us out of the hand of this Enemy; for my part I see no reason why we should distrust our God any more, since he has now, and in such a place as this, given us such testimony of his love as this.*

Old Honest asleep under an Oak.

Then they got up and went forward; now a little before them stood an Oak, and under it when they came to it, they found an old

Pilgrim fast asleep; they knew that he was a *Pilgrim* by his *Cloths*, and his *Staff*, and his *Girdle*.

So the *Guide* Mr. *Great-heart* awaked him, and the old Gentleman, as he lift up his eyes cried out; What's the matter? who are you? and what is your business here?

Great. Come man be not so hot, here is none but Friends; yet the old man gets up and stands upon his guard, and will know of them what they were. Then said the *Guide*, My name is *Great-heart*, I am the guide of these Pilgrims which are going to the Celestial Countrey.

Honest. Then said Mr. *Honest*, I cry you mercy; I feared that you had been of the Company of those that some time ago did rob *Little-faith* of his money; but now I look better about me, I perceive you are honester People.

One Saint sometimes takes another for his Enemy.

Greath. Why what would, or could you adone, to a helped your self, if we indeed had been of that Company?

Talk between Great-heart and he.

Hon. Done! Why I would a fought as long as breath had been in me; and had I so done, I am sure you could never have given me the worst on't, for a *Christian* can never be overcome, unless he shall yield of himself.

Greath. Well said, Father Honest, *quoth the Guide, for by this I know that thou art a Cock of the right kind, for thou hast said the Truth.*

Hon. And by this also I know that thou knowest what true Pilgrimage is; for all others do think that we are the soonest overcome of any.

Greath. Well, now we are so happily met, pray let me crave your Name, and the name of the Place you came from?

Whence Mr. Honest came.

Hon. My Name I cannot, but I came from the Town of *Stupidity;* it lieth about four Degrees beyond the City of *Destruction.*

Greath. Oh! *Are you that Country-man then? I deem I have half a guess of you, your Name is old* Honesty, *is it not?* So the old Gentleman blushed, and said, Not Honesty in the *Abstract*, but *Honest* is my Name, and I wish that my *Nature* shall agree to what I am called.

Hon. But Sir, said the old Gentleman, how could you guess that I am such a Man, since I came from such a place?

Greath. I had heard of you before, by my Master, for he knows all things that are done on the Earth: But I have often wondered that any should come from your place; for your Town is worse then is the City of Destruction it self.

Hon. Yes, we lie more off from the Sun, and so are more Cold and Sensless; but as a Man in a Mountain of Ice, yet if the Sun of Righteousness will arise upon him, his frozen Heart shall feel a Thaw; and thus it hath been with me.

Greath. I believe it, Father *Honest,* I believe it, for I know the thing is true.

Then the old Gentleman saluted all the Pilgrims with a holy Kiss of Charity,†and asked them of their Names, and how they had fared since they set out on their Pilgrimage.

Christ. Then said *Christiana,* My name I suppose you have heard of, good *Christian* was my Husband, and these four were his Children. But can you think how the old Gentleman was taken, when she told him who she was! He skip'd, he smiled, and blessed them with a thousand good Wishes, saying,

Hon. *I have heard much of your Husband, and of his Travels and Wars which he underwent in his days. Be it spoken to your Comfort, the Name of your Husband rings all over these parts of the World; His Faith, his Courage,* *his Enduring, and his Sincerity under all, has made his name Famous.* Then he turned him to the Boys, and asked them of their names, which they told him: And then said he unto them, *Mathew,* be thou like *Mathew* the Publican, not in Vice, but Virtue. *Samuel,* said he, be thou like *Samuel* the Prophet, a Man of Faith and Prayer. *Joseph,* said he, be thou like *Joseph* in *Potiphar's* House, Chast, and one that flies from Temptation. And, *James,* be thou like *James* the *Just,* and like *James* the brother of our Lord.

Then they told him of *Mercie,* and how she had left her Town and her Kindred to come along with *Christiana,* and with her Sons. At that the old *Honest* man said, *Mercie,* is thy Name? by *Mercie* shalt thou be sustained, and carried thorough all those Difficulties that shall assault thee in thy way; till thou come thither where thou shalt look the Fountain of Mercie in the Face with Comfort.

All this while the Guide Mr. *Great-heart,* was very much pleased, and smiled upon his Companion.

Now as they walked along together, the Guide asked the old *Talk of one Mr.* Fearing Gentleman, *if he did not know one Mr.* Fearing, *that came on Pilgrimage out of his Parts.*

Hon. Yes, very well, said he; he was a Man that had the Root of the Matter in him, but he was one of the most troublesome Pilgrims that ever I met with in all my days.

Greath. I *perceive you knew him, for you have given a very right Character of him.*

Hon. Knew him! I was a great Companion of his, I was with him most an end; when he first began to think of what would come upon us hereafter, I was with him.

Greath. I *was his Guide from my Master's House, to the Gates of the Celestial City.*

Hon. Then you knew him to be a troublesom one?

Greath. I *did so, but I could very well bear it: for Men of my Calling are often times intrusted with the Conduct of such as he was.*

Hon. Well then, pray let us hear a little of him, and how he managed himself under your Conduct.

Greath. Why he was always afraid that he should come short of *Mr.* Fearing's troublesom Pilgrimage.* whither he had a desire to go. Every thing frightned him that he heard any body speak of, that had the least appearance of Opposition in it. I heard that he lay roaring at the *Slow of Dispond,* for above *His behaviour at the Slow of Dispond.* a Month together, nor durst he, for all he saw several go over before him, venture, tho they, many of them, offered to lend him their Hand. *He would not go back again neither.* The Celestial City, he said he should die if he came not to it, and yet was dejected at every Difficulty, and stumbled at every Straw that any body cast in his way. Well, after he had layn at the *Slow of Dispond* a great while, as I have told you; one sunshine Morning, I do not know how, he ventured, and so got over. But when he was over, he would scarce believe it. He had, I think, a *Slow of Dispond* in his Mind, a *Slow* that he carried every where with him, or else he could never have been as he was. So he came up to the Gate, you know what I mean, that stands at the head of this way, and there also he stood a good while before he would adventure to knock. When the Gate was opened he *His behavior at the Gate.* would give back, and give place to others, and say that he was not worthy. For, for all he gat before some to the Gate, yet many of

them went in before him. There the poor man would stand shaking and shrinking; I dare say it would have pitied ones Heart to have seen him: *Nor would he go back again.* At last he took the Hammer that hanged on the Gate in his hand, and gave a small Rapp or two; then one opened to him, but he shrunk back as before. He that opened, stept out after him, and said, Thou trembling one, what wantest thou? with that he fell to the Ground. He that spoke to him wondered to see him so faint. So he said to him, Peace be to thee; up, for I have set open the Door to thee; come in, for thou art blest. With that he gat up, and went in trembling, and when he was in, he was ashamed to show his Face. Well, after he had been entertained there a while, as you know how the manner is, he was bid go on his way, and also told the way he should take. So he came till he came to our House, but as he behaved himself at the Gate, so he
His did at my master the *Interpreters* Door. He lay thereabout in the Cold
behavior a good while, before he would adventure to call; *Yet he would not go*
at the Inter- *back.* And the Nights were long and cold then. Nay he had a Note
preters *Door.* of *Necessity* in his Bosom to my Master, to receive him, and grant him the Comfort of his House, and also to allow him a stout and valiant Conduct, because he was himself so *Chickin-hearted* a Man; and yet for all that he was afraid to call at the Door. So he lay up and down there abouts, till, poor man, he was almost starved; yea so great was his Dejection, that tho he saw several others for knocking got in, yet he was afraid to venture. At last, I think I looked out of the Window, and perceiving a man to be up and down about the Door, I went out to him, and asked what he was; but poor man, the water stood in his Eyes. So I perceived what he wanted. I went therefore in, and told it in the House, and we shewed the thing to our Lord; So he sent me out again, to entreat him to come in, but I dare say I had hard work to do it. At last he came in, and I will say that for my Lord, he carried it wonderful lovingly to him. There
How he were but a few good bits at the Table, but some of it was laid upon
was enter- his Trencher. Then he presented the *Note*, and my Lord looked
tained there. thereon and said, His desire should be granted. So when he had bin
He is a little
encouraged there a good while, he seemed to get some Heart, and to be a little
at the more Comfortable. For my Master, you must know, is one of very
Interpreters tender Bowels, especially to them that are afraid, wherefore he
house.

carried it so towards him, as might tend most to his Incouragement. Well, when he had had a sight of the things of the place, and was ready to take his Journey to go to the City, my Lord, as he did to *Christian* before, gave him a Bottle of Spirits, and some comfortable things to eat. Thus we set forward, and I went before him; but the man was but of few Words, only he would sigh aloud.

When we were come to where the three Fellows were hanged, he said, that he doubted that that would be his end also. Only he seemed glad when he saw the Cross and the Sepulcher. There I confess he desired to stay a little, to look; and he seemed for a while after to be a little *Cheary*. When we came at the Hill *Difficulty*, he made no stick at that, nor did he much fear the Lyons. For you must know that his Trouble *was not about such things as those*, his Fear was about his Acceptance at last. *He was greatly afraid when he saw the Gibbit, Cheary when he saw the Cross.*

I got him in at the House *Beautiful*, I think before he was willing; also when he was in, I brought him acquainted with the Damsels that were of the Place, but he was ashamed to make himself much for Company, he desired much to be alone, yet he always loved good talk, and often would get behind the *Screen* to hear it; he also loved much to see *antient* things, and to be *pondering* them in his Mind. He told me afterwards, that he loved to be in those two Houses from which he came last, to wit, at the Gate, and that of the *Interpreters*, but that he durst not be so bold to ask. *Dumpish at the house Beautiful.*

When we went also from the House *Beautiful*, down the Hill, into the Valley of *Humiliation*, he went down as well as ever I saw man in my Life, for he cared not how mean he was, so he might be happy at last. Yea, I think there was a kind of a Sympathy betwixt that Valley and him. For I never saw him better in all his Pilgrimage, then when he was in that Valley. *He went down into, and was very Pleasant in the Valley of Humiliat.*

Here he would lie down, embrace the Ground, and kiss the very Flowers that grew in this Valley. He would now be up every Morning by break of Day, tracing, and walking to and fro in this Valley. *Lam. 3. 27, 28, 29.*

But when he was come to the entrance of the Valley of the Shadow of Death, I thought I should have lost my Man; not for that he had any Inclination to go back, that he alwayes abhorred, but he was ready to dy for Fear. O, the *Hobgoblins* will have me, the *Hobgoblins* will have me, cried he; and I could not beat him out on't. He made *Much perplexed in the Valley of the Shadow of Death.*

such a noyse, and such an outcry here, that, had they but heard him, 'twas enough to encourage them to come and fall upon us.

But this I took very great notice of, that this Valley was as quiet while he went thorow it, as ever I knew it before or since. I suppose, those Enemies here, had now a special Check from our Lord, and a Command not to meddle until Mr. *Fearing* was pass'd over it.

It would be too tedious to tell you of all; we will therefore only *His* mention a Passage or two more. When he was come at *Vanity Fair*, *behaviour* I thought he would have fought with all the men in the Fair; I *at Vanity-* *Fair.* feared there we should both have been knock'd o'th Head, so hot was he against their Fooleries; upon the inchanted Ground, he also was very wakeful. But when he was come at the *River* where was no Bridge, there again he was in a heavy Case; now, now he said he should be drowned for ever, and so never see that Face with Comfort, that he had come so many miles to behold.

And here also I took notice of what was very remarkable, the Water of that River was lower at this time, than ever I saw it in all my Life; so he went over at last, not much above wet-shod. When he was going up to the Gate, Mr. *Great-heart* began to take his Leave *His Bold-* of him, and to wish him a good Reception above; So he said, *I shall,* *ness at last.* *I shall.* Then parted we asunder, and I saw him no more.

Honest. *Then it seems he was well at last.*

Greath. Yes, yes, I never had doubt about him, he was a man of a choice Spirit, only he was always kept very low, and that made *Psal. 88.* his Life so burthensom to himself, and so troublesom to others. He *Rom. 14.* was above many, tender of Sin; he was so afraid of doing Injuries to *21.* *1 Cor. 8.* others, that he often would deny himself of that which was lawful, *13.* because he would not offend.

Hon. *But what should be the reason that such a good Man should be all his days so much in the dark?*

Reason why *Greath.* There are two sorts of Reasons for it; one is, The wise *good men* *are so in* God will have it so. Some must *Pipe,* and some must *Weep:* Now *the dark.* Mr. *Fearing* was one that play'd upon *this Base.* He and his fellows *Mat. 11.* *16, 17, 18.* sound the *Sackbut,* whose Notes are more doleful than the Notes of other Musick. Tho indeed some say, the Base is the ground of Musick. And for my part, I care not at all for that Profession that begins not in heaviness of Mind. The first string that the Musitian usually touches, *is the Base*, when he intends to put all in tune; God also plays upon this string first, when he sets the Soul in tune for himself. Only here was the imperfection of Mr. *Fearing*, he could play upon no other Musick but this, till towards his latter end.

I make bold to talk thus Metaphorically, for the ripening of the Wits of young Readers, and because in the Book of the Revelations, the Saved are compared to a company of Musicians that play upon their *Trumpets* and Harps, and sing their Songs before the Throne.† Revel. 8. 2. chap. 14. 2, 3.

Hon. *He was a very zealous man, as one may see by what Relation you have given of him. Difficulties, Lyons, or Vanity-Fair, he feared not at all: 'Twas only Sin, Death and Hell, that was to him a Terror; because he had some Doubts about his Interest in that Celestial Country.*

Greath. You say right. *Those were the things that were his Troub-* *A Close* lers, and they, as you have well observed, arose from the weakness of *about him.* his Mind there about, not from weakness of Spirit as to the practical part of a Pilgrims Life. I dare believe, that as the Proverb is, he could have bit a Firebrand, had it stood in his way: But the things with which he was oppressed, no man ever yet could shake off with ease.

Christiana. *Then said* Christiana, *This Relation of Mr.* Fearing *has* Chris- *done me good. I thought no body had been like me, but I see there was some* *tiana's Sentence.* *Semblance 'twixt this good man and I, only we differed in two things. His Troubles were so great they brake out, but mine I kept within. His also lay so hard upon him, they made him that he could not knock at the Houses provided for Entertainment; but my Trouble was always such, as made me knock the lowder.*

Mer. If I might also speak my Heart, I must say that something Mercie's of him has also dwelt in me. For I have ever been more afraid of the *Sentence.* Lake†and the loss of a place in *Paradice,* then I have been of the loss of other things. Oh, thought I, may I have the Happiness to have a Habitation *there,* 'tis enough, though I part with all the World to win it.

Mat. *Then said* Mathew, *Fear was one thing that made me think that* Mathew's *I was far from having that within me that accompanies Salvation, but if it* *Sentence.* *was so with such a good man as he, why may it not also go well with me?*

Jam. No fears, no Grace, said *James.* Though there is not always James's Grace where there is the fear of Hell; yet to be sure there is no *Sentence.* Grace where there is no fear of God.

Greath. *Well said,* James, *thou hast hit the Mark, for the fear of God is the beginning of Wisdom;*†*and to be sure they that want the beginning, have neither* middle *nor* end. *But we will here conclude our Discourse of Mr.* Fearing, *after we have sent after him this Farewel.*

Their
Farewell
about him.

> *Well, Master* Fearing, *thou didst fear*
> *Thy God, and wast afraid*
> *Of doing any thing, while here,*
> *That would have thee betray'd.*
> *And didst thou fear the Lake and Pit?*
> *Would others did so too;*
> *For, as for them that want thy Wit,*
> *They do themselves undo.*

Now I saw, that they still went on in their Talk. For after Mr. *Great-heart* had made an end with Mr. *Fearing,* Mr. *Honest* began to tell them of another, but his Name was Mr. *Selfwil.* He pretended himself to be a *Pilgrim,* said Mr. *Honest*; But I perswade my self, he never came in at the Gate that stands at the head of the way.

Of Mr. Self-will.

Greath. *Had you ever any talk with him about it?*

Old Honest had talked with him.

Hon. Yes, more then once or twice; but he would always be like himself, *self-willed.* He neither cared for man, nor Argument, nor yet Example; what his Mind prompted him to, that he would do, and nothing else could he be got to.

Greath. *Pray what Principles did he hold, for I suppose you can tell?*

Self-will's Opinions.

Hon. He held that a man might follow the Vices as well as the Virtues of the Pilgrims, and that if he did both, he should be certainly saved.

Greath. How! *If he had said, 'tis possible for the best to be guilty of the Vices, as well as to partake of the Virtues of Pilgrims, he could not much a been blamed: For indeed we are exempted from no Vice absolutely, but on condition that we Watch and Strive. But this I perceive is not the thing: But if I understand you right, your meaning is, that he was of that Opinion, that it was allowable so to be.*

Hon. Ai, ai, so I mean, and so he believed and practised.

Greath. *But what Ground had he for his so saying?*

Hon. Why, he said he had the Scripture for his Warrant.

Greath. *Prethee, Mr.* Honest, *present us with a few particulars.*

Hon. So I will. He said, to have to do with other mens Wives, had been practised by *David*, Gods Beloved, and therefore he could do it. He said, to have more Women then one, was a thing that *Solomon* practised, and therefore he could do it. He said, that *Sarah* and the godly Midwives of *Egypt* lyed, and so did saved *Rahab*, and therefore he could do it. He said, that the Disciples went at the bidding of their Master, and took away the Owners *Ass*, and therefore he could do so too. He said, that *Jacob* got the Inheritance of his Father in a way of Guile and Dissimulation, and therefore he could do so too.[†]

Greath. High base! indeed, and you are sure he was of this Opinion?

Hon. I have heard him plead for it, bring Scripture for it, bring Argument for it, *&c.*

Greath. An Opinion that is not fit to be, with any Allowance, in the World.

Hon. You must understand me rightly: He did not say that any man might do this; but, that those that had the Virtues of those that did such things, might also do the same.

Greath. But what more false then such a Conclusion? For this is as much as to say, that because good men heretofore have sinned of Infirmity, therefore he had allowance to do it of a presumptuous Mind. Or if because a Child, by the blast of the Wind, or for that it stumbled at a stone, fell down and so defiled it self in Myre, therefore he might wilfully ly down and wallow like a Bore therein. Who could a thought that any one could so far a bin blinded by the power of Lust? But what is written must be true: They stumble at the Word, being disobedient, whereunto also they were appointed. 1 Pet. 2. 8.

His supposing that such may have the godly Mans Virtues, who addict themselves to their Vices, is also a Delusion as strong as the other. 'Tis just as if the Dog *should say, I have, or may have the* Qualities *of the* Child, *because I lick up its stinking Excrements. To eat up the Sin of Gods People,* Hos. 4. 8. *is no sign of one that is possessed with their Virtues. Nor can I believe that one that is of this Opinion, can at present have Faith or Love in him. But I know you have made strong Objections against him, prethee what can he say for himself?*

Hon. Why, he says, To do this by way of Opinion, seems abundance more honest, then to do it, and yet hold contrary to it in Opinion.

Greath. *A very wicked Answer, for tho to let loose the Bridle to Lusts, while our Opinions are against such things, is bad; yet to sin, and plead a Toleration so to do, is worse; the one stumbles Beholders accidentally, the other pleads them into the Snare.*

Hon. There are many of this man's mind, that have not this mans mouth, and *that* makes going on Pilgrimage of so little esteem as it is.

Greath. *You have said the Truth, and it is to be lamented. But he that feareth the King of Paradice, shall come out of them all.*

Christiana. There are strange Opinions in the World. I know one that said 'twas time enough to repent when they came to die.

Greath. *Such are not over wise. That man would a bin loath, might he have had a week to run twenty mile in for his Life, to have deferred that Journey to the last hour of that Week.*

Hon. You say right, and yet the generality of them that count themselves Pilgrims, do indeed do thus. I am, as you see, an old Man, and have bin a Traveller in this Rode many a day; and I have taken notice of many things.

I have seen some that have set out as if they would drive all the World afore them, who yet have in few days, dyed as they in the Wilderness,†and so never gat sight of the promised Land.

I have seen some that have promised nothing at first setting out to be Pilgrims, and that one would a thought could not have lived a day, that have yet proved very good Pilgrims.

I have seen some that have run hastily forward, that again have after a little time, run just as fast back again.

I have seen some who have spoke very well of a Pilgrim's Life at first, that after a while, have spoken as much against it.

I have heard some, when they first set out for Paradice, say positively, there is such a place, who when they have been almost there, have come back again, and said there is none.

I have heard some vaunt what they would do in case they should be opposed, that have even at a false Alarm fled Faith, the Pilgrims way, and all.

Fresh News of trouble.
† *Part. pag.* 102-3. Now as they were thus in their way, there came one running to meet them, and said, Gentlemen, and you of the weaker sort, if you love Life, shift for your selves, for the Robbers are before you.

Greath. Then said Mr. *Greatheart,* They be the three that set upon Littlefaith heretofore. Well, said he, we are ready for them; so they went on their way. Now they looked at every Turning when they should a met with the Villains. But whether they heard of Mr. *Greatheart,* or whether they had some other Game, they came not up to the Pilgrims.

Great-heart's Resolution.

Chris. Christiana then wished for an Inn for her self and her Children, because they were weary. Then said Mr. *Honest,* There is one a little before us, where a very honourable Disciple, one *Gaius†* dwells. So they all concluded to turn in thither; and the rather, because the old Gentleman gave him so good a Report. So when they came to the Door, they went in, not knocking, for folks use not to knock at the Door of an Inn. Then they called for the Master of the House, and he came to them. *So they asked if they might lie there that Night?*

Christiana wisheth for an Inn. Rom. 16. 23. *Gaius.*

They enter into his House.

Gaius. Yes Gentlemen, if you be true Men, for my House is for none but Pilgrims. Then was *Christiana, Mercie,* and the *Boys,* the more glad, for that the Inn-keeper was a lover of Pilgrims. So they called for Rooms; and he shewed them one for *Christiana,* and her Children, and *Mercy,* and another for Mr. *Greatheart* and the old Gentleman.

Gaius Entertains them, and how.

Greath. Then said Mr. Great-heart, *good* Gaius, *what hast thou for Supper? for these Pilgrims have come far to day, and are weary.*

Gaius. It is late, said *Gaius;* so we cannot conveniently go out to seek Food; but such as we have you shall be welcome to, if that will content.

Greath. We will be content with what thou hast in the House, for as much as I have proved thee; thou art never destitute of that which is convenient.

Then he went down, and spake to the Cook, whose Name was Taste-that-which-is-good, to get ready Supper for so many Pilgrims. This done, he comes up again, saying, come my good Friends, you are welcome to me, and I am glad that I have an House to entertain you; and while Supper is making ready, if you please, let us entertain one another with some good Discourse. So they all said, content.

Gaius his Cook.

Talk between Gaius and his Guests.   Gaius. *Then said* Gaius, *Whose Wife is this aged Matron? and whose Daughter is this young Damsel?*

Greath. The Woman is the Wife of one *Christian,* a Pilgrim of former times, and these are his four Children. The Maid is one of her Acquaintance; one that she hath perswaded to come with her on Pilgrimage. *Mark this.* The Boys take all after their Father, and covet to tread in his Steps. Yea, if they do but see any place where the old Pilgrim hath lain, or any print of his Foot, it ministreth Joy to their Hearts, and they covet to lie, or tread in the same.

Act. 11. 26. Of Christian's Ancestors.   Gaius. *Then said* Gaius, is this *Christian's* Wife, and are these *Christian's* Children? I knew your Husband's Father, yea, also, his Fathers Father. Many have been good of this stock, their Ancestors dwelt first at *Antioch.*† *Christian's* Progenitors (I suppose you have heard your Husband talk of them) were very worthy men. They have above any that I know, shewed themselves men of great Virtue and Courage, for the Lord of the Pilgrims, his ways, and them that loved him. I have heard of many of your Husbands Relations that *Acts 7. 59, 60. chap. 12. 2.* have stood all Tryals for the sake of the Truth. *Stephen* that was one of the first of the Family from whence your Husband sprang, was knocked o'th' Head with Stones. *James,* an other of this Generation, was slain with the edge of the Sword. To say nothing of *Paul* and *Peter,*† men antiently of the Family from whence your Husband came. There was *Ignatius,* who was cast to the Lyons, *Romanus,* whose Flesh was cut by pieces from his Bones; and *Policarp,* that played the man in the Fire. There was he that was hanged up in a Basket in the Sun, for the Wasps to eat; and he who they put into a Sack and cast him into the Sea, to be drowned.† 'Twould be impossible, utterly to count up all of that Family that have suffered Injuries and Death, for the love of a Pilgrims Life. Nor can I, but be glad, to see that thy Husband has left behind him four such Boys as these. I hope they will bear up their Fathers Name, and tread in their Fathers Steps, and come to their Fathers End.

Greath. *Indeed Sir, they are likely Lads, they seem to chuse heartily their Fathers Ways.*

Advice to Christiana about her Boys.   Gaius. That is it that I said, wherefore *Christians* Family is like still to spread abroad upon the face of the Ground, and yet to be numerous upon the Face of the Earth. Wherefore let *Christiana* look out some Damsels for her Sons, to whom they may be Betroathed, *&c.* that the Name of their Father, and the House of his Progenitors may never be forgotten in the World.

Hon. '*Tis pity this Family should fall and be extinct.*

Gaius. Fall it cannot, but be diminished it may; but let *Christiana* take my Advice, and that's the way to uphold it.

And *Christiana*, said *This* Inn-keeper, I am glad to see thee and thy Friend *Mercie* together here, a lovely Couple. And may I advise, take *Mercie* into a nearer Relation to thee. If she will, let her be given to *Mathew* thy eldest Son. 'Tis the way to preserve you a posterity in the Earth. So this match was concluded, and in process of time they were married. But more of that hereafter. *Mercie and Matthew Marry.*

Gaius also proceeded, and said, I will now speak on the behalf of Women, to take away their Reproach. For as Death and the Curse came into the World by a Woman, so also did Life and Health; *God sent forth his Son, made of a Woman.* Yea, to shew how much those that came after did abhor the Act of their Mother, this Sex, in the old Testament, coveted Children, if happily this or that Woman might be the Mother of the Saviour of the World. I will say again, that when the Saviour was come, Women rejoyced in him, before either Man or Angel. I read not that ever any man did give unto Christ so much as one *Groat*, but the Women followed him, and ministred to him of their Substance. 'Twas a Woman that washed his Feet with Tears, and a Woman that anointed his Body to the Burial. They were Women that wept when he was going to the Cross; And Women that followed him from the Cross, and that sat by his Sepulcher when he was buried. They were Women that was first with him at his Resurrection *morn*, and Women that brought Tidings first to his Disciples that he was risen from the Dead. Women therefore are highly favoured, and shew by these things that they are sharers with us in the Grace of Life. Gen. 3.
Gal. 4.
Why Women of old so much desired Children.
Luke 2.
chap. 8. 2.
3.
chap. 7.
37, 50.
Joh. 11. 2.
chap. 12.
3.
Luke 23.
27.
Matt. 27.
55, 56, 61.
Luke 24.
22, 23.

Now the Cook sent up to signifie that Supper was almost ready, and sent one to lay the Cloath, the Trenshers, and to set the Salt and Bread in order. *Supper ready.*

Then said *Mathew*, *The sight of this Cloath, and of this Forerunner of the Supper, begetteth in me a greater Appetite to my Food then I had before.*

What to be gathered from laying of the Board with the Cloath and Trenshers. *Gaius.* So let all ministring Doctrines *to* thee in this Life, beget *in* thee a greater desire to sit at the Supper of the great King in his Kingdom; for all Preaching, Books, and Ordinances here, are but as the laying of the Trenshers, and as setting of Salt upon the Board, when compared with the Feast that our Lord will make for us when we come to his House.

Levit. 7. 32, 33, 34. chap. 10. 14, 15. Psal. 25. 1. Heb. 13. 15. So Supper came up, and first a *Heave-shoulder*, and a *Wave-breast* was set on the Table before them: To shew that they must begin their *Meal* with Prayer and Praise to God. The *Heave-shoulder David* lifted his Heart up to God with, and with the *Wave-breast, where his heart lay*, with that he used to lean upon his Harp when he played. These two Dishes were very fresh and good, and they all eat heartily-well thereof.

Deut. 32. 14. Judg. 9. 13. Joh. 15. 1. The next they brought up, was a Bottle of Wine, red as Blood. So *Gaius* said to them, Drink freely, this is the Juice of the true Vine, that makes glad the Heart of God and Man. So they drank and were merry.

The next was a Dish of Milk well crumbed. But *Gaius* said, *Let the Boys have that, that they may grow thereby.*

1 Pet. 2. 1, 2. A Dish of Milk. Of Honey and Butter. Isa. 7. 15. Then they brought up in course a Dish of *Butter* and *Hony.* Then said *Gaius*, Eat freely of *this*, for this is good to chear up, and strengthen your Judgments and Understandings: This was our Lords Dish when he was a Child; *Butter and Hony shall he eat, that he may know to refuse the Evil, and choose the Good.*

A Dish of Apples. Then they brought them up a Dish of Apples, and they were very good tasted Fruit. Then said *Mathew*, May we eat Apples, since they were such, by, and with which the Serpent beguiled our first Mother?

Then said *Gaius*,

> *Apples were they* with *which we were beguil'd,*
> *Yet Sin, not Apples hath our Souls defil'd.*
> *Apples forbid, if eat, corrupts the Blood:*
> *To eat such, when commanded, does us good.*
> *Drink of his Flagons then, thou, Church, his Dove,*
> *And eat* his *Apples, who art sick of Love.*†

Then said *Mathew*, I made the Scruple, because I *a while since was sick with eating of Fruit.*

Gaius. Forbidden Fruit will make you sick, but not what our Lord has tolerated.

While they were thus talking, they were presented with an other Dish; and 'twas a dish of *Nuts*. Then said some at the Table, *Nuts* spoyl tender Teeth; especially the Teeth of Children. Which when *Gaius* heard, he said, Song 6. 11. A dish of Nuts.

> *Hard* Texts *are* Nuts (*I will not call them* Cheaters,)
> *Whose* Shells *do keep their* Kirnels *from the* Eaters.
> *Ope then the Shells, and you shall have the* Meat,
> *They here are brought, for you to crack and eat.*

Then were they very Merry, and sate at the Table a long time, talking of many things. Then said the old Gentleman, My good Landlord, while we are cracking your *Nuts*, if you please, do you open this Riddle.

> *A man there was, tho some did count him mad,*
> *The more he cast away, the more he had.* A Riddle put forth by old Honest.

Then they all gave good heed, wondring what good *Gaius* would say, so he sat still a while, and then thus replyed:

> *He that bestows his Goods upon the Poor,*
> *Shall have as much again, and ten times more.* Gaius opens it.

Then said *Joseph*, I dare say Sir, I did not think you could a found it out. Joseph wonders.

Oh! said *Gaius*, I have bin trained up in this way a great while: Nothing teaches like Experience; I have learned of my Lord to be kind, & have found by experience that I have gained thereby: *There is that scattereth, yet increaseth, and there is that withholdeth more then is meet, but it tendeth to Poverty. There is that maketh himself Rich, yet hath nothing; there is that maketh himself poor, yet hath great Riches.* Prov. 11. 24. chap. 13. 7.

Then *Samuel* whispered to *Christiana* his Mother, and said, Mother, this is a very good mans House, let us stay here a good while, and let my Brother *Matthew* be married here to *Mercy*, before we go any further.

The which *Gaius* the Host overhearing, said, *With a very good Will my Child.*

So they stayed there more then a Month, and *Mercie* was given to *Mathew* to Wife.

While they stayed here, *Mercy* as her Custom was, would be making Coats and Garments to give to the Poor, by which she brought up a very good Report upon the Pilgrims.

But to return again to our Story. After Supper, the *Lads* desired a Bed, for that they were weary with Travelling. Then *Gaius* called to shew them their Chamber, but said *Mercy*, I will have them to Bed. So she had them to Bed, and they slept well, but the rest sat up all Night. For *Gaius* and they were such sutable Company, that they could not tell how to part. Then after much talk of their Lord, themselves, and their Journey, Old Mr. *Honest*, he that put forth the Riddle to *Gaius*, began to nod. Then said *Great-heart*, What Sir, you begin to be drouzy, come rub up, now here's a *Riddle* for you. Then said Mr. *Honest*, let's hear it.

Then said Mr. *Great-heart*,

> *He that will kill, must first be overcome:*
> *Who live abroad would, first must die at home.*

Huh, said Mr. *Honest*, it is a hard one, hard to expound, and harder to practise. But come Landlord, said he, I will, if you please, leave my part to you, do you expound it, and I will hear what you say.

No, said *Gaius*, 'twas put to you, and 'tis expected that you should answer it.

Then said the old Gentleman,

> *He first by Grace must conquer'd be,*
> *That Sin would mortifie.*
> *And who, that lives, would convince me,*
> *Unto himself must die.*

It is right, said *Gaius*; good Doctrine, and Experience teaches this. For first, until Grace displays it self, and overcomes the Soul with its Glory, it is altogether without Heart to oppose Sin. Besides, if Sin is Satan's Cords, by which the Soul lies bound, how should it make Resistance, before it is loosed from that Infirmity?

Secondly, Nor will any that knows either Reason or Grace, believe that such a man can be a living Monument of Grace, that is a Slave to his own Corruptions.

And now it comes in my mind, I will tell you a Story, worth the hearing. There were two Men that went on Pilgrimage, the one began when he was young, the other when he was old. The young man had strong Corruptions to grapple with, the old mans were decayed with the decays of Nature. The young man trod his steps as even as did the old one, and was every way as light as he; who now, or which of them had their Graces shining clearest, since both seemed to be alike? *A Question worth the minding.*

Honest. *The young mans doubtless. For that which heads it against the greatest Opposition, gives best demonstration that it is strongest. Specially when it also holdeth pace with that that meets not with half so much: as to be sure old Age does not.* *A Comparison.*

Besides, I have observed that old men have blessed themselves with this mistake; Namely, taking the decays of Nature for a gracious Conquest over Corruptions, and so have been apt to beguile themselves. Indeed old men that are gracious, are best able to give Advice to them that are young, because they have seen most of the emptiness of things. But yet, for an old and a young to set out both together, the young one has the advantage of the fairest discovery of a work of Grace within him, tho the old mans Corruptions are naturally the weakest. *A Mistake.*

Thus they sat talking till break of Day. Now when the Family was up, *Christiana* bid her Son *James* that he should read a Chapter; so he read the 53 of *Isaiah*. When he had done, Mr. *Honest* asked why it was said, *That the Savior is said to come out of a dry ground, and also that he had no Form nor Comliness in him?* *Another Question*

Greath. Then said Mr. *Great-heart*, To the first I answer, Because, the Church of the Jews, of which Christ came, had then lost almost all the Sap and Spirit of Religion. To the Second I say, The Words are spoken in the Person of the Unbelievers, who because they want *that* Eye that can see into our Princes Heart, therefore they judg of him by the meanness of his Outside.

Just like those that know not that precious Stones are covered over with a homely *Crust*; who when they have found one, because they know not what they have found, cast it again away as men do a common Stone.

Well, said *Gaius*, Now you are here, and since, as I know, Mr. *Great-heart* is good at his Weapons, if you please, after we have refreshed our selves, we will walk into the Fields, to see if we can do any good. About a mile from hence, there is one *Slaygood*,[†] a *Gyant*, that doth much annoy the Kings High-way in these parts. And I know whereabout his Haunt is, he is Master of a number of Thieves; 'twould be well if we could clear these Parts of him.

So they consented and went, Mr. *Great-heart* with his *Sword*, *Helmet* and *Shield*, and the rest with Spears and Staves.

When they came to the place where he was, they found him with one *Feeble-mind* in his Hands, whom his Servants had brought unto him, having taken him in the Way; now the Gyant was rifling of him, with a purpose after that to pick his Bones. For he was of the nature of *Flesh-eaters*.

Well, so soon as he saw Mr. *Great-heart*, and his Friends, at the mouth of his Cave with their Weapons, he demanded what they wanted?

Greath. We want thee; for we are come to revenge the Quarrel of the many that thou hast slain of the Pilgrims, when thou hast dragged them out of the Kings High-way; wherefore come out of thy Cave. So he armed himself and came out, and to a Battle they went, and fought for above an Hour, and then stood still to take Wind.

Slaygood. *Then said the Gyant, Why are you here on my Ground?*

Greath. To revenge the Blood of Pilgrims, as I also told thee before; so they went to it again, and the Gyant made Mr. *Great-heart* give back, but he came up again, and in the greatness of his Mind, he let fly with such stoutness at the Gyants Head and Sides, that he made him let his Weapon fall out of his Hand. So he smote him, and slew him, and cut off his Head, and brought it away to the *Inn*. He also took *Feeble-mind* the Pilgrim, and brought him with him to his Lodgings. When they were come home, they shewed his Head to the Family, and then set it up as they had done others before, for a Terror to those that should attempt to do as he, hereafter.

Then they asked Mr. *Feeble-mind* how he fell into his hands?

Gyant Slay-good assaulted and slain.

He is found with one Feeble-mind in his hands.

One Feeble-mind rescued from the Gyant.

Feeblem. Then said the poor man, I am a sickly man, as you see, and because *Death* did usually once a day *knock at my Door*, I thought I should never be well at home. So I betook my self to a Pilgrims life; and have travelled hither from the Town of *Uncertain*, where I and my Father were born. I am a man of no strength at all, of Body, nor yet of Mind, but would, if I could, tho I can but *craul*, spend my Life in the Pilgrims way. When I came at the Gate that is at the head of the Way, the Lord of that place did entertain me freely. Neither objected he against my weakly Looks, nor against my *feeble Mind*; but gave me such things that were necessary for my Journey, and bid me hope to the end. When I came to the House of the *Interpreter*, I received much Kindness there, and because the Hill *Difficulty* was judged too hard for me, I was carried up that by one of his Servants. Indeed I have found much Relief from Pilgrims, tho none was willing to go so softly as I am forced to do. Yet still as they came on, they bid me be of good Chear, and said that it was the will of their Lord, that Comfort should be given to the *feeble minded*, and so went on their *own* pace. When I was come up to *Assault-Lane*, then this *Gyant* met with me, and bid me prepare for an *Incounter*; but alas, feeble one that I was, I had more need of a *Cordial*. So he came up and took me, I conceited he should not kill me; also when he had got me into his Den, since I went not with him *willingly*, I believed I should come out alive again. For I have heard, that not any Pilgrim that is taken Captive by violent Hands, if he keeps Heart-whole towards his Master, is by the Laws of Providence to die by the Hand of the Enemy. *Robbed*, I looked to be, and Robbed to be sure I am; but I am as you see escaped with Life, for the which I thank my King as Author, and you as the Means. Other Brunts I also look for, but this I have resolved on, to wit, to *run* when I can, to *go* when I cannot *run*, and to *creep* when I cannot *go*. As to the main, I thank him that loves me, I am fixed; my way is before me, my Mind is beyond the *River* that has no Bridg, tho I am as you see, but of a *feeble Mind*.

How Feeble-mind came to be a Pilgrim.

1 Thess. 5. 14.

Mark this.

Mark this.

Hon. Then said old *Mr.* Honest, *Have not you some time ago, been acquainted with one Mr. Fearing, a Pilgrim?*

Mr. Fearing. Mr. Feeble-mind's Uncle.

Feeble. Acquainted with him? Yes. He came from the Town of *Stupidity*, which lieth *four Degrees* to the Northward of the City of

Destruction, and as many off, of where I was born; Yet we were well acquainted, for indeed he was mine Uncle, my Fathers Brother; he and I have been much of a Temper, he was a little shorter then I, but yet we were much of a Complexion.

Feeble-mind has some of Mr. Fearing's Features.

Hon. *I perceive you knew him, and I am apt to believe also that you were related one to another; for you have his whitely Look, a Cast like his with your Eye, and your Speech is much alike.*

Feebl. Most have said so, that have known us both, and besides, what I have read in him, I have for the most part found in my self.

Gaius Comforts him.

Gaius. *Come Sir, said good Gaius, be of good Chear, you are welcome to me, and to my House; and what thou hast a mind to, call for freely; and what thou would'st have my Servants do for thee, they will do it with a ready Mind.*

Notice to be taken of Providence.

Feebl. Then said Mr. *Feeble-mind*, This is unexpected Favour, and as the Sun shining out of a very dark Cloud. Did Gyant *Slay-good* intend me this Favour when he stop'd me, and resolved to let me go no further? Did he intend that after he had rifled my Pockets, I should go to *Gaius mine Host?* Yet so it is.

Tidings how one Not-right was slain with a Thunder-bolt, and Mr. Feeble-mind's Comment upon it.

Now, just as Mr. *Feeble-mind*, and *Gaius* was thus in talk; there comes one running, and called at the Door, and told, That about a Mile and an half off, there was one Mr. *Not-right* a Pilgrim, struck dead upon the place where he was, with a *Thunder-bolt*.

Feebl. Alas! said Mr. *Feeble-mind*, is he slain? he overtook me some days before I came so far as hither, and would be my Company-keeper. He also was with me when *Slay-good* the Gyant took me, but he was nimble of his Heels, and escaped. But it seems, he escaped to die, and I was took to live.

> *What, one would think, doth seek to slay* outright,
> *Ofttimes, delivers from the saddest* Plight.
> *That very* Providence, *whose Face is* Death,
> *Doth ofttimes, to the lowly,* Life bequeath.
> *I taken was, he did escape and flee,*
> *Hands Crost, gives Death to him, and Life to me.*

Now about this time *Mathew* and *Mercie* was Married; also *Gaius* gave his Daughter *Phebe* to *James*, *Mathew*'s Brother, to Wife; after which time, they yet stayed above ten days at *Gaius*'s House, spending their time, and the Seasons, like as Pilgrims use to do.

When they were to depart, *Gaius* made them a Feast, and they did eat and drink, and were merry. Now the Hour was come that they must be gon, wherefore Mr. *Great-heart* called for a Reckoning. But *Gaius* told him, that at his House, it was not the Custom for *Pilgrims* to pay for their Entertainment. He boarded them by the year, but looked for his pay from the good *Samaritan*, who had promised him at his return, whatsoever Charge he was at with them, faithfully to repay him. Then said Mr. *Great-heart* to him,

The Pilgrims prepare to go forward.

Luke 10. 33, 34, 35. How they greet one another at parting. 3 John 6.

Greath. Beloved, thou dost faithfully, whatsoever thou dost, to the Brethren and to Strangers, which have born Witness of thy Charity before the Church. Whom if thou (yet) bring forward on their Journey after a Godly sort, thou shalt do well.

Then *Gaius* took his leave of them all, and of his Children, and particularly of Mr. *Feeble-mind.* He also gave him something to drink by the way.

Gaius his last kindness to Feeble-mind.

Now Mr. *Feeble-mind*, when they were going out of the Door, made as if he intended to linger. The which, when Mr. *Great-heart* espied, he said, come Mr. *Feeble-mind*, pray do you go along with us, I will be your *Conductor*, and you shall fare as the rest.

Feebl. Alas, I want a sutable Companion, you are all lusty and strong, but I, as you see, am weak; I chuse therefore rather to come behind, lest, by reason of my many Infirmities, I should be both a Burthen to my self, and to you. I am, as I said, a man of a weak and feeble Mind, and shall be offended and made weak at that which others can bear. I shall like no Laughing, I shall like no gay Attire, I shall like no unprofitable Questions. Nay, I am so weak a Man, as to be offended with that which others have a liberty to do: I do not yet know all the Truth; I am a very ignorant Christian-man; *sometimes if I hear some rejoyce in the Lord, it troubles me because I cannot do so too. It is with me, as it is with a weak Man among the strong, or as with a sick Man among the healthy, or as a Lamp despised. (He that is ready to slip with his Feet, is as a Lamp despised, in the Thought of him that is at ease.) So that I know not what to do.*

Feeble-mind for going behind.

His Excuse for it.

Job 12. 5.

Greath. But Brother, said Mr. *Great-heart.* I have it in Commission, to comfort the *feeble minded*, and to support the weak. You must needs go along with us; we will wait for you, we will lend you our

Great-heart's Commission. 1 Thess. 5. 14.

Rom. 14. help, we will deny our selves of some things, both *Opinionative* and
1 Cor. 8. Practical, for your sake; we will not enter into doubtful Disputations
chap. 9. 22. before you, we will be made all things to you, rather then you shall
A Christian be left behind.
Spirit.

Now, all this while they were at *Gaius*'s Door; and behold as they
Psa. 38. 17. were thus in the heat of their Discourse, Mr. *Ready-to-hault* came
Promises. by, with his *Crutches* in his hand, and he also was going on Pil-
grimage.

Feeble-   Feebl. *Then said Mr.* Feeble-mind *to him, Man! how camest thou*
mind glad *hither? I was but just now complaining that I had not a sutable Companion,*
to see
Ready-to- *but thou art according to my Wish. Welcome, welcome, good Mr.* Ready-
hault *to-hault, I hope thee and I may be some help.*
come by.

Ready-to. I shall be glad of thy Company, said the other; and good
Mr. *Feeble-mind*, rather then we will part, since we are thus happily
met, I will lend thee one of my Crutches.

Feebl. *Nay, said he, tho I thank thee for thy good Will, I am not inclined
to hault before I am Lame. How be it, I think when occasion is, it may help
me against a Dog.*

Ready-to. If either my *self*, or my *Crutches*, can do thee a pleasure,
we are both at thy Command, good Mr. *Feeble-mind.*

Thus therefore they went on, Mr. *Great-heart* and Mr. *Honest* went
before, *Christiana* and her Children went next, and Mr. *Feeble-mind*
and Mr. *Ready-to-hault* came behind with his Crutches. Then said
Mr. *Honest*,

New Talk. Hon. *Pray Sir, now we are upon the Road, tell us some profitable things
of some that have gon on Pilgrimage before us.*

Greath. With a good Will. I suppose you have heard how *Christian*
of old, did meet with *Apollyon* in the Valley of *Humiliation*, and also
what hard work he had to go thorow the Valley of the Shadow of
Death. Also I think you cannot but have heard how *Faithful* was
1 Part, pag. put to it with *Madam Wanton*, with *Adam* the first, with one *Dis-*
56–61. *content*, and *Shame*; four as deceitful Villains, as a man can meet with
upon the Road.

Hon. *Yes, I have heard of all this; but indeed, good* Faithful *was hardest
put to it with* Shame, *he was an unwearied one.*

Greath. Ai, for as the Pilgrim well said, He of all men had the
wrong Name.

Hon. But pray Sir, where was it that Christian *and* Faithful *met* Talkative? *that same was also a notable one.*

Greath. He was a confident Fool, yet many follow his wayes.

Hon. He had like to a beguiled Faithful?

Greath. Ai, But *Christian* put him into a way quickly to find him out. Thus they went on till they came at the place where *Evangelist* met with *Christian* and *Faithful*, and prophecyed to them of what should befall them at Vanity-Fair. ⟨*1 Part, pag. 70–2.*⟩

Greath. Then said their Guide, Hereabouts did *Christian* and *Faithful* meet with *Evangelist*, who Prophesied to them of what Troubles they should meet with at *Vanity-Fair.*

Hon. Say you so! I dare say it was a hard Chapter that then he did read unto them.

Greath. 'Twas so, but he gave them Incouragement withall. But what do we talk of them, they were a couple of Lyon-like Men; they had set their Faces like Flint. Don't you remember how undaunted they were when they stood before the Judg? ⟨*1 Part, pag. 76–80.*⟩

Hon. Well Faithful, *bravely suffered!*

Greath. So he did, and as brave things came on't: For *Hopeful* and some others, as the Story relates it, were Converted by his Death.

Hon. Well, but pray go on; for you are well acquainted with things.

Greath. Above all that *Christian* met with after he had passed throw *Vanity-Fair*, one *By-ends* was the arch one. ⟨*1 Part, pag. 80.*⟩

Hon. By-ends; what was he?

Greath. A very arch Fellow, a downright Hypocrite; one that would be Religious, which way ever the World went, but so cunning, that he would be sure neither to lose, nor suffer for it.

He had his *Mode* of Religion for every fresh occasion, and his Wife was as good at it as he. He would turn and change from Opinion to Opinion; yea, and plead for so doing too. But so far as I could learn, he came to an ill End with his *By-ends*, nor did I ever hear that any of his Children were ever of any Esteem with any that truly feared God.

Now by this time, they were come within sight of the Town of *Vanity*, where Vanity Fair is kept. So when they saw that they were so near the Town, they consulted with one another how they should pass thorow the Town, and some said one thing, and some another. At last Mr. *Greatheart* said, I have, as you may understand, often ⟨*They are come within sight of Vanity. Psa. 12. 2.*⟩

They enter into one Mr. Mnasons to Lodg. been a *Conductor* of Pilgrims thorow *this* Town; Now I am acquainted with one Mr. *Mnason*,† a *Cyprusian* by Nation, an old Disciple, at whose House we may Lodg. If you think good, said he, we will turn in there.

Content, said old *Honest*; Content, said *Christiana*; Content, said Mr. *Feeble-mind*; and so they said all. Now you must think it was *Even-tide*, by that they got to the outside of the Town, but Mr. *Great-heart* knew the way to the Old man's House. So thither they came; and he called at the Door, and the old Man within knew his Tongue so soon as ever he heard it; so he opened, and they all came in. Then said *Mnason* their Host, How far have ye come to day? So they said, From the House of *Gaius* our Friend. I promise you, said he, you have gone a good stitch, you may well be a-weary; sit down. So they sat down.

Greath. Then said their Guide, Come what Chear Sirs, I dare say you are welcome to my Friend.

They are glad of entertain-ment. *Mna.* I also, said Mr. *Mnason*, do bid you Welcome; and whatever you want, do but say, and we will do what we can to get it for you.

Hon. Our great Want, a while since, was Harbor, and good Company, and now I hope we have both.

Mna. For Harbour, you see what it is, but for good Company, that will appear in the Tryal.

Greath. Well, said Mr. Great-heart, will you have the Pilgrims up into their Lodging?

Mna. I will, said Mr. *Mnason*. So he had them to their respective Places; and also shewed them a very fair Dining-Room, where they might be and sup together, until time was come to go to Rest.

Now when they were set in their places, and were a little cheary after their Journey, Mr. *Honest* asked his Landlord if there were any store of good People in the Town?

Mna. We have a few, for indeed they are but a few, when compared with them on the other side.

They desire to see some of the good People in the Town. *Hon.* But how shall we do to see some of them? for the sight of good men to them that are going on Pilgrimage, is like to the appearing of the Moon and the Stars to them that are sailing upon the Seas.

Mna. Then Mr. *Mnason* stamped with his Foot, and his Daughter *Grace* came up; so he said unto her, *Grace*, go you, tell my Friends, Mr. *Contrite*, Mr. *Holy-man*, Mr. *Love-saint*, Mr. *Dare-not-ly*, and Mr. *Penitent*; that I have a Friend or two at my House, that have a mind this Evening to see them. *Some sent for.*

So *Grace* went to call them, and they came, and after Salutation made, they sat down together at the Table.

Then said Mr. *Mnason* their Landlord, My Neighbours, I have, as you see, a company of *Strangers* come to my House, they are *Pilgrims*: They come from afar, and are going to Mount *Sion*. But who, quoth he, do you think this is? pointing with his Finger to *Christiana*. It is *Christiana*, the Wife of *Christian*, that famous Pilgrim, who with *Faithful* his brother were so shamefully handled in our Town. At that they stood amazed, saying, We little thought to see *Christiana*, when *Grace* came to call us, wherefore this is a very comfortable Surprize. Then they asked her of her welfare, and if these young men were her Husbands Sons. And when she had told them they were; they said, The King whom you love, and serve, make you as your Father, and bring you where he is in Peace.

Hon. Then Mr. *Honest* (*when they were all sat down*) asked Mr. *Contrite and the rest, in what posture their Town was at present?* *Some Talk betwixt Mr. Honest and Contrite.*

Cont. You may be sure we are full of Hurry, in Fair time. *'Tis hard keeping our Hearts and Spirits in any good Order, when we are in a cumbred condition. He that lives in such a place as this is, and that has to do with such as we have, has need of an Item to caution him to take heed, every moment of the Day. *• The Fruit of Watchfulness.*

Hon. But how are your Neighbors for quietness?

Cont. They are much more moderate now then formerly. You know how *Christian* and *Faithful* were used at our Town; but of late, I say, they have been far more moderate. I think the Blood of *Faithful* lieth with load upon them till now; for since they burned him, they have been ashamed to burn any more: In *those* days we were afraid to walk the Streets, but *now* we can shew our Heads. *Then* the Name of a Professor was odious, *now*, specially in some parts of our Town (for you know our Town is large) Religion is counted Honourable. *Persecution not so hot at Vanity Fair as formerly.*

Then said Mr. Contrite *to them, Pray how faireth it with you in your Pilgrimage, how stands the Country affected towards you?*

Hon. It happens to us, as it happeneth to Way-fairing men; sometimes our way is clean, sometimes foul; sometimes up-hill, sometimes down-hill; We are seldom at a Certainty. The Wind is not alwayes on our Backs, nor is every one a Friend that we meet with in the Way. We have met with some notable Rubs already; and what are yet behind we know not, but for the most part we find it true, that has been talked of of old, *A good Man must suffer Trouble.*

Contrit. *You talk of Rubs, what Rubs have you met withal?*

Hon. Nay, ask Mr. *Great-heart* our Guide, for he can give the best Account of that.

Greath. We have been beset three or four times already: First *Christiana* and her Children were beset with two Ruffians, that they feared would a took away their Lives; We was beset with Gyant *Bloody-man,* Gyant *Maul,* and Gyant *Slay-good.* Indeed we did rather beset the last, then were beset of him: And thus it was. After we had been some time at the House of *Gaius, mine Host, and of the whole Church,* we were minded upon a time to take our Weapons with us, and go see if we could light upon any of those that were Enemies to Pilgrims; (for we heard that there was a notable one thereabouts.) Now *Gaius* knew his *Haunt* better than I, because he dwelt thereabout, so we looked and looked, till at last we discerned the mouth of his Cave; then we were glad and pluck'd up our Spirits. So we approached up to his *Den,* and lo when we came there, he had dragged by meer force into his Net, this *poor man,* Mr. *Feeble-mind,* and was about to bring him to his End. But when he saw us, supposing as we thought, he had had an other Prey, he left the poor man in his Hole, and came out. So we fell to it full sore, and he lustily laid about him; but in conclusion, he was brought down to the Ground, and his Head cut off, and set up by the Way side for a Terror to such as should after practise such Ungodliness. That I tell you the Truth, here is the man himself to affirm it, who was as a Lamb taken out of the Mouth of the Lyon.

Feebl. Then said Mr. Feeble-mind, *I found this true to my Cost, and Comfort; to my Cost, when he threatned to pick my Bones every moment; and to my Comfort, when I saw Mr. Great-heart and his Friends with their Weapons approach so near for my Deliverance.*

Holym. Then said Mr. *Holy-man,* There are two things that they have need to be possessed with that go on Pilgrimage, *Courage* and an *unspotted Life*. If they have not *Courage,* they can never hold on their way; and if their Lives be *loose,* they will make the very Name of a *Pilgrim* stink.

Mr. Holy-man's Speech.

Loves. Then said Mr. *Love-saint;* I hope this Caution is not needful amongst you. But truly there are many that go upon the Road, that rather declare themselves Strangers to Pilgrimage, then Strangers and Pilgrims in the Earth.[†]

Mr. Love-saint's Speech.

Darenot. Then said Mr. *Dare-not-ly,* '*Tis true; they neither have the Pilgrims Weed, nor the Pilgrims Courage; they go not uprightly, but all awrie with their Feet, one Shoo goes* inward, *an other* outward, *and their Hosen out behind;* there a *Rag,* and there a *Rent,* to the Disparagement of their Lord.

Mr. Dare-not-ly his Speech.

Penit. These things, said Mr. *Penitent,* they ought to be troubled for, nor are the Pilgrims like to have that Grace put upon them and their pilgrims Progress, as they desire, until the way is cleared of such Spots and Blemishes.

Mr. Penitent his Speech.

Thus they sat talking and spending the time, until Supper was set upon the Table. Unto which they went and refreshed their weary Bodys, so they went to Rest. Now they stayed in this Fair a great while, at the House of this Mr. *Mnason,* who in process of time gave his Daughter *Grace* unto *Samuel, Christiana's* Son, to Wife, and his Daughter *Martha* to *Joseph.*

The time, as I said, that they lay here, was long (for it was not now as in former times). Wherefore the *Pilgrims* grew acquainted with many of the good people of the Town, and did them what Service they could. *Mercie,* as she was wont, laboured much for the Poor, wherefore their Bellys and Backs blessed her, and she was there an Ornament to her Profession. And to say the truth, for *Grace, Phebe,* and *Martha,* they were all of a very good Nature, and did much good in their place. They were also all of them very Fruitful, so that *Christian's* Name, as was said before, was like to live in the World.

While they lay here, there came a *Monster*[†] out of the Woods, and

A Monster.

slew many of the People of the Town. It would also carry away their Children, and teach them to suck its Whelps. Now no man in the Town durst so much as Face this *Monster;* but all Men fled when they heard of the noise of his coming.

Rev. 17. 3. *His Shape. His Nature.* The *Monster* was like unto no one Beast upon the Earth. Its Body was like a Dragon, and it had seven Heads and ten Horns, *It made great havock of Children, and yet it was governed by a Woman.* This *Monster* propounded Conditions to men; and such men as loved their Lives more then their Souls, accepted of those Conditions. So they came under.

Now this Mr. *Great-heart*, together with these that came to visit the Pilgrims at Mr. *Mnason*'s House, entred into a Covenant to go and ingage this Beast, if perhaps they might deliver the People of this Town, from the Paws and Mouths of this so devouring a Serpent.

How he is ingaged. Then did Mr. *Great-heart*, Mr. *Contrite*, Mr. *Holyman*, Mr. *Dare-not-ly*, and Mr. *Penitent*, with their Weapons go forth to meet him. Now the *Monster* at first was very Rampant, and looked upon these Enemies with great Disdain, but they so be-labored him, being sturdy men at Arms, that they made him make a Retreat: so they came home to Mr. *Mnasons* House again.

The *Monster*, you must know, had his certain Seasons to come out in, and to make his Attempts upon the Children of the People of the Town, also these Seasons did these valiant Worthies watch him in, and did still continually assault him, in so much, that in process of time, he became not only wounded, but lame; also he has not made that havock of the Towns mens Children, as formerly he has done. And it is verily believed by some, that this Beast will die of his Wounds.

This therefore made Mr. *Great-heart* and his Fellows, of great Fame in this Town, so that many of the People that wanted their tast of things, yet had a Reverend Esteem and Respect for them. Upon this account therefore it was that these Pilgrims got not much hurt here. True, there were some of the baser sort that could see no more then a *Mole*, nor understand more then a Beast, these had no reverence for these men, nor took they notice of their Valour or Adventures.

Well, the time drew on that the Pilgrims must go on their way, wherefore they prepared for their Journey. They sent for their Friends, they conferred with them, they had some time set apart therein to commit each other to the Protection of their Prince. There was again, that brought them of such things as they had, that was fit for the weak, and the strong, for the Women, and the men; and so *laded* them with such things as was necessary.

Act. 28. 10.

Then they set forwards on their way, and their Friends accompanying them so far as was convenient; they again committed each other to the Protection of their King, and parted.

They therefore that were of the Pilgrims Company went on, and Mr. *Great-heart* went before them; now the Women and Children being weakly, they were forced to go as they could bear, by this means Mr. *Ready-to-hault* and Mr. *Feeble-mind* had more to sympathize with their Condition.

When they were gone from the Towns-men, and when their Friends had bid them farewel, they quickly came to the place where *Faithful* was put to Death: There therefore they made a stand, and thanked him that had enabled him to bear his Cross so well, and the rather, because they now found that they had a benefit by such a manly Suffering as his was.

They went on therefore after this, a good way further, talking of *Christian* and *Faithful*, and how *Hopeful* joyned himself to *Christian* after that *Faithful* was dead.

Now they were come up with the *Hill Lucre*, where the *Silver-mine* was, which took *Demas* off from his Pilgrimage, and into which, as some think, *By-ends* fell and perished; wherefore they considered that. But when they were come to the old Monument that stood over against the *Hill Lucre*, to wit, to the Pillar of Salt that stood also within view of *Sodom*, and its stinking Lake; they marvelled, as did *Christian* before, that men of that Knowledg and ripeness of Wit as they was, should be so blinded as to turn aside here. Only they considered again, that Nature is not affected with the Harms that others have met with, specially if that thing upon which they look, has an attracting Virtue upon the foolish Eye.

1 *Part, pag.* 88.

I saw now that they went on till they came at the River that was on this side of the Delectable Mountains. To the River where the

1 *Part, pag.* 90.

fine Trees grow on both sides, and whose Leaves, if taken inwardly,
are good against Surfits; where the Medows are green all the year
long, and where they might lie down safely.

By this River side in the Medow, there were Cotes and Folds for
Sheep, an House built for the *nourishing* and bringing up of those
Lambs, the Babes of those Women that go on Pilgrimage. Also there
was here one that was intrusted with them, who could have com-
passion, and that could gather these Lambs with his Arm, and carry
them in his Bosom, and that could gently lead those that were with
young. Now to the Care of *this Man, Christiana* admonished her four
Daughters to commit their little ones; that by these Waters they
might be housed, harbored, suckered and nourished, and that none
of them might *be lacking in time to come.* This man, if any of them go
astray, or be lost, he will bring them again, he will also bind up that
which was broken, and will strengthen them that are sick. Here
they will never want Meat, and Drink and Cloathing, here they
will be kept from Thieves and Robbers, for this man will dye before
one of those committed to his Trust, shall be lost. Besides, here they
shall be sure to have good *Nurture* and Admonition, and shall be
taught to walk in right Paths, and that you know is a Favour of no
small account. Also here, as you see, are delicate *Waters,* pleasant
Medows, dainty *Flowers,* variety of *Trees,* and such as bear *wholsom
Fruit.* Fruit, not like that that *Matthew* eat of, that fell over the Wall
out of *Belzebubs* Garden, but Fruit that procureth Health where
there is none, and that continueth and increaseth it where it is.

So they were content to commit their little Ones to him; and that
which was also an Incouragement to them so to do, was, for that
all this was to be at the Charge of the King, and so was an Hospital
to young Children, and *Orphans.*

Now they went on: And when they were come to *By-path* Medow,
to the Stile over which *Christian* went with his Fellow *Hopeful,* when
they were taken by *Gyant-dispair,* and put into *doubting*-Castle, they
sat down and consulted what was best to be done, to wit, now they
were so strong, and had got such a man as Mr. *Great-heart* for their
Conductor; whether they had not best to make an Attempt upon
the Gyant, demolish his Castle, and if there were any Pilgrims in it,
to set them at liberty before they went any further. So one said one

---

Psal. 23.

Heb. 5. 2.
Isa. 40. 11.

Jer. 23. 4.
Ezek. 34.
11,12,13,
14,15,16.

John 10.
16.

They being
come to By-
path Stile,
have a mind
to have a
pluck with
Gyant
Dispair.
1 Part, pag.
93-7.

thing, and an other said the contrary. One questioned if it was lawful to go upon *unconsecrated* Ground, an other said they might, provided their end was good; but Mr. *Great-heart* said, Though that Assertion offered last, cannot be universally true, yet I have a Comandment to resist Sin, to overcome Evil, to fight the good Fight of Faith. And I pray, with whom should I fight this good Fight, if not with *Gyant-dispair*? I will therefore attempt the taking away of his Life, and the demolishing of *Doubting* Castle. Then said he, who will go with me? Then said old *Honest*, I will, And so will we too, said *Christian*'s four Sons, *Mathew*, *Samuel*, *James* and *Joseph*, for they were young men and strong.

1 John 2. 13, 14.

So they left the Women in the Road, and with them Mr. *Feeble-mind*, and Mr. *Ready-to-halt*, with his Crutches, to be their *Guard*, until they came back, for in that place tho *Gyant-Dispair* dwelt so near, they keeping in the Road, *A little Child might lead them.*

Isa. 11. 6.

So Mr. *Great-heart*, old *Honest*, and the four young men, went to go up to *Doubting* Castle, to look for *Gyant-Dispair*. When they came at the Castle Gate, they knocked for Entrance with an unusual Noyse. At that the old Gyant comes to the Gate, and *Diffidence* his Wife follows. Then said he, Who, and what is he, that is so hardy, as after this manner to molest the *Gyant-Dispair*? Mr. *Great-heart* replyed, It is I, *Great-heart*, one of the King of the Celestial Countries Conductors of Pilgrims to their Place. And I demand of thee that thou open thy Gates for my Entrance, prepare thy self also to Fight, for I am come to take away thy Head, and to demolish *Doubting* Castle.

Now *Gyant-Dispair*, because he was a *Gyant*, thought no man could overcome him, and again, thought he, since heretofore I have made a Conquest of Angels, shall *Great-heart* make me afraid? So he harnessed himself and went out. He had a Cap of Steel upon his Head, a Brestplate of Fire girded to him, and he came out in Iron Shoos, with a great Club in his Hand. Then these six men made up to him, and beset him behind and before; also when *Diffidence*, the *Gyantess*, came up to help him, old Mr. *Honest* cut her down at one Blow. Then they fought for their Lives, and *Gyant-Dispair* was brought down to the Ground, *but was very loth to die.* He strugled hard, and had, as they say, as many Lives as a Cat, but *Great-heart*

Despair has overcome Angels.

Despair is loth to die.

was his death, for he left him not till he had severed his head from his shoulders.

Then they fell to demolishing *Doubting* Castle, and that you know might with ease be done, since *Gyant-Dispair* was dead. They were seven Days in destroying of that; and in it of Pilgrims, they found one Mr. *Dispondencie*, almost starved to Death, and one *Much-afraid* his Daughter; these two they saved alive. But it would a made you a wondered to have seen the dead Bodies that lay here and there in the Castle Yard, and how full of dead mens Bones the Dungeon was.

When Mr. *Great-heart* and his Companions had performed this Exploit, they took Mr. *Despondencie*, and his Daughter *Much-afraid*, into their Protection, for they were honest People, tho they were Prisoners in *Doubting* Castle, to that Tyrant *Gyant-Dispair*. They therefore I say, took with them the Head of the *Gyant* (for his Body they had buried under a heap of Stones) and down to the Road and to their Companions they came, and shewed them what they had done. Now when *Feeble-mind*, and *Ready-to-hault* saw that it was the Head of *Gyant-Dispair* indeed, they were very jocond and merry. Now *Christiana*, if need was, could play upon the *Vial*, and her Daughter *Mercie* upon the *Lute:* So, since they were so merry disposed, she plaid them a Lesson, and *Ready-to-halt* would Dance. So he took *Dispondencie*'s Daughter, named *Much-afraid*, by the Hand, and to dancing they went in the Road. True, he could not Dance without one Crutch in his Hand, but I promise you, he footed it well; also the Girl was to be commended, for she answered the Musick handsomely.

As for Mr. *Despondencie*, the Musick was not much to him, he was for feeding rather then dancing, for that he was almost starved. So *Christiana* gave him some of her bottle of Spirits for present Relief, and then prepared him something to eat; and in little time the old Gentleman came to himself, and began to be finely revived.

Now I saw in my Dream, when all these things were finished, Mr. *Great-heart* took the Head of *Gyant-Dispair*, and set it upon a Pole by the Highway side, right over against the Piller that *Christian* erected for a *Caution* to Pilgrims that came after, to take heed of entering into his Grounds.

Doubting-Castle demolished.

They have Musick and dancing for joy.

Then he writ under it upon a *Marble* stone, these Verses following.

> *This is the* Head *of him, whose* Name *only,*
> *In former times, did* Pilgrims *terrify.*
> *His* Castle's *down, and* Diffidence *his Wife,*
> *Brave Master* Great-heart *has bereft of Life.*
> Dispondencie, *his Daughter* Much-afraid,
> Great-heart, *for them, also the Man has plaid.*
> *Who hereof doubts, if he'l but cast his Eye,*
> *Up hither, may his Scruples satisfy,*
> *This* Head, *also when doubting Cripples dance,*
> *Doth shew from Fears they have Deliverance.*

A Monument of Deliverance.

When these men had thus bravely shewed themselves against *Doubting-Castle*, and had slain *Gyant-Dispair*, they went forward, and went on till they came to the *Delectable* Mountains, where *Christian* and *Hopeful* refreshed themselves with the Varieties of the Place. They also acquainted themselves with the Shepherds there, who welcomed them as they had done *Christian* before, unto the Delectable Mountains.

Now the Shepherds seeing so great a train follow Mr. *Great-heart* (for with him they were well acquainted;) they said unto him, Good Sir, you have got a goodly Company here; pray where did you find all these?

Then Mr. *Great-heart* replyed,

> *First here's* Christiana *and her train,*
> *Her Sons, and her Sons Wives, who like the Wain*
> *Keep by the Pole, and do by Compass stere,*
> *From Sin to Grace, else they had not been here.*
> *Next here's old* Honest *come on Pilgrimage,*
> Ready-to-halt *too, who I dare ingage,*
> *True hearted is, and so is* Feeble-mind,
> *Who willing was, not to be left behind.*
> Dispondencie, *good-man, is coming after,*
> *And so also is* Much-afraid, *his Daughter.*
> *May we have Entertainment here, or must*
> *We further go? let's know whereon to trust.*

The Guides Speech to the Shepherds.

*. Their
Entertain-
ment.*
Then said the Shepherds; This is a comfortable Company; you are welcome to us, for we have for the *Feeble*, as for the *Strong*; our
Matt. 25.
40.
Prince has an Eye to what is done to the least of these. Therefore Infirmity must not be a block to our Entertainment. So they had them to the Palace Door, and then said unto them, Come in Mr. *Feeble-mind*, come in Mr. *Ready-to-halt*, come in Mr. *Dispondencie*, and Mrs. *Much-afraid* his Daughter. *These* Mr. *Great-heart*, said the Shepherds to the Guide, we call in by Name, for that they are most subject to draw back; but as for you, and the rest that are *strong*, we leave you to your wonted Liberty. Then said Mr. *Great-heart*,
*A Descrip-
tion of false
Shepherds.*
Ezek. 34.
21.
This day I see that Grace doth shine in your Faces, and that you are my Lords Shepherds indeed; for that you have not *pushed* these Diseased neither with Side nor Shoulder, but have rather strewed their way into the Palace with Flowers, as you should.

So the Feeble and Weak went in, and Mr. *Great-heart*, and the rest did follow. When they were also set down, the Shepherds said to those of the weakest sort, What is it that you would have? For said they, all things must be managed here, to the supporting of the weak, as well as to the warning of the Unruly.

So they made them a Feast of things easy of Digestion, and that were pleasant to the Palate, and nourishing; the which when they had received, they went to their rest, each one respectively unto his proper place. When Morning was come, because the Mountains were high, and the day clear; and because it was the Custom of the Shepherds to shew to the Pilgrims, before their Departure, some Rarities; therefore after they were ready, and had refreshed themselves, the Shepherds took them out into the Fields, and shewed them first, what they had shewed to *Christian* before.

*Mount-
Marvel.*
1 Part, pag.
98.
Then they had them to some new places. The first was to *Mount-Marvel*, where they looked, and behold a man at a Distance, *that tumbled the Hills about with Words.* Then they asked the Shepherds what that should mean? So they told him, that that man was the Son of one *Great-grace*, of whom you read in the first part of the Records of the *Pilgrims Progress.* And he is set there to teach Pil-
Mar. 11.
23, 24.
grims how to believe down, or to tumble out of their ways, what Difficulties they shall meet with, by faith. Then said Mr. *Great-heart*, I know him, he is a man above many.

Then they had them to another place, called *Mount-Innocent*. And *Mount-Innocent.*
there they saw a man cloathed all in White; and two men, *Prejudice,*
and *Ill-will*, continually casting Dirt upon him. Now behold the
Dirt, whatsoever they cast at him, would in little time fall off
again, and his Garment would look as clear as if no Dirt had been
cast thereat.

Then said the Pilgrims what means this? The Shepherds answered, This man is named *Godly-man*, and this Garment is to shew
the Innocency of his Life. Now those that throw Dirt at him, are
such as hate his *Well-doing*, but as you see the Dirt will not stick
upon his Clothes, so it shall be with him that liveth truly Innocently
in the World. Whoever they be that would make such men dirty,
they labor all in vain; for God, by that a little time is spent, will
cause that their *Innocence* shall break forth as the Light, and their
Righteousness as the Noon day.

Then they took them, and had them to *Mount-Charity*, where *Mount-*
they shewed them a man that had a bundle of Cloth lying before *Charity.*
him, out of which he cut Coats and Garments, for the Poor that
stood about him; yet his Bundle or Role of Cloth was never the less.

Then said they, what should this be? This is, said the Shepherds,
to shew you, That he that has a Heart to give of his Labor to the
Poor, shall never want wherewithal. He that watereth shall be
watered himself. And the Cake† that the Widdow gave to the
Prophet, did not cause that she had ever the less in her Barrel.

They had them also to a place where they saw one *Fool*, and one *The Work*
Want-wit, washing of an *Ethiopian* with intention to make him white, *of one Fool,*
but the more they washed him, the blacker he was. They then *Want-witt*
asked the Shepherds what that should mean. So they told them,
saying, Thus shall it be with the vile Person; all means used to get
such an one a good Name, shall in Conclusion tend but to make him
more abominable. Thus it was with the *Pharises*, and so shall it be
with all Hypocrites.

Then said *Mercie* the Wife of *Mathew* to *Christiana* her Mother, *1 Part, pag*
Mother, I would, if it might be, see the Hole in the Hill; or that, *99-100.*
commonly called, the *By-way* to Hell. So her Mother brake her *a mind to*
mind to the Shepherds. Then they went to the Door; it was in the *see the hole*
side of an Hill, and they opened it, and bid *Mercie* harken awhile. *in the Hill.*

So she harkened, and heard one saying, *Cursed be my Father for holding of my Feet back from the way of Peace and Life*; and another said, *O that I had been torn in pieces before I had, to save my life, lost my Soul*; and another said, *If I were to live again, how would I deny my self rather then come to this place*. Then there was as if the very Earth had groaned, and quaked under the Feet of this young Woman for fear; so she looked white, and came trembling away, saying, Blessed be he and she that is delivered from this Place.

Now when the Shepherds had shewed them all these things, then they had them back to the Palace, and entertained them with what the House would afford; But *Mercie* being a young, and breeding Woman, Longed for something which she saw there, but was ashamed to ask. Her Mother-in-law then asked her what she ailed, for she looked as one not well. Then said *Mercy*, *There is a Looking-glass hangs up in the Dining-room*, off of which I cannot take my mind; if therefore I have it not, I think I shall Miscarry. Then said her Mother, I will mention thy Wants to the Shepherds, and they will not deny it thee. But she said, I am ashamed that these men should know that I longed. Nay my Daughter, said she, it is no Shame, but a Virtue, to long for such a thing as that; so *Mercie* said, Then Mother, if you please, ask the Shepherds if they are willing to sell it.

Mercie longeth, and for what.

Now the Glass was one of a thousand. It would present a man, one way, with his own Feature exactly, and turn it but an other way, and it would shew one the very Face and Similitude of the Prince of Pilgrims himself. Yea I have talked with them that can tell, and they have said, that they have seen the very Crown of Thorns upon his Head, by looking in that Glass, they have therein also seen the holes in his Hands, in his Feet, and his Side. Yea such an excellency is there in that Glass, that it will shew him to one where they have a mind to see him; whether living or dead, whether in Earth or Heaven, whether in a State of Humiliation, or in his Exaltation, whether coming to Suffer, or coming to Reign.

It was the Word of God.

Jam. 1. 23.

1 Cor. 13. 12.

2 Cor. 3. 18.

Christiana therefore went to the Shepherds apart. (Now the Names of the Shepherds are *Knowledge*, *Experience*, *Watchful*, and *Sincere*), and said unto them, There is one of my Daughters a breeding Woman, that, I think doth long for some thing that she hath seen

1 Part, pag. 98.

in this House, and she thinks she shall miscarry if she should by you be denied.

Experience. Call her, call her, She shall assuredly have what we can help her to. So they called her, and said to her, *Mercie*, what is that thing thou wouldest have? Then she blushed and said, The great Glass that hangs up in the Dining-room: So *Sincere* ran and fetched it, and with a joyful Consent it was given her. Then she bowed her Head, and gave Thanks, and said, By this I know that I have obtained Favour in your Eyes. *She doth not lose her Longing.*

They also gave to the other young Women such things as they desired, and to their Husbands great Commendations, for that they joyned with Mr. *Great-heart* to the slaying of *Gyant-Dispair*, and the demolishing of *Doubting-Castle.*

About *Christiana*'s Neck, the Shepherds put a Bracelet, and so they did about the Necks of her four Daughters, also they put Ear-rings in their Ears, and Jewels on their Fore-heads. *How the Shepherds adorn the Pilgrims.*

When they were minded to go hence, they let them go in Peace, but gave not to them those certain Cautions which before was given to *Christian* and his Companion. The Reason was, for that these had *Great-heart* to be their Guide, who was one that was well acquainted with things, and so could give them their Cautions more seasonably, to wit, even then when the Danger was nigh the approaching. *1 Part, pag. 100-1.*

What Cautions *Christian* and his Companions had received of the Shepherds, they had also lost, by that the time was come that they had need to put them in practice. Wherefore here was the Advantage that this Company had over the other. *1 Part, pag. 109.*

From hence they went on Singing, and they said,

> *Behold, how fitly are the Stages set !*
> *For their Relief, that Pilgrims are become;*
> *And how they us receive without one let,*
> *That make the other Life our Mark and Home.*
> *What Novelties they have, to us they give,*
> *That we, tho Pilgrims, joyful Lives may live.*
> *They do upon us too such things bestow,*
> *That shew we Pilgrims are, where ere we go.*

When they were gone from the Shepherds, they quickly came to the Place where *Christian* met with one *Turn-a-way*, that dwelt in the Town of *Apostacy*. Wherefore of him Mr. *Great-heart* their Guide did now put them in mind; saying, This is the place where *Christian* met

1 Part, pag. 102. with one *Turn-a-way*, who carried with him the Character of his Rebellion at his Back. And this I have to say concerning this man,

How one Turn-a-way man-aged his Apostacy. Heb. 10. 26, 27, 28, 29. He would harken to no Counsel, but once a falling, perswasion could not stop him. When he came to the place where the Cross and the Sepulcher was, he did meet with one that did bid him *look there*, but he gnashed with his Teeth, and stamped, and said, he was re-solved to go back to his own Town. Before he came to the Gate, he met with *Evangelist*, who offered to lay Hands on him, to turn him into the way again. But this *Turn-a-way resisted him*, and having done much *despite* unto him, he got away over the Wall, and so escaped his Hand.

Then they went on, and just at the place where *Little-faith* formerly was Robbed, there stood a man with his Sword drawn, and his Face all bloody. Then said Mr. *Great-heart*, What art thou? The

One Valiant-for-truth beset with Thieves. man made Answer, saying, I am one whose Name is *Valiant-for-Truth*, I am a Pilgrim, and am going to the Celestial City. Now as I was in my way, there was three men did beset me, and propounded unto me these three things. 1. Whether I would become one of them? Or go back from whence I came? Or die upon the Place? To

Prov. 1. 10, 11, 12, 13, 14. the first I answered, I had been a true Man a long Season, and there-fore, it could not be expected that I now should cast in my Lot with Thieves. Then they demanded what I would say to the Second. So I told them that the Place from whence I came, had I not found Incommodity there, I had not forsaken it at all, but finding it alto-gether unsutable to me, and very unprofitable for me, I forsook it for this Way. Then they asked me what I said to the third. And I told them, my Life cost more dear far, then that I should lightly give it away. Besides, you have nothing to do thus to put things to my Choice; wherefore at your Peril be it, if you meddle. Then these three, to wit, *Wild-head*, *Inconsiderate*, and *Pragmatick*, drew upon me, and I also drew upon them.

How he behaved himself, and put them to flight. So we fell to it, one against three, for the space of above three Hours. They have left upon me, as you see, some of the Marks of their Valour, and have also carried away with them some of mine.

They are but just now gone. I suppose they might, as the saying is, hear your Horse dash, and so they betook them to flight.

Greath. *But here was great Odds, three against one.*

Valiant. 'Tis true, but *little* and *more*, are nothing to him that has the Truth on his side. *Though an Host should encamp against me, said* one, *My Heart shall not fear. Tho War should rise against me, in this will I be Confident,* etc. Besides, said he, I have read in some Records, that one man has fought an Army; and how many did *Sampson* slay† with the Jaw Bone of an Ass? [margin: Psal. 27. 3. Great-heart *wonders at his Valour.*]

Greath. *Then said the Guide, Why did you not cry out, that some might a came in for your Succour?*

Valiant. So I did, to my King, who I knew could hear, and afford invisible Help, and that was sufficient for me.

Greath. *Then said* Great-heart *to Mr.* Valiant-for-Truth, *Thou hast worthily behaved thy self; Let me see thy Sword; so he shewed it him.* [margin: *Has a mind to see his* Sword, *and spends his Judgment on it.*]

When he had taken it in his Hand, and looked thereon a while, he said, Ha! *It is a right* Jerusalem *Blade*. [margin: Isa. 2. 3.]

Valiant. It is so. Let a man have one of *these Blades*, with a Hand to wield it, and skill to use it, and he may venture upon an Angel with it. He need not fear its holding, if he can but tell how to lay on. Its Edges will never blunt. It will cut *Flesh,* and *Bones,* and *Soul,* and *Spirit,* and all. [margin: Ephes. 6. 12, 13, 14, 15, 16, 17. Heb. 4. 12.]

Greath. *But you fought a great while, I wonder you was not weary?*

Valiant. I fought till my Sword did cleave to my Hand, and when they were joyned together, as if a Sword grew out of my Arm,† and when the Blood run thorow my Fingers, then I fought with most Courage. [margin: 2 Sam. 23. 10. *The Word. The Faith. Blood.*]

Greath. *Thou hast done well, thou hast resisted unto Blood, striving against Sin. Thou shalt abide by us, come in, and go out with us; for we are thy Companions.*

Then they took him and washed his Wounds, and gave him of what they had, to refresh him, and so they went on together. Now as they went on, because Mr. *Great-heart* was delighted in him (for he loved one greatly that he found to be a man of his Hands) and because there was with his Company, them that was feeble and

weak, therefore he questioned with him about many things; as
first, *What Country-man he was?*

What
Countrey
man Mr.
Valiant
was.

Valiant. I am of *Dark-land*, for there I was born, and there my
Father and Mother are still.

Great. *Dark-land*, said the Guide, *Doth not that ly upon the same
Coast with the City of* Destruction?

How Mr.
Valiant
came to
go on Pil-
grimage.

Valiant. Yes it doth. Now that which caused me to come on Pilgrimage, was this: We had one Mr. *Tell-true* came into our parts, and he told it about, what *Christian* had done, that went from the City of *Destruction*. Namely, how he had forsaken his *Wife* and *Children*, and had betaken himself to a *Pilgrims* Life. It was also confidently reported how he had killed a *Serpent*†that did come out to resist him in his Journey, and how he got thorow to whither he intended. It was also told what Welcome he had at all his Lords Lodgings; specially when he came to the Gates of the Celestial City. For there, said the man, He was received with sound of Trumpet, by a company of shining ones. He told it also, how all the Bells in the City did ring for Joy at his Reception, and what Golden Garments he was cloathed with; with many other things that now I shall forbear to relate. In a word, that man so told the Story of *Christian* and his Travels, that my Heart fell into a burning hast to be gone after him, nor could Father or Mother stay me, so I got from them, and am come thus far on my Way.

Great. *You came in at the Gate, did you not?*

He begins
right.

Valiant. Yes, yes. For the same man also told us, that all would be nothing if we did not begin to enter this way at the Gate.

Christian's
Name
famous.

Great. *Look you, said the Guide to* Christiana, *The Pilgrimage of your Husband, and what he has gotten thereby, is spread abroad far and near.*

Valiant. Why, is this *Christian*'s Wife?

Great. *Yes, that it is, and these are also her four Sons.*

Valiant. What! and going on Pilgrimage too?

Great. *Yes verily, they are following after.*

He is much
rejoyced
to see
Christian's
Wife.

Valiant. It glads me at the Heart! Good man! How Joyful will he be, when he shall see them that would not go with him, yet to enter after him, in at the Gates into the City!

Great. *Without doubt it will be a Comfort to him; for next to the Joy of seeing himself there, it will be a Joy to meet there his Wife and his Children.*

Valiant. But now you are upon that, pray let me see your Opinion about it. Some make a question whether we shall know one another when we are there.

Whether we shall know one another when we come to Heaven.

Greath. Do they think they shall know themselves then? Or that they shall rejoyce to see themselves in that Bliss? and if they think they shall know and do these; Why not know others, and rejoyce in their Welfare also?

Again, Since Relations are our second self, tho that State will be dissolved there, yet why may it not be rationally concluded that we shall be more glad to see them there, then to see they are wanting?

Valiant. Well, I perceive whereabouts you are as to this. Have you any more things to ask me about my beginning to come on Pilgrimage?

Greath. Yes, Was your Father and Mother willing that you should become a Pilgrim?

Valiant. Oh, no. They used all means imaginable to perswade me to stay at Home.

Greath. Why, what could they say against it?

Valiant. They said it was an idle Life, and if I my self were not inclined to Sloath and Laziness, I would never countenance a Pilgrim's Condition.

The great Stumbling-Blocks that by his Friends were laid in his way.

Greath. And what did they say else?

Valiant. Why, They told me that it was a dangerous Way, yea the most dangerous Way in the World, said they, is that which the Pilgrims go.

Greath. Did they shew wherein this Way is so dangerous?

Valiant. Yes. And that in many Particulars.

Greath. Name some of them.

Valiant. They told me of the Slow of *Dispond*, where *Christian* was well nigh Smuthered. They told me that there were Archers standing ready in *Belzebub-Castle*, to shoot them that should knock at the *Wicket* Gate for Entrance. They told me also of the Wood, and dark Mountains, of the Hill *Difficulty*, of the Lyons, and also of the three Gyants, *Bloodyman, Maul*, and *Slay-good*. They said moreover, That there was a foul *Fiend* haunted the Valley of *Humiliation*, and that *Christian* was, by him, almost bereft of Life. Besides, said they, You must go over the *Valley of the Shadow of Death*, where the *Hobgoblins* are, where the Light is Darkness, where the Way is full of Snares,

The first Stumbling-Block.

Pits, Traps and Ginns. They told me also of *Gyant Dispair*, of *Doubting Castle*, and of the *Ruins* that the Pilgrims met with there. Further, They said, I must go over the enchanted Ground, which was dangerous. And that after all this I should find a River, over which I should find no Bridg, and that that River did lie betwixt me and the Celestial Countrey.

Greath. *And was this all?*

The Second Valiant. No, They also told me that this way was full of *Deceivers*, and of Persons that laid await there, to turn good men out of the Path.

Greath. *But how did they make that out?*

The Third. Valiant. They told me that Mr. *Worldly-wise-man* did there lie in wait to deceive. They also said that there was *Formality* and *Hypocrisie* continually on the Road. They said also that *By-ends*, *Talkative*, or *Demas*, would go near to gather me up; that the Flatterer would catch me in his Net, or that with greenheaded *Ignorance* I would presume to go on to the Gate, from whence he always was sent back to the Hole that was in the side of the Hill, and made to go the By-way to Hell.

Greath. *I promise you, This was enough to discourage. But did they make an end here?*

Fourth. Valiant. No, stay. They told me also of many that had tryed that way of old, and that had gone a great way therein, to see if they could find something of the Glory there, that so many had so much talked of from time to time; and how they came back again, and befooled themselves for setting a Foot out of Doors in that Path, to the Satisfaction of all the Country. And they named several that did so, as *Obstinate*, and *Plyable*, *Mistrust*, and *Timerous*, *Turn-a-way*, and old *Atheist*, with several more; who, they said, had, some of them, gone far to see if they could find, but not one of them found so much Advantage by going, as amounted *to the weight of a Fether*.

Greath. *Said they any thing more to discourage you?*

The Fifth. Valiant. Yes, They told me of one Mr. *Fearing*, who was a Pilgrim, and how *he* found this way so Solitary, that he never had comfortable Hour therein, also that Mr. *Dispondency* had like to been starved therein; Yea, and also, which I had almost forgot, that *Christian* himself, about whom there has been such a Noise, after all his

Ventures for a Celestial Crown, was certainly drowned in the black
River, and never went foot further, however it was smuthered up.

Greath. *And did none of these things discourage you?*

Valiant. No. They seemed but as so many Nothings to me.

Greath. *How came that about?*

Valiant. Why, I still believed what Mr. *Tell-True* had said, and that carried me beyond them all. *How he got over these Stumbling-Blocks.*

Greath. *Then this was your Victory, even your Faith?*

Valiant. It was so, I believed and therefore came out, got into the Way, fought all that set themselves against me, and by believing am come to this Place.

> *Who would true Valour see*
> *Let him come hither;*
> *One here will Constant be,*
> *Come Wind, come Weather.*
> *There's no Discouragement,*
> *Shall make him once Relent,*
> *His first avow'd Intent,*
> *To be a Pilgrim.*
>
> *Who so beset him round,*
> *With dismal Storys,*
> *Do but themselves Confound;*
> *His Strength the more is.*
> *No Lyon can him fright,*
> *He'l with a Gyant Fight,*
> *But he will have a right,*
> *To be a Pilgrim.*
>
> Hobgoblin, *nor foul* Fiend,
> *Can daunt his Spirit:*
> *He knows, he at the end,*
> *Shall Life Inherit.*
> *Then Fancies fly away,*
> *He'l fear not what men say,*
> *He'l labour Night and Day,*
> *To be a Pilgrim.*

1 *Part, pag.*
III.
By this time they were got to the *enchanted Ground*, where the Air naturally tended to make one *Drowzy*. And that place was all grown over with Bryers and Thorns; excepting *here* and *there*, where was an *inchanted Arbor*, upon which, if a Man sits, or in which if a man sleeps, 'tis a question, say some, whether ever they shall rise or wake again in this World. Over this Forrest therefore they went, both one with an other, and Mr. *Great-heart* went before, for that he was the Guide, and Mr. *Valiant-for-truth*, he came behind, being there a Guard, for fear lest paradventure some *Fiend*, or *Dragon*, or *Gyant*, or *Thief*, should fall upon their Rere, and so do Mischief. They went on here each man with his Sword drawn in his Hand; for they knew it was a dangerous place. Also they cheared up one another as well as they could. *Feeble-mind*, Mr. *Great-heart* commanded should come up after him, and Mr. *Dispondency* was under the Eye of Mr. *Valiant*.

Now they had not gone far, but a great Mist and a darkness fell upon them all; so that they could scarce, for a great while, see the one the other. Wherefore they were forced for some time, to feel for one another, by Words; for they walked not by Sight.[†]

But any one must think, that here was but sorry going for the best of them all, but how much worse for the Women and Children, who both of *Feet* and *Heart* were but tender. Yet so it was, that, thorow the incouraging Words of he that led in the Front, and of him that brought them up behind, they made a pretty good shift to wagg along.

The Way also was here very wearysom, thorow Dirt and Slabbiness. Nor was there on *all* this Ground, so much as one *Inn*, or *Victualling-House*, therein to refresh the feebler sort. Here therefore was *grunting*, and *puffing*, and *sighing*: While one tumbleth over a Bush, another sticks fast in the Dirt, and the Children, some of them, lost their Shoos in the Mire. While one crys out, I am down, and another, Ho, Where are you? and a third, The Bushes have got such fast hold on me, I think I cannot get away from them.

An Arbor
on the
Inchanting
Ground.
Then they came at an *Arbor*, warm, and promising much Refreshing to the Pilgrims; for it was finely wrought above-head, beautified with *Greens*, furnished with *Benches*, and *Settles*. It also had in it a soft Couch whereon the weary might lean. This, you must think, all things considered, was tempting; for the Pilgrims already

began to be foyled with the badness of the way; but there was not one of them that made so much as a motion to stop there. Yea, for ought I could perceive, they continually gave so good heed to the Advice of their Guide, and he did so faithfully tell them of *Dangers*, and of the *Nature* of Dangers when they were at them, that usually when they were nearest to them, they did most pluck up their Spirits, and hearten one another to deny the Flesh. This *Arbor* was called *The sloathfuls Friend*, on purpose to allure, if it might be, some of the Pilgrims there, to take up their Rest when weary. *The Name of the Arbor.*

I saw then in my Dream, that they went on in this their *solitary* Ground, till they came to a place at which a man is apt to lose his Way. *Now*, tho when it was light, their Guide could well enough tell how to miss those ways that led wrong, yet in the dark he was put to a stand: But he had in his Pocket a Map of all ways leading to, or from the Celestial City; wherefore he strook a Light (for he never goes also without his Tinder-box) and takes a view of his Book or Map; which bids him be careful in that place to turn to the right-hand-way. And had he not here been careful to look in his Map, they had all, in probability, been smuthered in the Mud, for just a little before them, and that at the end of the cleanest Way too, was a Pit, none knows how deep, full of nothing but Mud; there made on purpose to destroy the Pilgrims in. *The way difficult to find.* *The Guide has a Map of all ways leading to or from the City.*

Then thought I with my self, who, that goeth on Pilgrimage, but would have one of these Maps about him, that he may look when he is at a *stand*, which is the way he must take? *God's Book.*

They went on then in this *inchanted* Ground, till they came to where was an other *Arbor*, and it was built by the High-way-side. And in that *Arbor* there lay two men whose Names were *Heedless* and *Too-bold*. These two went thus far on Pilgrimage, but here being wearied with their Journy, they sat down to rest themselves, and so fell fast asleep. When the Pilgrims saw them, they stood still and shook their Heads; for they knew that the Sleepers were in a pitiful Case. Then they consulted what to do; whether to go on and leave them in their Sleep, or to step to them and try to awake them. So they concluded to go to them and wake them; that is, if they could; but with this Caution, namely, to take heed that themselves did not sit down, nor imbrace the offered Benefit of that *Arbor*. *An Arbor and two asleep therein.* *The Pilgrims try to wake them.*

So they went in and spake to the men, and called each by his Name, (for the Guide, it seems, did know them) but there was no Voice nor Answer. Then the Guide did shake them, and do what he could to disturb them. Then said one of them, *I will pay you when I take my Mony*; At which the Guide shook his Head. *I will fight so long as I can hold my Sword in my Hand*, said the other. At that, one of the Children laughed.

Their Endeavour is fruitless.

Then said *Christiana*, What is the meaning of this? The Guide said, *They talk in their Sleep.* If you strike them, beat them, or whatever else you do to them, they will answer you after this fashion; or as one of them said in old time, when the Waves of the Sea did beat upon him, and he slept as one upon the Mast of a Ship, *When I awake I will seek it again.* You know when men talk in their Sleeps, they say any thing; but their Words are not governed, either by Faith or Reason. There is an *Incoherencie* in their Words *now*, as there was before betwixt their going on Pilgrimage, and sitting down here. This then is the Mischief on't, when *heedless* ones go on Pilgrimage, 'tis twenty to one, but they are served thus. For this *inchanted* Ground is one of the last Refuges that the Enemy to Pilgrims has; wherefore it is as you see, placed almost at the end of the Way, and so it standeth against us with the more advantage. For when, thinks the Enemy, will these Fools be so desirous to sit down, as when they are weary; and when so like to be weary, as when almost at their Journys end? Therefore it is, I say, that the *inchanted* Ground is placed so nigh to the Land *Beulah*, and so neer the end of their Race. Wherefore let Pilgrims look to themselves, lest it happen to them as it has done to these, that, as you see, are fallen asleep, and none can wake them.

Prov. 23. 34, 35.

The light of the Word.

2 Pet. 1. 19.

Then the Pilgrims desired with trembling to go forward, only they prayed their Guide to strike a Light, that they might go the rest of their way by the help of the light of a Lanthorn. So he strook a light, and they went by the help of that thorow the rest of this way, tho the Darkness was very great.

The Children cry for weariness.

But the Children began to be sorely weary, and they cryed out unto him that loveth Pilgrims, to make their way more Comfortable. So by that they had gone a little further, a Wind arose that drove away the Fog, so the Air became more clear.

Yet they were not off (by much) of the *inchanted* Ground; only now they could see one an other better, and the way wherein they should walk.

Now when they were almost at the end of this Ground, they perceived that a little before them, was a *solemn* Noise, as of one that was much concerned. So they went on and looked before them, and behold, they saw, as they thought, *a Man upon his Knees*, with Hands and Eyes lift up, and speaking, as they thought, earnestly to one that was above. They drew nigh, but could not tell what he said; so they went softly till he had done. When he had done, he got up and began to run towards the Celestial City. Then Mr. *Great-heart* called after him, saying, So-ho, Friend, let us have your Company, if you go, as I suppose you do, to the Celestial City. So the man stoped, and they came up to him. But as soon as Mr. *Honest* saw him, he said, I know this man. Then said Mr. *Valiant-for-truth*, Prethee who is it? 'Tis one, said he, that comes from whereabouts I dwelt, his Name is *Stand-fast*, he is certainly a right good Pilgrim. *Standfast upon his Knees in the Inchanted Ground.* *The Story of Stand-fast.*

So they came up one to another, and presently *Stand-fast* said to old *Honest*, Ho, Father *Honest*, are you there? Ai, said he, that I am, as sure as you are there. Right glad am I, said Mr. *Stand-fast*, that I have found you on this Road. And as glad am I, said the other, that I espied you upon your Knees. Then Mr. *Stand-fast* blushed, and said, But why, did you see me? Yes, that I did, quoth the other, and with my Heart was glad at the Sight. Why, what did you think, said *Stand-fast*? Think, said old *Honest*, what should I think? I thought we had an honest Man upon the Road, and therefore should have his Company by and by. If you thought not amiss, how happy am I! But if I be not as I should, I alone must bear it. That is true, said the other; but your fear doth further confirm me that things are right betwixt the Prince of Pilgrims and your Soul. For he saith, *Blessed is the Man that feareth always.*† *Talk betwixt him and Mr. Honest.*

Valiant. Well, But Brother, I pray thee tell us what was it that was the cause of thy being upon thy Knees, even now? Was it for that some special Mercy laid Obligations upon thee, or how? *They found him at Prayer.*

Stand. Why we are as you see, upon the *inchanted* Ground, and as I was coming along, I was musing with my self of what a dangerous Road, the Road in this place was, and how many that had come *What it was that fetched him upon his Knees.*

even thus far on Pilgrimage, had here been stopt, and been destroyed. I thought also of the manner of the Death with which this place destroyeth Men. Those that die here, die of no violent Distemper; the Death which such die, is not grievous to them. For he that goeth away in a *Sleep*, begins that Journey with Desire and Pleasure. Yea, such acquiesce in the Will of that Disease.

Hon. Then Mr. Honest *Interrupting of him said, Did you see the two Men asleep in the Arbor?*

Stand. Ai, ai, I saw *Heedless*, and *Too-bold* there; and for ought I know, there they will ly till they Rot. But let me go on in my Tale: As I was thus Musing, as I said, there was one in very pleasant Attire, *but old*, that presented her self unto me, and offered me three things, to wit, her *Body*, her *Purse*, and her *Bed*. Now the Truth is, I was both a weary, and sleepy, I am also as poor as a *Howlet*, and that, perhaps, the *Witch* knew. Well, I repulsed her once and twice, but she put by my Repulses, and smiled. Then I began to be angry, but she mattered that nothing at all. Then she made Offers again, and said, If I would be ruled by her, she would make me great and happy. For, said she, I am the Mistriss of the World, and men are made happy by me. Then I asked her Name, and she told me it was *Madam Bubble*. This set me further from her; but she still followed me with Inticements. Then I betook me, as you see, to my Knees, and with Hands lift up, and crys, I pray'd to him that had said, he would help. So just as you came up, the Gentlewoman went her way. Then I continued to give thanks for this my great Deliverance; for I verily believe she intended no good, but rather sought to make stop of me in my Journey.

Hon. Without doubt her Designs were bad. But stay, now you talk of her, methinks I either have seen her, or have read some story of her.

Standf. Perhaps you have done both.

Hon. Madam Buble! Is she not a tall comely Dame, something of a Swarthy Complexion?

Standf. Right, you hit it, she is just such an one.

Hon. Doth she not speak very smoothly, and give you a Smile at the end of a Sentence?

Standf. You fall right upon it again, for these are her very Actions.

Prov. 10. 7.

Madam Buble, or this vain World.

Hon. Doth she not wear a great Purse by her Side, and is not her Hand often in it fingering her Mony, as if that was her Hearts delight?

Standf. 'Tis just so. Had she stood by all this while, you could not more amply set her forth before me, nor have better described her Features.

Hon. Then he that drew her Picture was a good *Limner*, and he that wrote of her, said true.

Greath. This Woman is a *Witch*, and it is by Virtue of her *Sorceries* *The World.* that this Ground is *enchanted*; whoever doth lay their Head down in her *Lap*, had as good lay it down upon that Block over which the Ax doth hang; and whoever lay their Eyes upon her Beauty, are counted the Enemies of God. This is she that maintaineth in their *Jam. 4. 4.* Splendor, all those that are the Enemies of Pilgrims. Yea, This is she *1 John 2.* that has bought off many a man from a Pilgrims Life. She is a great *15.* *Gossiper*, she is always, both she and her Daughters, at one Pilgrim's Heels or other, now Commending, and then preferring the excellencies of this Life. She is a bold and impudent Slut; She will talk with any Man. She always laugheth Poor Pilgrims to scorn, but highly commends the Rich. If there be one cunning to get Mony in a Place, she will speak well of him, from House to House. She loveth Banqueting, and Feasting, mainly well; she is always at one full Table or another. She has given it out in some places, that she is a Goddess, and therefore some do Worship her. She has her times and open places of Feasting, and she will say and avow it, that none can shew a Food comparable to hers. She promiseth to dwell with Childrens Children, if they will but love and make much of her. She will cast out of her Purse, Gold like Dust, in some places, and to some Persons. She loves to be sought after, spoken well of, and to ly in the Bosoms of Men. She is never weary of commending of her Commodities, and she loves them most that think best of her. She will promise to some Crowns, and Kingdoms, if they will but take her Advice, yet many has she brought to the Halter, and ten thousand times more to Hell.

Standf. *O! Said* Stand-fast, *What a Mercy is it that I did resist her; for whither might she a drawn me?*

Greath. Whither! Nay, none but God knows whither. But in

· Tim. 6. 9. general to be sure, she would a drawn thee *into many foolish and hurtful Lusts, which drown men in Destruction and Perdition.*

'Twas she that set *Absalom* against his Father, and *Jeroboam* against his Master. 'Twas she that perswaded *Judas* to sell his Lord, and that prevailed with *Demas* to forsake the godly Pilgrims Life;† none can tell of the Mischief that she doth. She makes Variance betwixt Rulers and Subjects, betwixt Parents and Children, 'twixt Neighbor and Neighbor, 'twixt a Man and his Wife, 'twixt a Man and himself, 'twixt the Flesh and the Heart.

Wherefore good Master *Stand-fast*, be as your Name is, and when you have done all, *stand*.

At this Discourse there was among the Pilgrims a mixture of Joy and Trembling, but at length *they brake* out and Sang:

> *What Danger is the Pilgrim in,*
> *How many are his Foes,*
> *How many ways there are to Sin,*
> *No living Mortal knows.*
> *Some of the Ditch, shy are, yet can*
> *Lie tumbling in the Myre:*
> *Some tho they shun the Frying-pan,*
> *Do leap into the Fire.*

1 Part, pag. 126. After this I beheld, until they were come into the Land of *Beulah*, where the Sun shineth Night and Day. Here, because they was weary, they betook themselves a while to Rest. And because this Country was common for Pilgrims, and because the Orchards and Vineyards that were here, belonged to the King of the Celestial Country; therefore they were licensed to make bold with any of his things.

But a little while soon refreshed them here, for the Bells did so ring, and the Trumpets continually sound so Melodiously, that they could not sleep, and yet they received as much refreshing, as if they had slept their Sleep never so soundly. Here also all the noise of them that walked the Streets, was, *More Pilgrims are come to Town.* And an other would answer, saying, And so many went over the Water, and were let in at the Golden Gates to Day. They would cry again, There is now a Legion of Shining ones, just come to Town;

by which we know that there are more Pilgrims upon the Road, for here they come to wait for them and to comfort them after all their Sorrow. Then the Pilgrims got up and walked to and fro: But how were their Ears now filled with heavenly Noises, and their Eyes delighted with Celestial Visions! In this Land, they *heard* nothing, *saw* nothing, *felt* nothing, *smelt* nothing, *tasted* nothing, that was offensive to their Stomach or Mind; only when they tasted of the Water of the River, over which they were to go, they thought that tasted a little Bitterish to the Palate, but it proved sweeter when 'twas down. *Death bitter to the Flesh, but sweet to the Soul.*

In this place there was a Record kept of the Names of them that had been Pilgrims of old, and a History of all the famous Acts that they had done. It was here also much discoursed how the *River* to some had had its *flowings*, and what *ebbings* it has had while others have gone over. It has been in a manner *dry* for some, while it has overflowed its Banks for others. *Death has its Ebbings and Flowings like the Tide.*

In this place, the Children of the Town would go into the Kings Gardens and gather Nose-gaies for the Pilgrims, and bring them to them with much affection. Here also grew *Camphire*, with *Spicknard*, and *Saffron*, *Calamus*, and *Cinamon*, with all its Trees of *Frankincense*, *Myrrhe*, and *Aloes*, with all *chief* Spices.† With these the Pilgrims Chambers were perfumed, while they stayed here; and with these were their Bodys anointed to prepare them to go over the *River* when the time appointed was come.

Now, while they lay here, and waited for the good Hour; there was a Noyse in the Town, that there was a *Post* come from the Celestial City, with Matter of great Importance, to one *Christiana*, the Wife of *Christian* the Pilgrim. So Enquiry was made for her, and the House was found out where she was, so the Post presented her with a Letter; the Contents whereof was, *Hail, Good Woman, I bring thee Tidings that the Master calleth for thee, and expecteth that thou shouldest stand in his Presence, in Cloaths of Immortality, within this ten Days.* *A Messenger of Death sent to Christiana. His Message.*

When he had read this Letter to her, he gave her therewith a *sure* Token that he was a true Messenger, and was come to bid her make haste to be gone. The Token was, *An Arrow with a Point sharpened with Love, let easily into her Heart, which by degrees wrought so effectually with her, that at the time appointed she must be gone.* *How welcome is Death to them that have nothing to do but to dy.*

Her Speech to her Guide. When *Christiana* saw that her time was come, and that she was the first of this Company that was to go over: She called for Mr. *Great-heart* her Guide, and told him how Matters were. So he told him he was heartily glad of the News, and could a been glad had the Post came for him. Then she bid that he should give Advice, how all things should be prepared for her Journey.

So he told her, saying, Thus and thus it must be, and we that Survive will accompany you to the Riverside.

To her Children. Then she called for her Children, and gave them *her Blessing*; and told them that she yet read with Comfort the Mark that was set in their Foreheads, and was glad to see them with her there, and that they had kept their Garments so white. Lastly, She bequeathed to the Poor that little she had, and commanded her Sons and her Daughters to be ready against the Messenger should come for them.

To Mr. Valiant. When she had spoken these Words to her Guide and to her Children, she called for Mr. *Valiant-for-truth*, and said unto him, Sir, You have in all places shewed your self true-hearted, be Faithful unto Death, and my King will give you a Crown of Life. I would also intreat you to have an Eye to my Children, and if at any time you see them faint, speak comfortably to them. For my Daughters, my Sons Wives, they have been Faithful, and a fulfilling of the Promise upon them, will be their end. But she gave Mr. *Stand-fast* a Ring.

To Mr. Stand-fast. To old Honest. Then she called for old Mr. *Honest*, and said of him, Behold an Israelite indeed, in whom is no Guile. Then said *he*, I wish you a fair Day when you set out for Mount *Sion*, and shall be glad to see that you go over the River dry-shod. But she answered, Come *Wet*, come *Dry*, I long to be gone; for however the Weather is in my Journey, I shall have time enough when I come there to sit down and rest me, and dry me.

To Mr. Ready-to-halt. Then came in that good Man Mr. *Ready-to-halt* to see her. So she said to him, Thy Travel hither has been with Difficulty, but that will make thy Rest the sweeter. But watch, and be ready, for at an Hour when you think not, the Messenger may come.

To Dis-pondencie, and his Daughter. After him, came in Mr. *Dispondencie*, and his Daughter *Much-a-fraid*. To whom she said, You ought with Thankfulness for ever, to remember your Deliverance from the Hands of Gyant *Dispair*, and out of *Doubting-Castle*. The effect of that Mercy is, that you are brought with Safety hither. Be ye watchful, and cast away Fear; be sober, and hope to the End.

Then she said to Mr. *Feeble-Mind*, Thou was delivered from the Mouth of Gyant *Slay-good*, that thou mightest live in the Light of the Living for ever, and see thy King with Comfort. Only I advise thee to repent thee of thy aptness to fear and doubt of his Goodness before he sends for thee, lest thou shouldest when he comes, be forced to stand before him for that Fault with Blushing. {.marginnote} *To Feeble-mind.*

Now the Day drew on that *Christiana* must be gone. So the Road was full of People to see her take her Journey. But behold all the Banks beyond the River were full of Horses and Chariots, which were come down from above to accompany her to the City-Gate. So she came forth and entered the *River* with a *Beck'n* of Fare well, to those that followed her to the River side. The last word she was heard to say here was, *I come Lord, to be with thee and bless thee.* {.marginnote} *Her last Day, and manner of Departure.*

So her Children and Friends returned to their Place, for that those that waited for *Christiana*, had carried her out of their Sight. So she went, and called, and entered in at the Gate with all the Ceremonies of Joy that her Husband *Christian* had done before her.

At her Departure her Children wept, but Mr. *Great-heart*, and Mr. *Valiant*, played upon the well tuned Cymbal and Harp for Joy. So all departed to their respective Places.

In process of time there came a *Post* to the Town again, and his Business was with Mr. *Ready-to-halt*. So he enquired him out, and said to him, I am come to thee in the Name of him whom thou hast Loved and Followed, tho upon *Crutches*. And my Message is to tell thee, that he expects thee at his Table to Sup with him in his Kingdom the next Day after *Easter*. Wherefore prepare thy self for this Journey. {.marginnote} *Ready-to-halt Summoned.*

Then he also gave him a Token that he was a true Messenger, saying, *I have broken thy golden Bowl, and loosed thy silver Cord.* {.marginnote} *Eccles. 12. 6.*

After this Mr. *Ready-to-halt* called for his Fellow Pilgrims, and told them, saying, I am sent for, and God shall surely visit you also. So he desired Mr. *Valiant* to make his *Will*. And because he had nothing to bequeath to them that should Survive him, but his *Crutches*, and his good *Wishes*, therefore thus he said: *These Crutches,* {.marginnote} *Promises.*

His Will. I bequeath to my Son that shall tread in my Steps with an hundred warm Wishes that he may prove better then I have done.

Then he thanked Mr. *Great-heart*, for his Conduct, and Kindness, and so addressed himself to his Journey. When he came at the brink of the River, he said, Now I shall have no more need of these *Crutches*, since yonder are Chariots and Horses for me to ride on. *His last* The last Words he was heard to say, was, *Welcome Life*. So he went *words.* his Way.

Feeble- After this, Mr. *Feeble-mind* had Tidings brought him, that the *mind Sum-* Post sounded his Horn at his Chamber Door. Then he came in and *moned.* told him, saying, I am come to tell thee that the Master has need of thee, and that in very little time thou must behold his Face in Brightness. And take this as a Token of the Truth of my Message. *Eccles. 12.* *Those that look out at the Windows shall be darkned.*

3. Then Mr. *Feeble-mind* called for his Friends, and told them what Errand had been brought unto him, and what Token he had received of the truth of the Message. Then he said, Since I have *He makes* nothing to bequeath to any, to what purpose should I make a Will? *no Will.* As for my *feeble Mind*, that I will leave behind me, for that I shall have no need of that in the place whither I go; nor is it worth bestowing upon the poorest Pilgrim: Wherefore when I am gon, I desire, that you Mr. *Valiant*, would bury it in a Dunghil. This done, and the Day being come, in which he was to depart; he entered the *His last* River as the rest. His last Words were, *Hold out Faith and Patience.* *words.* So he went over to the other Side.

Mr. Dis- When Days, had many of them passed away: Mr. *Dispondencie* was *pondencie's* sent for. For a *Post* was come, and brought this Message to him; *Summons.* *Trembling Man, These are to summon thee to be ready with thy King, by the next Lords Day, to shout for Joy for thy Deliverance from all thy Doubtings.*

And said the Messenger, That my Message is true, take this for *Eccl. 12. 5.* a Proof. So he gave him *The Grashopper to be a Burthen unto him.* Now *His Daugh-* Mr. *Dispondencie*'s Daughter, whose Name was *Much-a-fraid*, said, *ter goes too.* when she heard what was done, that she would go with her Father. Then Mr. *Dispondencie* said to his Friends; My self and my Daughter, you know what we have been, and how troublesomly we have *His Will.* behaved our selves in every Company. My Will and my Daughters is, that our *Disponds*, and slavish Fears, be by no man ever received, from the day of our Departure, for ever; For I know that after my Death they will offer themselves to others. For, to be plain with you, they are *Ghosts*, the which we entertained when we first began to be Pilgrims, and could never shake them off after. And they will walk about and seek Entertainment of the Pilgrims, but for our Sakes, shut ye the Doors upon them.

When the time was come for them to depart, they went to the Brink of the *River*. The last Words of Mr. *Dispondencie*, were, *Farewel Night, welcome Day*. His Daughter went thorow the River singing, but none could understand what she said. *His last words.*

Then it came to pass, a while after, that there was a *Post* in the Town that enquired for Mr. *Honest*. So he came to the House where he was, and delivered to his Hand these Lines: *Thou art Commanded to be ready against this Day seven Night, to present thy self before thy Lord, at his Fathers House*. And for a Token that my Message is true, *All thy Daughters of Musick shall be brought low*. Then Mr. *Honest* called for his Friends, and said unto them, I Die, but shall make no Will. As for my Honesty, it shall go with me; let him that comes after be told of this. When the Day that he was to be gone, was come, he addressed himself to go over the *River*. Now the *River* at that time overflowed the Banks in some places. But Mr. *Honest* in his Life time had spoken to one *Good-conscience* to meet him there, the which he also did, and lent him his Hand, and so helped him over. The last Words of Mr. *Honest* were, *Grace Reigns*. So he left the World. *Mr. Honest Summoned. Eccl. 12. 4. He makes no Will. Good-conscience helps Mr. Honest over the River.*

After this it was noised abroad that Mr. *Valiant-for-truth* was taken with a Summons, by the same *Post* as the other; and had this for a Token that the Summons was true, *That his Pitcher was broken at the Fountain*. When he understood it, he called for his Friends, and told them of it. Then said he, I am going to my Fathers, and tho with great Difficulty I am got hither, yet now I do not repent me of all the Trouble I have been at to arrive where I am. *My Sword*, I give to him that shall succeed me in my Pilgrimage, and my *Courage* and *Skill*, to him that can get it. My *Marks* and *Scarrs* I carry with me, to be a witness for me, that I have fought his Battels, who now will be my Rewarder. When the Day that he must go hence, was come, many accompanied him to the River side, into which, *Mr. Valiant Summoned. Eccl. 12. 6. His Will.*

His last as he went, he said, *Death, where is thy Sting?* And as he went down
words. deeper, he said, *Grave where is thy Victory?* So he passed over, and the
Trumpets sounded for him on the other side.

Mr. Stand-   Then there came forth a Summons for Mr. *Stand-fast*, (This Mr.
fast is *Stand-fast*, was he that the rest of the Pilgrims found upon his Knees
summoned. in the *inchanted* Ground.) For the *Post* brought it him open in his
Hands. The Contents whereof were, *That he must prepare for a change
of Life, for his Master was not willing that he should be so far from him any
longer.* At this Mr. *Stand-fast* was put into a Muse; Nay, said the
Messenger, you need not doubt of the truth of my Message; for here
Eccl. 12. 6. is a Token of the Truth thereof, *Thy Wheel is broken at the Cistern.*
He calls for Then he called to him Mr. *Great-heart*, who was their Guide, and
Mr. Great-
Heart. said unto him, Sir, Altho it was not my hap to be much in your good
His Speech
to him. Company in the Days of my Pilgrimage, yet since the time I knew
you, you have been profitable to me. When I came from home, I left
behind me a Wife, and five small Children. Let me entreat you, at
your Return (for I know that you will go, and return to your
Masters House, in Hopes that you may yet be a Conductor to more
of the Holy Pilgrims,) that you send to my Family, and let them be
acquainted with all that hath, and shall happen unto me. Tell them
His Errand moreover, of my happy Arrival to this Place, and of the present late
to his blessed Condition that I am in. Tell them also of *Christian*, and of
Family.
Christiana his Wife, and how *She* and her Children came after her
Husband. Tell them also of what a happy End she made, and
whither she is gone. I have little or nothing to send to my Family,
except it be Prayers, and Tears for them; of which it will suffice, if
thou acquaint them, if paradventure they may prevail. When Mr.
Stand-fast had thus set things in order, and the time being come for
him to haste him away; he also went down to the River. Now there
was a great Calm at that time in the River, wherefore Mr. *Stand-
fast*, when he was about half way in, he stood a while and talked to
his Companions that had waited upon him thither. And he said,

His last   This River has been a Terror to many, yea the thoughts of it also
words. have often frighted me. But now methinks I stand easie, my Foot is
Jos. 3. 17. fixed upon that, upon which the Feet of the Priests that bare the
Ark of the Covenant stood while *Israel* went over this *Jordan*. The
Waters indeed are to the Palate bitter, and to the Stomack cold; yet

the thoughts of what I am going to, and of the Conduct that waits for me on the other side, doth lie as a glowing Coal at my Heart.

I see my self now at the *end* of my Journey, my *toilsom* Days are ended. I am going now to see *that* Head that was Crowned with Thorns, and *that* Face that was spit upon, for me.

I have formerly lived by Hear-say, and Faith, but now I go where I shall live by sight, and shall be with him, in whose Company I delight my self.

I have loved to hear my Lord spoken of, and wherever I have seen the print of his Shooe in the Earth, there I have coveted to set my Foot too.

His Name has been to me as a *Civit-Box*, yea sweeter then all Perfumes. His Voice to me has been most sweet, and his Countenance, I have more desired then they that have most desired the Light of the Sun. His Word I did use to gather for my Food, and for Antidotes against my Faintings. He has held me, and I have kept me from mine Iniquities: Yea, my Steps hath he strengthened in his Way.

Now while he was thus in Discourse his Countenance changed, his *strong men* bowed under him, and after he had said, *Take me, for I come unto thee*, he ceased to be seen of them.

But glorious it was, to see how the open Region was filled with Horses and Chariots, with Trumpeters and Pipers, with Singers, and Players on stringed Instruments, to welcome the Pilgrims as they went up and followed one another in at the beautiful Gate of the City.

As for *Christian*'s Children, the four Boys that *Christiana* brought with her, with their Wives and Children, I did not stay where I was, till they were gone over. Also since I came away, I heard one say, that they were yet alive, and so would be for the Increase of the Church in that Place where they were for a time.

Shall it be my Lot to go that way again, I may give those that desire it, an Account of what I here am silent about; mean time I bid my Reader *Adieu*.

FINIS

EXPLANATORY NOTES

The following abbreviations are used in the notes:

Brown   John Brown, *John Bunyan (1628–88): His Life, Times and Work*, tercentenary edn rev. Frank Mott Harrison (1928)

Casebook   Roger Sharrock (ed.), *The Pilgrim's Progress: a Casebook* (1976)

Coleridge   *Coleridge on the Seventeenth Century*, ed. Roberta Florence Brinkley (New York, 1968)

GA   John Bunyan, *Grace Abounding to the Chief of Sinners*, ed. Roger Sharrock (Oxford, 1962)

Greaves   Richard L. Greaves, *John Bunyan* (Appleford, Berks., 1969)

Kaufmann   U. Milo Kaufmann, *The Pilgrim's Progress and Traditions in Puritan Meditation* (New Haven, 1966)

Misc. Wks. John Bunyan, *The Miscellaneous Works*, gen. ed. Roger Sharrock, 12 vols. in progress (Oxford, 1976–)

Newey   Vincent Newey (ed.), *The Pilgrim's Progress: Critical and Historical Views* (Liverpool, 1980)

NT   New Testament

OET   *The Pilgrim's Progress*, ed. James Blanton Wharey, rev. Roger Sharrock, Oxford English Texts, 2nd edn (Oxford, 1960)

OT   Old Testament

PP   *The Pilgrim's Progress*

Sharrock   Roger Sharrock, *John Bunyan*, corrected reissue (1968)

Talon   Henri A. Talon, *John Bunyan: the Man and his Works*, transl. Barbara Wall (1951)

Works   *The Works of John Bunyan*, ed. George Offor, 3 vols. (1862)

THE FIRST PART

1 *Apology*: the prefatory verses constitute a defence (or *apologia*) of Bunyan's allegorical method against anticipated charges of being

obscure, fictional (that is, false), imaginatively fanciful and frivolous, on the grounds that metaphor may more compellingly and lastingly affect a reader than plain discourse and that the method has Biblical warrant (cf. the epigraph on the title page from Hos. 12:10, 'I have used similitudes'). These, we may suppose, were among the opinions of those friends who advised against publication (p. 2), but suspicion of fictional narratives was not confined to Puritans, as the general unwillingness of authors to admit their works were fictions testifies. In 1688 Aphra Behn claimed to have been an eye-witness of the events narrated in her novel *Oroonoko*, and in the early eighteenth century Defoe still presented his novels as authentic autobiographies. Bunyan's 'Apology' contributed significantly to the establishment of fiction as a serious literary mode, distinct from both lying and escapist romance. It is discussed in Kaufmann, pp. 3–60.

When at the first I took my Pen in hand: the view of Brown, pp. 239–47, and Talon, pp. 166–9, 317, that this was during Bunyan's second imprisonment, dated by Brown 1675–6, has been cogently challenged by OET, pp. xxi–xxxv, which, on the basis of Joyce Godber, 'The Imprisonments of John Bunyan', *Trans Cong. His. Soc.*, xvi (1949) 23–32, dates the second term in 1677 and, since this would have allowed barely a year for composition, criticism by friends and publication, argues for the second half of the twelve year term (1666–72) as the period of composition of *PP*.

the Way/ And Race of Saints: OET, pp. xxxi–iii, argues that this was *The Heavenly Footman* (posthumously published in 1698), a discourse on the Pauline image of the race for the prize of salvation, and not, as Brown, p. 247, suggested, *The Strait Gate* (1676). See also Sharrock, pp. 71–2; Talon, pp. 315–16.

Still as I pull'd, it came: the image is of a spinner pulling thread from the distaff.

3 *Pearl may in a Toads-head dwell*: according to contemporary lapidary lore, toadstones (various precious stones of supposedly therapeutic value) were formed in the heads of toads. cf. *As You Like It*, II. i. 12–14. 'Sweet are the uses of adversity,/ Which, like the toad, ugly and venemous,/ Wears yet a precious jewel in his head.' In 1646, however, Sir Thomas Browne had, in *Pseudodoxia Epidemica*, III. xiii, doubted their existence (*The Works*, ed. Geoffrey Keynes, 4 vols. (1928; rpt. 1964), ii. 199–200).

4 *Types*: typological exegesis interpreted characters and incidents in the OT as 'types' prefiguring or symbolizing something in the NT. In a developed form it could lead to allegorical readings of the OT, and so is thought by Bunyan to justify his literary method.

pins and loops...blood of Lambs: examples of details in the OT which Bunyan would interpret symbolically, from Exod. 26:5, 27:19 ('pins and loops'); Lev. 16:3,14,15, Heb. 9:12,19 ('Calves'); Lev. 1:10, 22:19 ('Sheep'); Deut. 21:3–9, Heb. 9:13 ('Heifers'); Exod. 29:15–32, Lev. 5:15–16 ('Rams'); Lev. 14:4–7 ('Birds'); Exod. 12:7,8 ('Herbs', 'blood'). For a sustained example of such exegesis see Bunyan's *Solomon's Temple Spiritualiz'd* (1688) in *Works*, iii. 460–509.

5 *Silver Shrines*: Acts 19:24.

Timothy: 1 Tim. 6:3, 4:7.

I find...Dialogue-wise: precedents for writings works of divinity in dialogue included Arthur Dent's *The Plain Mans Path-way to Heaven* (1601), one of the two books Bunyan's first wife brought with her (*GA*, §15), to which Bunyan's contemporary nonconformist Richard Baxter appealed to justify his own dialogue, *The Poor Man's Family Book* (1674).

6 *first to Plow*: Isa. 28:24–6; 1 Cor. 9:9–10.

the everlasting Prize: 1 Cor. 9:24.

8 *I saw...upon his Back*: the marginal references identify the Biblical sources for this striking image of spiritual desolation. The rags signify the inadequacy of man's own moral effort (Isa. 64:6) and the burden the guilt of sin which that effort cannot remove (Ps. 38:4, Matt. 11:28; 2 Cor. 5:1–4).

9 *Evangelist*: represents a minister of the gospel, a figure further exemplified in the emblem of 'a very grave Person' on p. 24 and in Great-heart in Part II. During his own spiritual crisis Bunyan himself received from John Gifford, pastor of the Bedford Church, just such guidance as Christian is given by Evangelist (*GA*, §§77, 117; Brown, pp. 77–90).

Wicket-gate: the 'strait gate' of Matt 7:13–14 and Luke 13:24, which was interpreted by Puritans as the experience of conversion and regeneration, the only possible beginning to an authentic Christian life. Christian's reception at the gate (pp. 21–3), whose keeper is later identified as Christ (p. 173; cf. p. 20), may therefore be taken to represent his conversion. OET, p. 315, suggests his 'admission into a community of believers', but this is more clearly represented in the Palace Beautiful (see below n. on p. 37).

10 *City of Destruction*: Isa. 19:18. The city stands for the fallen world and is opposed to the eternal Celestial City which is Christian's destination (Heb. 13:14). Brainerd P. Stranahan, 'Bunyan and the Epistle to the Hebrews', *Studies in Philology*, lxxix (1982), 284–6, points out that Bunyan's sense of its corruption, its impending doom and the details of Christian's flight derive from the account in Gen. 19 of Lot's departure prior to the destruction of Sodom and Gomorrah.

Christian: the protagonist is so named only after the process of conversion has begun; he was previously Graceless (p. 38).

12 *Dispond*: consciousness of sin, causing despair, was a recognized preliminary stage in the process of conversion, as Help now explains. Faithful later specifies its experience as a mark of grace (p. 68), Hopeful experiences it (p. 114), and Christian argues to Ignorance it is essential to true faith (pp. 119–21). Bunyan recounts his own struggle with the temptation 'to dispond' (*GA*, §195) in *GA*, esp. §§82–111, 132–228.

13 (marg.) *The Promises*: Biblical texts assuring salvation to the truly repentant faithful.

I stepped: on occasions intrusions by the narrator signify a point Bunyan especially wished the reader to notice, but generally the interruptions appear unsystematic; it is in the marginalia Bunyan consistently directs the reader's response.

13–14 *pleasure of the King...mended*: i.e. since the time of Christ ('this sixteen hundred years') God's ministers ('Labourers') have applied Biblical doctrine ('by direction of...Surveyors') in sermons and theological works ('wholesom Instructions') to try to prevent sinners from despairing ('mended').

14 *Mr. Worldly-Wiseman*: a type of the complacent believer who rests securely in his high opinion of his own moral behaviour without ever having experienced the desperate need for grace represented in the Slough of Despond. Such easy confidence as his, which contrasts with Christian's spiritual agony, was repeatedly denounced by Puritan divines as hypocrisy, 'formal' or 'opinionative' Christianity, as a reliance upon works rather than faith (see Evangelist's exposition on p. 18). That he 'looked like a Gentleman' (p. 17) reminds us of the generally lower-class social origins of the Baptists and their tendency to associate this kind of hypo-

crisy with the socially and politically acceptable established episcopal church, with its insistence upon uniformity of ceremonial observance amongst its members and its teaching which, as the century wore on, increasingly stressed morality at the expense of inner experience of the Spirit (represented in Legality and Civility (p. 16)).

15 *Wearisomness . . . death*: 2 Cor. 11:23-7.

16 *yonder high hill*: on Mt. Sinai Moses received the Ten Commandments (Exod. 19:1-3, 20:1-18). The hill represents this moral law, which inevitably condemns sinners since no mere man can fulfil it (Rom. 7; Heb. 12:18-21). On pp. 19-20, Evangelist explains that only Christ's grace answers sin, an idea embodied in the emblem of the parlour on pp. 24-5. For Bunyan's thinking on the law and grace, faith and works, see Greaves, esp. pp. 97-121.

19 *Now there are three things . . . utterly abhor*: Evangelist sets against Worldly-Wiseman's formal religion firstly, the necessity of faith, exemplified in the 'way' and the 'strait gate'; secondly, the inevitability of trial, self-denial and mortification, exemplified in the Israelites' wilderness wanderings to Canaan, with particular reference to Heb. 11:26 – 'the Cross' is used as a metonymy for such commitment in Matt. 10:38, 16:24 and Gal. 6:12,14; thirdly, the certainty of death, that is, the eternal death of damnation, as the consequence of a life of mere social rectitude – 'the ministration of death' is Paul's phrase in 2 Cor. 3:1-9, where he contrasts the old Mosaic law ('the letter killeth') with the new covenant of Christ ('the spirit giveth life').

21 *Good-will*: represents divine grace, Christ's mercy to sinners, in the words used by the angels at the nativity (Luke 2:14; cf. Eph. 2:4,7; 2 Thess. 2:16; 1 John 4:9).

Mount Zion: originally the citadel of Jerusalem captured by David (2 Sam. 5:6-7), the name came to stand for the city of Jerusalem (e.g. 1 Kings 8:1, Ps. 51:18, Isa. 24:23), and so to represent the 'New Jerusalem' (Rev. 3:12, 21:2), heavenly or celestial city (Heb. 12:22; Rev. 14:1).

Belzebub: the god of Ekron (2 Kings 1:2), who is identified in the NT as 'the prince of devils', that is, Satan (Matt. 12:24,26). Deut. 32:17 and 1 Cor. 10:20 encouraged the belief that the fallen angels came to earth and deceived men as pagan deities (cf. *Paradise Lost*, i. 364-75), but in Renaissance demonology the names of OT deities were variously applied. In the Bible itself, only in the later OT books and NT does Satan (Greek, from Hebrew: 'enemy', 'adversary', cf. *Paradise Lost*, iii. 156) emerge as the embodiment of evil, the fallen angel leader implacably hostile to God and man. The margin here so identifies Belzebub, although in *Paradise Lost*, i. 79, he is Satan's lieutenant, as he had been in Marlowe's *Dr. Faustus*, and on p. 72 he appears as but one of a company of devils, rather than their leader.

23 *Patriarchs*: the fathers of Israel, Abraham, Isaac, Jacob and his sons.

house of the Interpreter: the instruction of the new convert by the inner light of the Holy Spirit (the candle 'Illumination' on p. 24) takes the form of a series of emblems, some of which had appeared in such earlier collections as Francis Quarles's *Emblems* (1635). Emblem books, of which Bunyan's own *A Book for Boys and Girls* (1688, in *Misc. Wks.*, vi. 183–269) is a late example, consisted of a series of allegorical pictures each followed by explanatory verses (see Rosemary Freeman, *English Emblem Books* (1948)). This episode is discussed in Kaufmann, pp. 61–79; see also Roger Sharrock, 'Bunyan and the English Emblem Writers', *Review of English Studies* xxi (1945), 105–16.

25–6 *Then I saw...but Rags*: Coleridge, p. 470, objected to this as 'faulty allegory' since the time-scale is so foreshortened as to be 'not legitimately imaginable'. He is answered by Kaufmann, pp. 83–5.

27 (marg.) *The valiant man*: for his sword and helmet see Eph. 6:17, and with his bearing compare that of Great-heart and Valiant-for-Truth in Part II.

28 *the Three*: Enoch, Moses and Elijah, who 'looked over the gate' on p. 132.

Man in an Iron Cage: identified in *Works*, iii. 72, as John Child, who conformed to the Church of England in 1660, was visited after his lapse by Baptists who may have included Bunyan, and who in 1684 committed suicide (OET, p. 318).

Despair: in theological terms the sin of believing oneself beyond God's mercy (see Christian's comment to Piety on p. 40). With the man's words on p. 29, compare Marlowe's *Dr. Faustus*, II. i. 17–20: 'My heart's so hardened I cannot repent./Scarce can I name salvation, faith or heaven,/But fearful echoes thunder in mine ears/ "Faustus, thou art damned"' (in *The Complete Plays*, ed. J. B. Steane (Harmondsworth, 1969), p. 285).

31 *the Comforter*: the Holy Spirit (John 14:26).

a Cross: that Christian's burden of guilt and sin falls so suddenly and effortlessly from him represents the free imputation of Christ's righteousness to sinners. Christian has become sinless in the eyes of God through Christ's atoning sacrifice, though not, as he well knows (p. 41), sinless in fact.

a Roll: identified on p. 37 as assurance of salvation, which in Puritan thought was a mark of the elected saint. That Ignorance is subsequently found to have no 'roll' or 'certificate' proves he has not experienced regeneration through faith in Christ (p. 133), but for a criticism of this interpretation see Gordon Campbell in Newey, pp. 259–60.

33 *more then a thousand years*: the length of time Roman Catholicism had, in Puritan eyes, obscured the doctrine of justification by faith through its insistence on the merit of works and had substituted for inner Christian commitment a formal religion of external observance and ceremony such as the established Church of England continued to promote by its determination to enforce conformity to its liturgy.

if we are in, we are in: Coleridge, pp. 479–80, opined that here 'the allegory degenerates into a sort of pun' and 'is clearly defective, inasmuch as "the way" represents two diverse meanings; – 1. the outward profession of Christianity, and 2. the inward and spiritual grace', but, as Kaufmann, pp. 98–100, points out, these two senses of 'way' are in play throughout *PP*, reflecting a variety of forms of Christian behaviour variously motivated.

Coat: the white garment of the righteous (Rev. 19:8) Christian had been given on p. 31. On p. 34 Bunyan thinks of this as the livery a lord's retainer would wear.

35 *where he stumbled and fell*: the OT image of the wicked stumbling in the crooked and dark way of evil (e.g. Prov. 4:19, Isa. 59:7–10) is combined with 'dark mountains' in Jer. 13:16. The expostulation at the 'obscurity' of a field of mountains in Coleridge, p. 480, misses this Biblical allusion, but it is true that, with his geographical experience limited to South-East England, Bunyan seems to have had no conception of the scale of mountains (Talon, p. 147; Kaufmann, p. 16).

Lions: represent the persecution of nonconformists by the 'Clarendon Code' (*Works*, iii. 106, distinguishes the two as 'civil despotism' and 'ecclesiastical tyranny'). This persecution was intermittent; hence, the lions are at the moment chained; they are far more dangerous in Part II (p. 180). For this penal legislation and its effect see G. R. Cragg, *Puritanism in the Period of the Great Persecution 1660–88* (Cambridge, 1957) and Michael R. Watts, *The Dissenters* (Oxford, 1978), pp. 221–62.

37 *Palace...Beautiful*: the episode represents church fellowship. The Porter's promise of lodging to Christian only 'after discourse' with Discretion and 'if she likes your talk' (pp. 38–9) suggests further Christian's admission to a particular church, since, in Congregational practice, members were admitted into communion only after giving evidence of having experienced conversion (see Geoffrey F. Nuttall, *Visible Saints* (Oxford, 1957), pp. 109–16). In this respect, the episode properly follows the Wicket-gate and the Cross which represent this experience. From the fact the building stands 'just by the High-way side' OET, p. 321, infers church membership, though desirable, 'is not necessary to salvation', but this does not accord with the exposition of Bunyan's thinking on the significance of the church in Greaves, p. 125, and the position of the palace may be just a realistic detail – it could hardly stand *in* the road. Brown, pp. 18, 323, suggests the seat of Sir Thomas Hillersdon at Elstow, or (more probably) Houghton House, built by Inigo Jones in 1615 for Mary Sidney, Countess of Pembroke, near Ampthill, may have prompted Bunyan's depiction.

42 *Wife and four small Children*: in 1660 Bunyan's own surviving children were the four born to his first wife (Brown, pp. 388–9). In *GA*, §§327–9, he recalls how hard it was – 'as the pulling the flesh from my bones' – for him to resolve to leave them and undergo imprisonment, rather than to give over preaching.

43 *precise*: scrupulous. 'Precisian' was an alternative sobriquet to 'Puritan'.

stript himself of his glory: 2 Cor. 8:9, Phil. 2:7.

44 *Ancient of Days*: God (Dan. 7:9).

45 *Armory*: the Pauline imagery of 'the whole armour of God' inspired many Puritan works (see William Haller, *The Rise of Puritanism* (New York, 1957), pp. 142, 150–72).

Moses Rod...Man of Sin: respectively Exod. 4:2–5,17,20, 7:8–12; Judges 4:17–22, 7:16–23, 3:31, 15:15–17; 1 Sam. 17:38–51; 2 Thess. 2:3–8.

Delectable Mountains: for a discussion of Bunyan's responsiveness to the OT imagery of the fertile desert and the fruitfulness of Canaan in his landscapes of spiritual ease, see John R. Knott in *Casebook*, pp. 234–8.

Immanuel: Hebrew: 'God [is] with us.' Its OT use in Isa. 7:14 and 8:8 was interpreted in Matt. 1:23 as referring to the Messiah, that is, Jesus.

46 *Apollyon*: Greek: 'the Destroyer' (and so described on p. 48), used, after Rev. 9:11, of Satan, particularly in his capacity as tempter of mankind. In ch. xiii of *The Travels of True Godliness* (1683) Bunyan's fellow Baptist Benjamin Keach has Apollyon, the 'cursed prince of darkness', beset the potential convert Thoughtful with worldly temptation and spiritual depression, and in ch. i of its sequel, *The Progress of Sin* (1684), Apollyon is identified as 'the Old Serpent, the Devil and Satan'. On p. 50, however, Bunyan distinguishes Apollyon from '*Great* Beelzebub, *the Captain of this Fiend*', that is, Satan.

47 *the Monster*: the idea of the dragon derives from Rev. 12, with particular details from Job 41:15 and Rev. 13:2. The combat is modelled on the fights of popular romance, which Bunyan admitted having read in his youth (*Misc. Wks.* i. 332–3). For his indebtedness to romance and folktale, see the essays by Roger Sharrock, Nick Davis, Nick Shrimpton and S. J. Newman, in Newey, pp. 49–68, 182–250, and the articles by Harold Golder there cited; Charles Firth in *Casebook*, pp. 88–93; Talon, pp. 172–8.

51 *Satyrs*: Bunyan thinks of the hairy demons which inhabit the deserts of Biblical story (Isa. 13:21, 34:14), not the woodland companions of Bacchus in Classical myth.

blind have led the blind: the context of this dominical saying, Matt. 15:1–14, makes it clear the 'Ditch' represents doctrinal error; reliance upon works and presumptuous hopes have also been suggested (OET, p. 323; *Works*, iii. 114).

King David once did fall: the allusion to 2 Sam. 11:1–12:13 identifies the 'Quagg' as moral failing; antinomianism and despairing fears have also been suggested (OET, p. 323; *Works*, iii. 114).

54 *till more of you be burned*: an allusion to the 'fires of Smithfield', the execution of Protestants by burning during the reign of the Roman Catholic Mary Tudor (1553–8), which, through its commemoration in John Foxe's 'Book of Martyrs', *Acts and Monu-*

ments (1563), became for Puritans a potent image of Roman tyranny. Bunyan is known to have had the *Acts and Monuments* with him in prison (Brown, pp. 153–4) – what may have been his own copy is in the Pierpont Morgan Library (*GA*, p. 153) – and he refers to Foxe several times in his published works (e.g. *Misc. Wks*. viii. 383–4, *Works*, iii. 532, *GA*, §270).

55 *Avenger of Blood*: Deut. 19:5–6; Joshua 20:5,9.

   last was first: Matt. 19:30, 20:16.

57 *Adam the first*: the use of 'first' or 'old Adam' and 'second' or 'new Adam' for the carnal and regenerate nature of man respectively derives from Paul's presentation of Christ as the Second Adam (Rom. 5:12–14, 1 Cor. 15:45). As is appropriate to his allegorical significance, Faithful has experienced not the spiritual temptations of Christian (he escapes the Slough and does not meet Apollyon) but the lure of the world and the flesh (Madam Wanton, the Old Man, Discontent, and Shame (pp. 56–61; Talon, pp. 198–9).)

63 *knowledge of many things*: Talkative accurately reproduces the topics and phraseology of Puritan discourse in a parody of many of the conversations in *PP* itself. The inadequacy of such a merely opinionative faith as his was a constant theme of Puritan sermons. Richard Baxter's observation that 'It was never the will of God that bare *speculation* should be the end of his Revelations, or of our *belief*' (*Directions for Weak Christians* (1669), pt i, pp. 97–8) is exactly Faithful's point on p. 68, while the gist of Faithful's discountenancing of Talkative is contained in Thomas Adams' aphorism that we are 'to walke, not to talke of loue: One steppe of our feet, is worth ten words of our tongues' (*Works* (1630), p. 802).

67 *lie at the catch*: lie in wait to catch me out.

   Blessed are ye if ye do them: John 13:17.

69 *not he that commendeth . . .*: 2 Cor. 10:18.

70 *From such withdraw thy self*: 1 Tim. 6:5.

71 *peace be to your helpers*: 1 Chron. 12:18.

   I say, right glad am I . . .: this paragraph is a catena of Biblical texts. In addition to those cited in the margin, see Ps. 81:12; Isa, 50:7; Jer. 17:9; Matt. 28:18; Heb. 11:1, 12:4; Jas. 4:7.

72 *Vanity-Fair*: *Works*, iii. 127. notes that Isaac James in 1815 suggested the annual fair at Stourbridge, near Cambridge, as the

source of many details in Bunyan's depiction, a suggestion followed by Brown, pp. 254–5, which, however, also notes Bunyan's familiarity with the fair at Elstow (p. 17). Firth adds the possibility of Bartholomew Fair, held at Smithfield in London (*Casebook*, p. 94). The fair represents in a general way the vanity, that is, the inconsequential folly and futility, of the world and worldly concerns (Eccles. 1:2,14, 3:19), but it is also the most specifically topical and satirical episode in *PP*.

Almost five thousand years agone: the Biblical chronology of archbishop James Ussher (1581–1656) established the date of the Creation as 4004 BC. Bunyan refers, not quite accurately, to the age after Noah's flood (2348 BC), the period of the patriarchs, the first pilgrims (Heb. 11:13).

Legion: from Mark 5:9 Bunyan takes this to be the name of a devil (rather than 'the Devil', as OET, p. 327).

74 *Language of Canaan...Barbarians*: Isa. 19:18, 1 Cor. 14:11. Puritan habits of speech and dress were a frequent object of ridicule: see, for example, Ben Jonson's *Bartholomew Fair*, a play which enjoyed something of a vogue at the Restoration.

76 *by what he told them would happen to them*: Faithful's martyrdom is from Foxe rather than contemporary experience. Although death was a real possibility for nonconformists confined in overcrowded and insanitary jails, it was not a penalty prescribed by legislation against religious dissent. However, the trial which precedes the martyrdom accurately represents the kind of treatment which could be anticipated by those accused of holding or attending unlawful religious meetings or of publishing 'seditious libels' (that is, criticism of the established church). Firth (in *Casebook*, p. 95) compares the behaviour of Lord Chief Justice George Jeffreys at Richard Baxter's trial in 1685 with that of Lord Hategood (*The Autobiography of Richard Baxter*, ed. N. H. Keeble (1974), pp. 257–66), though Sir John Kelynge, before whom Bunyan appeared, may lie behind the portrait (OET, p. 327). For Bunyan's trial, see *GA*, pp. 104–31, and Brown, pp. 128–50.

80 *a Chariot*: 2 Kings 2:11. The allegorical mode is, of course, broken in Faithful's literal death.

By-ends: incidental or secondary considerations. The figure represents a worldliness which, through dubious casuistry, compromises Christian commitment. The stress on changeableness and the case

of 'a greater benefice' are a hit at those ministers who secured
their livelihoods by conforming to the established church in 1662.
Richard Baxter pointedly asked 'how came *Bartholomew-day*,
1662, to be so happy a day, as to bring 7000 to sudden understanding and repentance, that never shewed that repentance till
then…unless it was by changing with the times?' (*An Answer
to Mr. Dodwell and Dr. Sherlocke* (1682), p. 178).

81 *the stricter sort:* i.e. the Puritans.

82 *Mr. Save-all:* may, as OET, p. 329, be intended as an Arminian,
that is, a follower of the Dutch theologian Jacob Arminius (1560–
1609) who denied the Calvinist contention that the atonement was
limited to the predestined elect and held that Christ died for all
men who have the free will to respond to, or reject, divine grace.
However, the worldliness of his companions and their common
schooling by Mr. Gripe-man (a covetous or miserly man) suggest
that he is an ungenerous person who saves all his wealth.

83 *righteous over-much:* Eccles. 7:16. A feature of this conversation
is the interlocutors' ability to press Biblical texts to their purpose.
wise as Serpents: Matt. 10:16.

83–4 *sometimes Rain, and sometimes Sunshine:* Matt. 5:45.

84 *Job saies:* Job 22:24.
extraordinary Zealous: 'zealot' (after Titus 2:14) was another
nickname for Puritans.

85 *so makes him a better man:* a fundamental error, since, in Calvinist thought, works do not, in themselves, constitute goodness.
This will be Ignorance's error.

86 *the bag:* which contained the thirty pieces of silver paid Judas for
betraying Jesus (Matt. 26:14–16).

87 *Demas:* 2 Tim. 4:10, where Demas 'having loved this present
world' deserts his companion Paul, in contrast to Philem. 24 and
Col. 4:14.

88 *one of his Majesties Judges:* i.e. Paul.

93 *Doubting-Castle…Giant Despair:* after the period of spiritual ease
by the river the pilgrims are assailed by doubts as to their election.
Bunyan's own struggle 'with fears of my own damnation' forms
the major part of *GA*. The castle and giant are from popular story,
on which see above, n. on p. 47.

96 *Key…called Promise:* the Biblical assurances of salvation to the
faithful. Christian and Hopeful, mistaking a propensity to sin

(forsaking the way on p. 91) for evidence of unregeneracy, had forgotten that failings do not in themselves deny election: the Biblical promise of everlasting life is not to the perfect but to those who believe (John 3:15,36). Examples of such texts, particularly from what Bunyan, in *GA*, §250, describing his own efforts to believe they applied to him, called 'that blessed sixth of *John*', are given by Hopeful on pp. 116–17. The escape recalls that of Peter from prison in Acts 12:1–11, but the sudden discovery of the key has been thought fictionally implausible (e.g. Sharrock, pp. 85–6; Newey, p. 143).

damnable hard: Firth saw here 'a bad relapse' into the language of Bunyan's 'unregenerate days' (*Casebook*, p. 84), but OET, p. 332, suggests a 'grim theological punning'.

100 *Esau...Saphira*: respectively Gen. 25:29–34; Matt. 26:14–16, 21–5,47–50; 2 Tim. 4:14–15; Acts 5:1–10.

101 *So I awoke from my Dream*: usually taken to signify Bunyan's release from prison and hence to imply the remainder of Part I was written when he was at liberty (e.g. Brown, pp. 247–8; OET, pp. xxxi, 333). For the view that the interruption is 'artistically justifiable' see James F. Forrest, 'Bunyan's Ignorance and the Flatterer', *Studies in Philology*, lx (1963), 14–15. Brian Nellist in Newey, p. 148, sees it as a thematic pointer to Christian's 'climactic victory' over despair.

Ignorance: represents the complacent folly of reliance upon deeds (that is, the opposite temptation to that to despair because of moral failing just represented in Doubting Castle). Ignorance's self-righteousness is caught in words which echo the Pharisee in Luke 18:9–14. His misconceptions are later laid bare by Christian (pp. 118–22), particularly his supposition that through Christ he may earn salvation by 'obedience to his Law' (p. 120). For further discussion, see Maurice Hussey, 'Bunyan's Mr. Ignorance', in *Casebook*, pp. 128–38; James F. Forrest, 'Bunyan's Ignorance and the Flatterer', pp. 12–22; Kaufmann, pp. 113–14; Talon, pp. 209–12; Newey, pp. 40–4, 145–6, 149–50.

103 *Jewels*: his saving faith; the stolen money represents his assurance of election (OET, p. 334).

106 *I my self have been Ingaged*: i.e. in the contest with Apollyon.

Goliah as David did: 1 Sam. 17:4–51.

went to the walls: proverbial saying, deriving from the practice of setting pews for the infirm by the wall in medieval churches, which were otherwise without seating.

107 *David...Peter*: many of the psalms, traditionally attributed to David, express spiritual desolation and personal misgivings, as does Ps. 88, ascribed to Heman; the struggle of Hezekiah, king of Judah, to retain his faith in the Lord in the face of Assyrian aggression is recorded in 2 Kings 18–19; the collapse of Peter's bravado in Luke 22:31–4 follows in vv. 54–62.

108 *Leviathan*: Job 41, traditionally taken to signify Satan.

man black of flesh: the incident represents primarily the lure of false pastors (2 Cor. 11:13, which is its source), but, behind that, Satan (2 Cor. 11:14), whom Bunyan envisages spreading a net of temptation and who was frequently depicted as black.

112 *Where God began with us*: spiritual experience is proposed as the topic of conversation. Hopeful's following account corresponds to the essential pattern of Puritan autobiographies: his debt to other saints, his ignorance of the 'new birth', his adoption of formal Christianity, his growing sense of hypocrisy and sinfulness with consequent realization of the need for grace and his final assurance may all be paralleled not only in *GA*, which, as Talon, p. 201, points out, Hopeful actually quotes, but in the majority of Puritan autobiographies. They are precisely the experiences the ensuing discussion shows Ignorance has *not* had (pp. 118–22). See B. J. Mandel, 'Bunyan and the Autobiographer's Artistic Purpose', *Criticism*, x (1968), 225–43; Owen Watkins, *The Puritan Experience* (1972), pp. 9, 37–49, 101–20; G. A. Starr, *Defoe and Spiritual Autobiography* (Princeton, 1965), pp. 16–17, 36ff; Sharrock, pp. 55–6, 61–4; Roger Sharrock, 'Spiritual Autobiography in *The Pilgrim's Progress*', *Review of English Studies*, xxiv (1948), 102–20.

118 *The Soul of the Sluggard*: Prov. 13:4.

121 *What!...believe it*: Ignorance cogently points to antinomianism, the complete separation of the moral life from the scheme of salvation, as the apparent consequence of Calvinism's stress on the imputation of Christ's righteousness to the predetermined elect. The libertinism to which this could lead, and which Bunyan knew from the radical Ranters of the Interregnum (*GA*, §§44–5), is later represented in Mr. Selfwill (pp. 212–13). Christian's answer here was the standard one: faith, if true, has 'effects', namely, a desire to lead a godly life.

‹24 *not every one that cries, Lord, Lord*: Matt. 7:21.

‹26 *the City*: the New Jerusalem of Rev. 21:10–27, described more
fully on p. 130 and elucidated at large by Bunyan in *The Holy
City* (1665; *Works*, iii. 397–459).

‹27 *If you see . . . sick of love*: Song of Sol. 5:8.

‹28 *Enoch and Elijah*: were translated to heaven without the experience of death (Gen. 5:22–4; Heb. 11:5; 2 Kings 2:1,11).

I sink . . . Salah: Ps. 42:7, 88:7.

the land . . . Honey: the promised land of Canaan to which the
Israelites journeyed from Egypt under divine guidance through
the wilderness (Exod. 3:8,17, 13:5, 33:1–3), and so symbolically
the heaven to which saints journey through the wilderness of the
world (Heb. 11:14–16).

‹29 *Jesus Christ maketh thee whole*: Acts 9:34.

enemy was . . . as a stone: Exod. 15:16.

130 *Just Men*: i.e. those justified by faith (Rom. 1:17; Gal. 3:11;
Heb. 10:38).

THE SECOND PART

136 *as they did my firstling*: the demand for Part I had been extraordinary. It went through three editions within a year of publication, and had gone through thirteen by Bunyan's death in 1688
(OET, pp. xxxvi–lxxiii), earning its publisher, Nathaniel Ponder,
the nickname 'Bunyan Ponder' (F. M. Harrison, 'Nathaniel
Ponder: the Publisher of *The Pilgrim's Progress*', *The Library*,
4th series, xv (1934), 264).

Counterfeit/My Pilgrim: *The Second Part of The Pilgrim's
Progress* (1682) by T. S. (Thomas Sherman) had attempted to give
what the author took to be literary dignity and due seriousness to
Bunyan's allegory, and in the later 1680s ballads by 'J.B.' featur-
ing 'Pilgrim' in their titles were on sale in London (OET, p. 339;
Brown, p. 435). The vogue of *PP*, of which Bunyan both in this
preface and in Part II itself shows himself well aware (pp. 137–8,
144), is evidenced not only by the number of editions put out by
the licensed publisher Nathaniel Ponder, but by the many
translations (p. 137), piratical and composite editions (OET,
pp. lxxiii–lxxxvii; Harrison, 'Ponder', pp. 268–74, 282–6; Brown,
pp. 444–5), and the proliferation of imitations, adaptations and
supposititious works from the mid-1680s. These included the

spurious Part III of *PP* (1693), which was reprinted into the nineteenth century. For other, and later, examples, see Brown, pp. 456–61, and F. M. Harrison, *A Bibliography of...Bunyan* (Oxford, 1932), pp. 75–7.

137 *As to be Trim'd...Gems*: i.e. finely bound.

139 *Jacob...Sheep*: Gen. 29: 10:11.

What Christian left lock't up...Key: OET, p. 339, maintains 'The promise here held out that allegories concealed in the First Part will be revealed by the Second is not fulfilled'; for a contrary view, see N. H. Keeble, 'Christiana's Key', in Newey, pp. 1–20.

143 *Now it hath so happened...*: Bunyan's handling of the narrative point of view is rather uncertain. The remarks in this paragraph make little sense if the landscape of Part I is a dream landscape but, if it is no longer supposed to be, then this narrator need not dream. Mr. Sagacity is another false start. Bunyan soon realizes he is redundant and Mr. Sagacity leaves him 'to Dream out my Dream by my self' (p. 155). This gets rid of Mr. Sagacity, but is again inconsistent: since Bunyan's dream hitherto has not been of Christiana and her family but of Mr. Sagacity telling him about them, Mr. Sagacity *is* his dream. And, at the end of Part II, the narrator never awakens but talks of going 'that way again' (p. 261).

145 *Christiana*: it has been suggested that in Christiana, 'with her vigorous strength of character', Bunyan depicts his second wife who was able to confront the judges on his behalf in 1661, and that in Mercy is reflected his first wife (Brown, p. 260; Talon, pp. 202–3).

146 *Caul of her Heart*: Hos. 13:8.

146 *Wo, worth the day*: Ezek. 30:2.

147 *Secret*: the character represents divine knowledge of, and concern about, men's innermost longings and anxieties, and underlines the essentially private nature of the saints' experience of grace as they are let into the secret of God's mercy (Ps. 25:14; Prov. 3:32; Matt. 13:35; Rom. 16:25).

148 *by root-of-Heart*: a conflation of 'by rote' and 'by heart'.

150 *her Bowels yearned*: 1 Kings 3:26.

151 *Mrs. Bats-eyes*: the name ironically suggests that she is herself blind to the truth of what she regards as Christiana's blindness.

154 *many there be...Labourers*: OET, p. 341, suggests the point is aimed at the Strict Baptists who differed from Bunyan's open communion church in requiring believers' baptism of members, but it may refer more generally to ministers in other denominations whose views Bunyan regarded as erroneous.

155 *a Dog*: Ps. 22:20; the dog is explained on p. 158.

(marg.) *perplexed about Prayer*: the dominical saying 'knock, and it shall be opened unto you', which lies behind this incident, is an encouragement to fervent prayer (Luke 11:1-13; cf. pp. 21, 157).

Suffer the little Children...unto me: 'Luke 18:16.

156 *she comes...without sending for*: Bunyan's point is that, although Mercy fears she has not, like Christiana, received a call, her coming itself witnesses to a vocation whether or no she be conscious of spiritual regeneration (cf. pp. 170-71). It is the first of many examples in Part II of Bunyan's recognition that grace works in different ways, and to different degrees, upon different personalities, a point to which the margin calls attention at the foot of this page (Brown, p. 263). Mercy's name itself signifies not only her own charity but the divine willingness to accept all who come (Kaufmann, pp. 96-7).

158 *Let my Lord...Lips*: Hos. 14:2; Heb. 13:15.

purchased one: i.e. by Christ's redeeming blood (Acts 20:28).

164 *Daughter of Abraham*: Abraham, the father of the children of Israel (Gen. 17:5-8), was taken to represent Christ typologically, and so was symbolically the father of the faithful (Gal. 3:7-9). Christiana's visit to the Interpreter's House is discussed in Kaufmann, pp. 188-95.

165 *ugly Spider*: the verse cited from Proverbs offers the interpretative key: despite their sin (the spider) the faithful are acceptable to the divine mercy (the spacious room).

171 *the Bath*: believers' baptism by immersion. As their name indicates, admission of members by adult baptism (rather than by paedobaptism, the baptism of children) is the distinctive mark of Baptist congregations, though not insisted on by Bunyan's open-communion Bedford church (Greaves, p. 129; Brown, pp. 219-225).

fair as the Moon: Song of Sol. 6:10.

the Seal: the mark of election (Eph. 1:11-14), which could be

paralleled by the Passover since that marked the liberation of
Israel from Egypt by Moses (Exod. 12:1–17) which was taken to
prefigure the liberation of Christians from sin by Christ.

172 *Great-heart*: represents a minister in his armour of faith (Eph.
6:11–18). Macaulay and Firth both surmised he embodies
Bunyan's recollections of Parliamentarian officers, perhaps an
individual, he had known at Newport Pagnell (*Casebook*, pp. 75,
100).

173 (marg.) *discourse of our being justified by Christ*: for an exposition
of Bunyan's theology here see Greaves, pp. 27–96.

174 *two Coats*: Luke 3:11.

180 *Grim or Bloody man*: the 'civil authority responsible for the
persecution of nonconformists with renewed vigour during the
'Tory Revenge' of the early 1680s following the defeat of the
Whigs in the Exclusion Crisis. Offor identified him as the Lord
Chief Justice George Jeffreys (*Works*, iii. 106).

187 *Mr. Brisk*: is, in his breeding, his sexual motivation and his
worldliness, a poor cousin of the courtly gallant, a familiar figure
in the old days of Charles I who was again voguish after the
Restoration.

188 *I might a had Husbands...Person*: Firth commented: 'Here the
allegory disappears altogether. We have simply an incident in the
life of a fair Puritan described with absolute fidelity to nature:
the actors are ordinary men and women of the time, and the fact
that their names have a moral significance makes no difference to
the story. We are passing, in fact, from allegory to the novel with
an improving tendency' (*Casebook*, p. 102; cf. Sharrock, p. 145).

189 *cried her down at the Cross*: publicly to renounce a wife at a
market cross was, in rural areas, a traditional form of divorce
(*Works*, iii. 201).

190 *a Purge*: this incident has been often adduced by those who feel
the allegory in Part II descends to the merely ingenious and
decorative: see e.g. Sharrock, pp. 145–6; Talon, pp. 161–2, 202;
Firth in *Casebook*, p. 100; Dorothy Van Ghent, *The English
Novel* (New York, 1961), pp. 22–3.

192 *Why doth the Pelican...with her Bill*: it was believed the pelican
fed its young, and could revive them from death, with blood from
a self-inflicted wound (see e.g. T. H. White, *The Bestiary* (1954;
rpt. New York, 1960), pp. 132–3), but in *Pseudodoxia Epidemica*,

V. i, Sir Thomas Browne counted this a vulgar error (*The Works*, ed. Geoffrey Keynes, 4 vols. (1928; rpt. 1964), ii. 337–40).

193 *within the vail*: the 'veil' referred to in Heb. 6:19–20 is the curtain separating the outer sanctuary in the wilderness tabernacle and Solomon's Temple from the inner sanctuary, or Holy of Holies, which OT law permitted only a priest to enter (Exod. 26:31–5, 1 Kings 6:5, 2 Chron. 3:8–14). The sense of Bunyan's allusion is that by trusting in faith to the promises of salvation (the 'anchor') Christians may secure themselves against the adversities of this world (the 'outer sanctuary') and so pass through death (the 'vail') to heaven (the 'inner sanctuary'), which the death of Christ, the high priest who has preceded them (Gen. 14:18; Heb. 5,7), has made possible by rending the veil (Matt. 27:51; Heb. 10:19–22). This idea is developed more at large in *Solomon's Temple Spiritualiz'd* (1688) which distinguishes the two sanctuaries as 'a...shadow of the church both in heaven and earth' (*Works*, iii. 497–8).

194 *Let thy Garments...few*: Eccles. 9:8; Deut. 33:6.

195 *Through all my Life...*: the two stanzas are from Ps. 23:6, 100:5, in the version of Thomas Sternhold and John Hopkins (1562) (OET, p. 346). The Puritan tradition was generally suspicious of the singing of non-scriptural hymns. Bunyan's own responsiveness to music is evidenced throughout *PP* (see esp. p. 211), and Sharrock, p. 150, suggests its lyrics were included to promote congregational hymn singing.

199 *To this man ... Word*: Isa. 66:2.

Hercules: the only Classical reference in *PP*.

200 *Resist the devil...you*: Jas. 4:7.

201 *The heart knows...Joy*: Prov. 14:10.

the Earth...God: Jonah 2:6; Isa. 50:10.

202 *Snares*: Ps. 141:9.

Maull a Gyant: probably the Roman Catholic Church, whose Jesuits, particularly, were charged with tendentiousness and hypocritical sophistry by Protestants. Maul's claim to sovereignty over the land would then represent Rome's claim to a universal jurisdiction over all Christians and that he is a more threatening figure than Giant Pope in Part I (p. 54) would be appropriate since the Popish Plot of 1678–80 had reawakened English fears of Roman subversion. However, Talon, p. 163, noting that anti-

Catholicism plays a relatively small part in *PP*, prefers to see Maul as late seventeenth-century rationalism.

206 *holy Kiss of Charity*: Rom. 16:16; 1 Pet. 5:14.

211 *I make bold...the Throne*: if these words are Bunyan's, they come in oddly since he has (of course) been talking 'metaphorically' throughout *PP*, but Great-heart, to whom they are more appropriate, would not address himself to 'Readers'. Such narrative slips may be found elsewhere: on p. 210 Bunyan forgets Great-heart is narrating ('Mr. *Great-heart* began to take his Leave of him') and on p. 205 reference is made to 'Mr. *Honest*' before the pilgrim has been identified (*OET*, p. 348).

the Lake: Rev. 19:20, 20:10–15.

212 *fear of God...Wisdom*: Ps. 111:10; Prov. 9:10.

213 *He said...could do so too*: Selfwill's recollection of Biblical story, if not his theology, is reliable; see 2 Sam. 11; 1 Kings 11:1–3; Gen. 12:11–20; Exod. 1:16–17; Joshua 2:1–6; Matt. 21:1–7; Gen. 25:28–34, 27:1–36.

214 *as they in the Wilderness*: i.e. those Israelites whose waverings denied them entry to Canaan (Num. 14:11–33; Heb. 3:7–19).

215 *Gaius*: noted for his hospitality in Rom. 16:23 and 3 John 1–6.

216 *Antioch*: the Syrian city where Jesus' followers were first called Christians (Acts 11:26).

Paul and Peter: St. Paul, the authority most frequently cited by Bunyan (Greaves, pp. 154–5), was a persecutor of Christians who, converted within a few years of the Crucifixion (Acts 7:58, 9:1–19), became an indefatigable evangelist and the pre-eminent theologian of the Apostolic church. He was martyred at Rome during the reign of Nero (AD 54–68) *c.* AD 65, perhaps on the same day as St. Peter, chief of the Apostles during Jesus' ministry and their leader after the Ascension (Acts 1:15–22) who ended his life, according to tradition, as bishop of Rome, where he was crucified (John 21:18).

Ignatius...drowned: the only non-Biblical historical figures named in *PP*, whom Bunyan knew from the first book of Foxe's *Acts and Monuments* (see above, n. on p. 54) 'containing...the ten persecutions of the Primitive Church'. St. Ignatius, bishop of Antioch, was martyred at Rome, traditionally in the Colosseum, *c.* AD 107; St. Romanus, a deacon at Caesarea, was martyred at Antioch *c.* AD 304; St. Polycarp, bishop of Smyrna, was martyred by burning *c*. AD 155; Marcus, bishop of Arethusa, was smeared with honey and tortured as Bunyan describes during the reign of the apostate emperor Julian, AD 332–63 (John Coulson (ed.), *The Saints: a Concise Biographical Dictionary* (1958), s.vv.). Drowning is named by Foxe as one of the 'sundry torments of the holy martyrs', but to specifically which martyr Bunyan finally refers is uncertain.

218 *Drink of his Flagons...Love*: Song of Sol. 2:5; 5.1–2,8.

222 *Slaygood*: has not the precise significance of giants encountered earlier in *PP*, and the casualness of the episode has persuaded some critics of Bunyan's declining interest in his allegory in Part II (see e.g. Sharrock, p. 151). However, that Great-heart should rescue Feeble-mind is the allegorically appropriate point: ministerial care, tactfully bestowed in a liberal spirit, can preserve even such tender spirits as he (pp. 225–6, 238). Slaygood must necessarily be of vague import and not too fearsome character since the scrupulous are susceptible not to a specific temptation but to a variety of vague fears exaggerated out of proportion (p. 225).

227 *like Flint*: Isa. 1:7.

228 *Mnason*: Paul's host in Acts 21:16.

229 *stamped with his Foot*: to summon his daughter from below (OET, p. 350).

231 *Strangers and Pilgrims in the Earth*: Heb. 11:13, the essential imaginative idea of *PP*.

Monster: the beast of Rev. 17:3–18, the Antichrist, was taken by Protestants to be manifested in the Church of Rome. Hence, though defeated, the monster is not killed.

236 *demolishing Doubting Castle*: the allegory appears unsound at this point. While it is quite proper that Great-heart should prevent his followers from falling into the giant's clutches, it is difficult to understand how he could ensure other believers should not be subject to the temptation to despair. Sharrock, p. 152, sees in the incident evidence of the more secure mood of Part II.

238 *tumbled the Hills about with Words*: a literal presentation of the text from Mark cited in the margin.

239 *the Cake*: 1 Kings 17:8–16.

243 *how many did Sampson slay*: 1,000 (Judges 15:15).

243 *I fought . . . out of my Arm*: an image C. S. Lewis considered 'Any of the epic poets would be glad to have thought of' (*Casebook*, p. 200).

244 *killed a Serpent*: 'Presumably Apollyon, though he is not slain by Christian' (OET, p. 351).

248 *walked not by Sight*: 2 Cor. 5:7.

251 *Blessed is the Man . . .*: Prov. 28:14.

254 *'Twas she . . . Demas . . .*: 2 Sam. 15:1–18:33; 1 Kings 12: 25–33 and 2 Chron. 13:1–20; Matt. 26:14–16; 2 Tim. 4–10.

255 *Camphire . . . Aloes, . . . chief Spices*: Biblical spices, all mentioned in the Song of Solomon, whose love poetry was taken to express Christ's love for his church. With fine tact, Bunyan avoids having to rival his own depiction of the triumphal entry into the City which concluded Part I by focusing the end of Part II upon the final hours and deaths of the pilgrims. He draws upon the Song and the imagery of death in Ecclesiastes 12 to create the hieratic and ritualistic texture of the modulated repetitions which constitute the most carefully structured episode in *PP*, an episode which has been much admired: see e.g. Sharrock, pp. 152–3, F. R. Leavis in *Casebook*, p. 216 and *The Common Pursuit* (1952), p. 210.

260 *Death . . . where is thy Victory?*: 1 Cor. 15:55.

GLOSSARY

[colloq.] = colloquialism; [dial.] = dialectal usage; [var.] = variant form; a page reference indicates that a specific usage is glossed.

a	[colloq.] have (e.g. p. 22)
accoutred	equipped
affront	attack, assault (p. 50)
amain	with full force, vehemently
angel	a gold coin, worth 6s. 8d., named from its device of the archangel St. Michael and the dragon (p. 194)
apes	fools (p. 73)
arch	(i) notable, chief; (ii) clever, cunning (p. 227)
a toside	[var.] atoneside, on one side
back	support, encourage (p. 180)
beck'n	gesture, wave
bedabled	spattered
bedlams	madmen
bemired	soiled, smeared with mud
beshrow	[var.] beshrew, curse, blame
betterment	difference for the better
bin	[dial.] been
blank	disappointment, nonplus (p. 162)
boot, to	into the bargain, besides
bottom	valley floor, stretch of low-lying land (p. 32)
bowels	(i) the seat of tender feeling (p. 150); (ii) pity, sympathy, compassion (p. 153)
brast	[dial.] burst
brave	(i) fashionable, dashing (p. 59); (ii) worthy, noble (p. 137)
bravadoes	boastings, defiant assertions
breathing	exhausting (p. 178)
breeding	pregnant (p. 240)
brest	[var.] breast

brisk	cheerful, sprightly, lively (p. 101)
broken	distressed by remorse (p. 146)
bruit	[var.] brute
but	except (p. 1)
butt	[var.] abut, join
by-blows	blows which miss their mark
carriages	conduct, behaviour
caytiff	[var.] caitiff, wretch
checkle	[dial.] chuckle
churl	boorish, insensitive person
civit-box	[var.] civet box, box of perfume
clap, at a	in one go, at once (p. 83)
clogged	burdened (p. 146)
commixes	mingles, mixes up
compassed	surrounded, encircled
conceited	supposed, imagined (p. 223)
conditions	habits, personal qualities (p. 188)
condole	lament
convenient	suitable, appropriate (p. 30)
conversation	dealings with others, behaviour (p. 43)
courage	ardour (p. 188)
covet	desire eagerly
coxcombs	fools
crumbed	covered with crumbs
cut	steal by cutting a purse from a belt (p. 90)
damp	stifle (p. 42)
dash	splash through puddles (p. 243)
discretion	discernment, good judgement
disserting	[var.] deserting
distemper	disorder, derangement
dumpish	sad, dejected
dumps	perplexity, bewilderment (p. 22)
engins	(i) contrivances (p. 3); (ii) implements (p. 45)

equipage	accoutrements and appurtenances appropriate to a person of high rank
experimental	founded on personal experience, experiential
fact	misdemeanour (p. 36)
fall	fade away, decline to extinction (p. 217)
fancies	delusions (p. 177)
fantastical	imaginary, illusory
fat	(i) rich, appetizing, nourishing (p. 43); (ii) fertile (p. 196)
fatt	tub
figures	(i) prefiguring types (p. 4); (ii) emblems, symbols (p. 26)
followed	prosecuted, went on with (p. 49)
fondness	foolishness
forrest	wilderness, waste land (p. 248)
foyled	discomfited, worn out
friend	relation, kin
gat	[var.] got
gins	traps, snares
Gospel-day	day of the Christian dispensation
greenheaded	inexperienced, foolish
handmaiden	servant
hands, man of his	courageous and dexterous swordsman
hap	luck, fortune
hazarding	running the risk of incurring (p. 18)
heady	passionate, violent
hectoring	swaggering, domineering
howlet	[dial.] owlet
humane	human
humours	disposition, moods
ill-favoured	ugly
imploy	employment, occupation
increase	prosper (p. 170)

item	admonition, reminder (p. 229)
journey-men	hirelings (p. 106)
kind, in his	naturally, in his way (p. 64)
leered	walked stealthily with averted look
lesson	musical piece, recital (p. 236)
let	hinder (pp. 74, 160)
lime-twigs	twigs smeared with a sticky preparation from holly bark (birdlime) to catch birds
limner	portrait painter
loose	[var.] lose
lumbring	rumbling
lusty	lively (p. 73)
mantles	cloaks
mattered	was concerned about, paid heed to
merchandizers	merchants, stall-keepers
mistrusting	being afraid, suspecting (p. 92)
most an end	[dial.] almost all the time
mystery, in a	mystically, symbolically (p. 19)
naughty	wicked
naughty-wise	wickedly, viciously
noddy	simpleton
noise	melodious sound of a company of musicians (p. 184)
noised	rumoured
notion	opinions, theoretical beliefs (p. 69)
ordinances	religious observances, particularly the sacraments (p. 33)
outlandish-men	strange-looking foreigners
over-run	overtake
pair	set (p. 193)
passengers	passers-by, travellers
peevish	ill-tempered, irritable, spiteful

pelting	sweating, excessively hot
perspective glass	spy-glass, telescope
pickthank	tell-tale, sycophant
plash	bend or break down
plat	[dial.] place, patch of ground
post	courier, messenger (p. 255)
pretty	(i) fine (p. 64); (ii) clever, ingenious (p. 138)
price	sum of money
professor	one who openly professes Christianity
proof	tried and tested strength (p. 46)
prove	test the genuineness of (p. 10)
purpose, to	thoroughly (p. 14)
rack	drive before the wind (p. 30)
rateing	chiding angrily, reproving vehemently
refined	clarified (p. 43)
renovation	spiritual renewal and purification (p. 185)
ridged	[var.] rigid, dogmatically strict (p. 83)
round	whisper (p. 111)
roundeth up	rebukes (p. 88)
roundly	bluntly, sharply, severely
rude	superficial, unskilled (p. 4)
ruffins	[var.] ruffians
runagate	renegade, villain
sackbut	bass trumpet
scheme	conspectus, summary (p. 195)
scrub	[colloq.] shabby wretch (p. 79)
settle	wooden bench with arms and back (p. 37)
shadows	images, symbols, OT foreshadowings of the NT (e.g. p. 4)
sight	awareness, consciousness (p. 68)
slabbiness	muddiness
slow	[var.] slough (p. 12)
sneaking	skulking, cringing (p. 14)
snibbeth	[dial.] rebukes, reproves

sober	temperate
stage	scaffold (p. 180)
stitch, a good	[dial.] a considerable distance to walk
straight	[var.] strait, dilemma (p. 2)
strodled	[dial.] straddled
strook	[var.] struck
stumble	form an obstacle to belief for (p. 70)
stounded	[dial.] bewildered, at a loss (p. 128)
stunds	[var. dial] stounds, astonishes, stupefies (p. 149)
surfeits	fevers (p. 90)
swound	[dial.] swoon
then	[var.] than
thorow	[var.] through
through	[var.] thorough (p. 64)
thorough	all the way, to the end (p. 182)
tinder-box	box containing inflammable material and steel and flint with which to strike a spark
tollerate	allow, permit (p. 121)
toull	[var.] toll
tracing	ranging over, making his away about (p. 209)
travailers	[var.] travellers (p. 35)
travel	[var.] travail, labour (p. 24)
trenshers	[var.] trenchers, platters
tro	[var.] trow, believe
Turk	barbarian, savage (p. 65)
types	emblems, symbols (e.g. p. 4), particularly of the NT in the OT
typical	figurative, symbolic (e.g. p. 105)
unwarrantable	not genuine, fraudulent (p. 136)
venturous	daring, bold
wagg	[colloq.] stagger
Wain	Great Bear constellation (p. 237)
walk	behave, conduct herself (p. 166)
wotted	knew

INDEX TO *THE PILGRIM'S PROGRESS*